THE FUNCTIONING OF ECOSYSTEMS

Edited by **Mahamane Ali**

The Functioning of Ecosystems
Edited by Mahamane Ali

Published by InTech

Janeza Trdine 9, 51000 Rijeka, Croatia

Publishing Process Manager Jana Sertic

Technical Editor Teodora Smiljanic

Cover Designer InTech Design Team

Additional hard copies can be obtained from orders@intechopen.com

The Functioning of Ecosystems, Edited by Mahamane Ali
p. cm.
ISBN 978-953-51-0573-2

For EU product safety concerns:
IN TECH d.o.o., Prolaz Marije Krucifikse Kozulić 3, 51000 Rijeka, Croatia,
info@intechopen.com or visit our website at intechopen.com.

Meet the editor

Professor Ali Mahamane is a Lecturer at Abdou Moumouni University (Niger). He was born in 1964 at Kendadji, Tillabéri, Niger. He got his first degree in Agricultural Sciences from the Abdou Moumouni University and later specialised in Arid Regions Forestry (ENGREF, Montpellier, France). He pursued his studies at the University of Ouagadougou, Burkina Faso where he obtained his M. Phil in 1997. In 2000, he got a tenure appointment at the Faculty of Sciences at Abdou Moumouni University. He registered for his Ph. D thesis at the Université Libre de Bruxelles, Belgium in April 2005. He published more than 38 Scientific articles both in national and international journals. He is National Coordinator of UNDESERT Project (Understanding and combating desertification to mitigate its impact on ecosystem services). Presently Ali Mahamane is Deputy Vice Chancellor and Dean of Faculty of Sciences and Technics at the University of Maradi (Niger).

Contents

Preface

If we unanimously agree that the survival of each living organism depends on the nature and ecological services rendered by ecosystems, we must equally agree unanimously that these ecosystems do not always benefit the required attention. Hence they are subjected to many treats. Indeed, the ecosystems are subjected to many pressing usages at unknown tolerable levels. It generally results to equilibrium breaks leading ineluctably to their degradation. This tendency has engendered many international initiatives to prevent ecosystems degradation. For instance, the Biodiversity Convention, United Nations Convention to Combact Desertification (UNCCD), etc.

Regarding to the magnitude and modification consequences of ecosystems, the United Nations have commended the study on Millennium Ecosystems Assessment (MA). In order to better conserve the ecosystems and their services, it will be better to understand these ecosystems in all their complexity. It is to this aim that this book suggests some case studies undertaking all continents.

Indeed we try to fill the gap on the knowledge on ecosystems diversity and functioning. Since we have started the book project, we were invaded by many chapters on current environmental issues from reputable international research teams and laboratories. This shows that this book has aroused many interests from international scientific community given the important number of chapters submitted form the beginning. Consequently, we were subjected to make selection. We use this opportunity to thank the Publisher for publishing this book to the benefit of the international scientific community.

This book is educational and useful to students, researchers and all those are interested in environmental issues.

This book offers a compilation of 16 chapters on the methods of study of dynamics of ecosystems, diverse methods on ecosystems biogeochemical cycle, impact evaluation methods of human activities on dynamics of exposed environments, ecological imprint, evaluation methods of toxicity determined by heavy metals, pollution impacts aquatic vegetation community equilibrium.

We invite students, researchers, teachers and all people interested in environmental issues to aread this book which is educational considering the different methods which are presented.

Dr. Mahamane Ali,
Deputy Vice Chancellor and Dean of Faculty of Sciences and Technics (FST),
University of Maradi,
Maradi, Niger

1

Autonomic Management of Networked Small-Medium Factories

Flavio Bonfatti, Matteo Berselli, Luca Martinelli and Federico Stradi
Softech-ICT Research Centre,
University of Modena and Reggio Emilia
Italy

1. Introduction

This Chapter addresses SMEs that are qualified to perform complementary manufacturing activities as producers, subcontractors or suppliers. They are used to participate in different supply chains taking the form of networks where every node manages in turn its own supply chain. More precisely, there are at least three main models (or viewpoints) to take into account:

- Medium-large company and its supply chain members. When issuing an order the medium-large company should select the most convenient supplier depending on factors like cost, lead time, capacity, supply conditions, distance. These factors are variable over time since every supply chain member has other customers and often its own supply chain to manage.
- Small-medium company and its scattered customers. Many manufacturing companies are moving still more from mass production to small customised batches frequently requested by their numerous and demanding customers. In order to preserve these customers it is necessary to assure faster and more accurate replies to their requests and orders.
- Cluster of small-medium companies behaving as a virtual factory. The cluster intends to show one face to its customers, and to this purpose it must behave as a single organisation. The critical point is actually assuring a fast reaction to customer requests and orders even though decisions are taken through intense communications between partners.

The three models share a number of features: steady composition of supply chains, customer-supplier relations based on consolidated rules, autonomy of collaborating parties. Moreover they suffer the same kinds of problems: time wasted on phone calls and distributed decisions, blind acceptance of unprofitable orders, limited capacity to react fast to unexpected events. In spite of these conditions successful small-medium distributed factories reach high levels of effectiveness. They are able to innovate products by adapting to the changing market demand, their products are normally of good quality, and a significant percentage of their production is sold in other countries thus showing a relevant attitude to stay actively on the global market.

What is often missing is a high level of efficiency. Efficiency relates to aspects such as (a) fast response to customer requests and orders, (b) shift from static to dynamic relations, (c) choice of the most convenient network configurations, (d) collaboration to damp down perturbations, (e) better logistics to minimise transports, and (f) interoperability of information systems. Once the mentioned objectives are achieved by the networked small-medium factories they result in higher productivity and competitiveness. This is of paramount relevance in present years, characterised by a global economic crisis and a very slow recovery path to previous production levels. Investing in the proposed direction is a condition for surviving or even growing and grasping new business opportunities.

It is worth observing that the identified problem affects different industrial sectors. The Authors have experience of networked factories in the mechanical, machinery, fluid power, fashion, building and construction, and automotive sectors. The common feature of these companies is participation in discrete manufacturing of one-of-a-kind products or small batches, which implies flexibility, adaptation and fast response. These are raising requirements since mass production itself is evolving towards small batches of ever more customised products while short and certain delivery time is becoming a fundamental success factor. And SME networks are particularly exposed to risks because of the limited resources their members can divert from the core business activities.

The Chapter reports the achievements of a research project that is developing a software platform with a suite of autonomic services enabling every company in the network to move from a situation where it wastes valuable resources in struggling with its customers and suppliers, towards a rational business environment where communication becomes faster, and operation and collaboration more efficient. The ultimate objective of the project is to set-up, develop, experiment and promote the adoption of a new collaboration practice within networked factories taking advantage of the autonomic model applied to a suite of support software services. This is done to help overcoming the present crisis and having in mind potential economic and industrial scenarios in the next ten years.

The proposed approach is autonomic since the planning, scheduling and decision-making steps as well as the implied data exchanges can be fully automated, and nonetheless each company maintains its autonomy by imposing its policy to the autonomic tools. Thus, once coded the desired behaviour the company is finally relieved from the daily manned interactions with customers and suppliers. Of course, in the analogy with other automation cases it is necessary that each company in the network can switch from the full automatic to the manual mode. This will slow down the distributed decision process but it could be very appreciated, at least during the initial stage, to overcome the expected distrust in such a new technology-supported collaboration model.

More in detail, the research is pursuing a number of operational objectives ranging from the detailed definition of the intended organisational model to the development and experimentation of the relative support functions. The critical research goals are:

- Study the present conditions of networked small-medium factories, taking advantage of the indications coming from the real-life cases examined, to define a new and more efficient collaboration framework highlighting policies and decision points that every single company can customise to express its autonomous behaviour.

- Design and develop an autonomic run-time support for every small-medium factory to efficiently manage its supply chain by fully automating the network planning activity. The resulting plan, obtained by an intense automatic interaction with the homologous function at suppliers, represents the most convenient network configuration for executing a given order.
- Design and develop an autonomic run-time support for every supplier node in the network to efficiently allocate its internal resources by fully automating the devised scheduling activity. The automatic scheduling function is needed to provide real-time estimation of the best execution times and costs for the tasks assigned by the upper-level planning function.
- Design and develop an autonomic run-time support for every small-medium factory to efficiently manage possible exceptions by automatically performing the needed re-planning and re-scheduling activities. The ultimate aim is damping down the perturbation in such a way to minimise its propagation to the other actors in the network.
- Study key performance indicators to measure the behaviour of every node in the network, and of the network itself as a whole, and derive general rules using them to influence decisions about network configuration and partner selection. This means adapting the autonomic decision points to include knowledge from past behaviour.
- Solve the interoperability problem of exchanging data and business documents across the network and between every node and its legacy ERP system, possibly in a multilingual environment, by means of a proper translation and document transformation service based on a reference ontology to annotate the involved information.

The rest of the Chapter is organised as follows. Section 2 justifies the research effort in the addressed field by examining the state-of-the-art to understand what has been done so far and what is still missing to meet the defined goals. Section 3 goes deeper into the research methodology by decomposing the faced problem into its components, namely the aspects and challenges that are taken into consideration to meet the requirements. Section 4 represents the algorithms of the autonomic services from the twofold viewpoint of network leader and network supplier. Finally, Section 5 draws the conclusions by highlighting benefits and possible limitations of the achieved results.

2. Research rationale and related work

In order to achieve the objectives addressed in the Introduction it is necessary to innovate in three main fields, namely collaborative networks, autonomic approach and semantic interoperability. This section describes the reasons behind the proposed approach and presents the state-of-the-art in the three fields together with the progress that our research is pursuing beyond that situation.

2.1 Collaborative networks

Networks in industry have existed for a long time (Dekkers, 2010). Particularly along the last decades, the shift from make-or-buy to co-makership and alliances, the search for flexibility, the emergence of concepts for computer integrated manufacturing, fractal company, holonic manufacturing systems, intelligent manufacturing systems, and balanced automation, all demonstrate a continuous move to more loosely connected industrial manufacturing entities. The industrial networks and concepts of distributed manufacturing are now perceived as potential solutions to the needed flexibility and agility in response to fast changes in market demands.

The advances in the ICT, and particularly the Internet and pervasive computing, have revolutionised virtual collaborations (Ommeren et al., 2009) and enabled and induced the emergence of new organisational paradigms leading to the establishment of the discipline of collaborative networks. This discipline covers the study of networks consisting of a variety of entities (e.g. organisations and individuals) that are largely autonomous, geographically distributed, and heterogeneous in terms of their operating environment, culture, social capital and goals, but that collaborate to better achieve common or compatible goals (e.g. problem solving, production, or innovation), and whose interactions are supported by a computer network.

Nowadays collaborative networks manifest in a large variety of forms. Moving from the classical supply chains format, characterised by relatively stable networks with well defined roles and requiring only minimal coordination and information exchange, more dynamic structures are emerging in industry, science, and services. With the development of new collaborative tools supported by Internet and a better understanding of the mechanisms of collaborative networks, new organisational forms are naturally emerging in manufacturing and services (Camarinha-Matos et al., 2008a). With the consolidation of collaborative networks as a new discipline, more emphasis is being put on the theoretical foundation for the area and reference models that form the basis for further sustainable developments. Projects such as ECOLEAD (http://www.ecolead.vtt.fi) are examples of precursors in this direction.

More recent initiatives, namely projects included in the FInES cluster, such as COIN (http://www.coin-ip.eu) and COMMIUS (http://www.commius.eu), have been contributing important elements for the consolidation and expansion of the area, including interoperability services, semantic mediation, service-oriented computing, security infrastructures, and so on. Nevertheless, the support for collaboration still lacks important elements namely in what regards the behavioural aspects and "soft issues" of collaboration, which are difficult to conceive with current approaches in spite of the potential of the semantic web. Some experiments have been tried with multi-agent systems but no mature solutions are yet made available. Other areas such as CSCW and VR have been developing complementary components (e.g. coordination, argumentation, avatars) but all these developments still lack a deeper understanding of the collaboration needs and important "soft" and "social" aspects of collaborative networks.

Sustainable development of collaborative networked organisations needs to be supported by stronger fundamental research combined with real-world applications. The ARCON reference modelling framework for collaborative networks (Camarinha-Matos & Afsarmanesh, 2008b) is a contribution in this direction. Some important results from this area that are relevant for our research are value systems and benefit analysis models for collaborative networks (Romero et al., 2009), soft modelling techniques applied to complex problems such as rational trust assessment and management (Msanjila & Afsarmanesh, 2008), value systems alignment, negotiation wizards, behavioural modelling, and so on.

The proposed approach takes into account the already extensive empirical knowledge and technological achievements on collaborative networks and known limitations of current ICT support in order to design a new organisational structure and collaboration infrastructure offering support for dynamic composition of supply chains. The collaborative models needed in innovative enterprise networks (Tidd, 2006; Arana et al., 2007) present distinctive characteristics e.g. in terms of duration of relationships (among stakeholders and between service providers and clients), nature of business relationships, scale and territorial coverage, types of participants, that require new organisational forms, contractual models, and new governance models.

2.2 Autonomic approach

The adjective "autonomic" was first introduced to denote the Autonomic Nervous System (ANS or visceral nervous system), that is, the part of the peripheral nervous system that acts as a control system functioning largely below the level of consciousness. Since then it was intended to express the idea of automatic behaviour of functions personalised and then delegated by humans.

The Autonomic Computing Initiative or ACI (IBM, 2010) was launched by IBM in 2001 to develop computer networks capable of self-management for facing the rapidly growing complexity of distributed computing. The essence of autonomic computing is automating low-level management tasks while assuring better performances at the network level. In a self-managing autonomic system, the human operator takes on a new role in fact he/she does not control the system directly but defines general policies and rules that serve as an input for the self-management process. For this process, IBM has defined the following four functional areas: (a) Self-Configuration, for automatic configuration of the network components; (b) Self-Optimisation, for automatic monitoring of resources to ensure their optimal functioning with respect to requirements; (c) Self-Healing, for automatic discovery and correction of faults; and (d) Self-Protection for proactive identification and protection from arbitrary attacks.

Successively the Autonomic Network Architecture (ANA) integrated project (http://www.ana-project.org) moved the focus to network organisation based on the application of the autonomic principle. More recently the CASCADAS integrated project (http://www.cascadas-project.org) developed an autonomic component-based framework to deploy distributed applications capable of coping well with uncertain environments. In general, studies moved from the original idea to the possibility of exploiting the autonomic model in different application directions (Deussen, 2007; Di Ferdinando et al., 2008). Our research identifies the efficient management of manufacturing networks as a promising application field for the autonomic approach. This seems to be in fact the only feasible way to overcome the organisational delays induced by the participation of different members to strictly successive business processes and thus assuring fast response to external and internal signals (Pouly & Huber, 2009).

In this context, as depicted in Figure 1, every company is subject in principle to a double autonomic process, respectively as leader of the supply chain (or network coordinator) and as subcontractor. If leader, whenever reached by a customer order it will perform a planning activity aimed at choosing the most suited network configuration (Self-Configuration) and resulting in task assignments to the selected suppliers; if supplier, possibly of itself, it will manage to optimally schedule the internal resources (Self-Optimisation) to meet the requirements of the tasks assigned by the network leader (Hülsmann & Grapp, 2006). Moreover, in either cases it will undertake the necessary recovery actions (Self-Healing) to damp down the perturbations crossing the network as consequence of exceptions raised by nodes. Finally, it will use performance indicators derived from the past experience (Self-Protection) to tune and correct its planning or internal scheduling decisions.

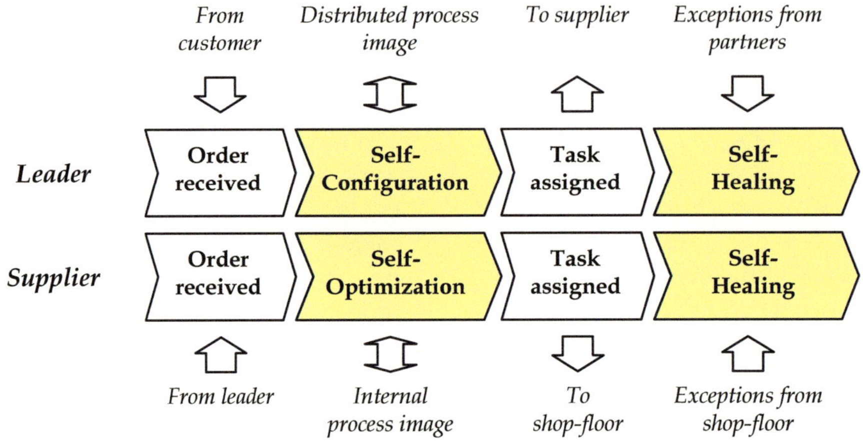

Fig. 1. Autonomic functions in leader-supplier relations.

According to the autonomic principle all these activities can be completely automated while preserving the autonomy of every network node to apply its own policies and habits. The progress brought by our research beyond the state-of-the-art is in the original adaptation of autonomic computation, a still young technology born in the world of technological infrastructures, to a practical and quite common problem of communication and coordination in manufacturing networks (Bonfatti et al., 2010a). If properly promoted and customised at the target companies it could have a dramatic impact similar to that produced by the first MRP systems in the far '60s and '70s.

More precisely, the proposed approach is inspired to some extent to the recent studies on autonomic service composition based on semantic, goal-oriented, pattern-matching (Fujii & Suda, 2006; Quitadamo et al., 2007). Their basic idea is that semantic description can be attached to services expressing what a service can provide to other services and what it requires from other services. In our case this is done by providing every node with the knowledge of distributed processes, if leader, and internal routings, if supplier. On this basis automatic propagation mechanisms for service composition can be enforced in an unsupervised way.

Besides relieving small-medium factories from burdensome and repetitive activities, thus saving resources that could be better employed in their core businesses, the autonomic approach introduces a clearer view of internal and distributed processes and a much deeper awareness of roles and decision policies. The increased effort required in the start-up phase will be paid back in a short time by the valuable benefits generated by the new organisational model. Therefore, the proposed collaboration model introduces as by-product a general cultural shift which is itself a breakthrough for moving the target companies towards a more efficient, profitable and competitive working environment.

2.3 Semantic interoperability

This third innovation area is justified by the intention to provide the autonomic platform with a technology assuring the possibility, for the user companies, to import/export business documents from/to their own legacy systems and to exchange documents with customer and partners adopting different data models and languages. It is well known that this requires the construction of a Reference Ontology - a data model and multilingual vocabulary for the specific application domain - as well as its use to "annotate" (or "map") the concepts of the legacy systems at the network companies and to cross-reference the terms from the origin language to English as "lingua franca" up to the destination language.

The generic WordNet (http://wordnet.princeton.edu) lexicon, the UNSPSC (http://www.unspsc.org) and eCl@ss (http://www.eclass-online.com) product/service standard taxonomies, the UBL (http://docs.oasis-open.org/ubl/os-UBL-2.0.zip) data model for business documents, all them represent good starting points for the construction of the intended reference ontology. What is needed is a strong simplification obtained by focusing on the only concepts and terms that are actually used for communication and document exchange in the devised collaboration environment (Bonfatti & Monari, 2007). This was done, for instance, with the domain ontology developed by the SEAMLESS project (http://www.seamless-eu.org) for the textile and building & construction sectors (Lima et al., 2006), the studies on semantic modelling in the logistics and transport sector (Brock et al., 2005), and the experience gained in the frame of the KASSETTS project (http://www.kassetts.eu) to support transnational collaboration in logistics (Bonfatti et al., 2010b).

Concerning the tools needed to define and manage the reference ontology (editor) and to map it versus legacy system data models (mapper) there are a couple of interesting candidate packages: PROTÉGÉ (http://protege.stanford.edu), the most known open-source editor of OWL ontologies exported in XML format, and MAPFORCE by Altova (http://www.altova.com/products/mapforce/xml_to_xml_mapping.html), a professional mapper of data models producing XSLT stylesheets for XML-to-XML transformations.

Although the document processing steps are well known, very few examples of business document transformations, including simultaneous structure conversion and contents translation, exist to be taken as reference. And the challenge is made even more critical by the need to hide the process complexity under an easy and simple user application. Our research is contributing to the advance of knowledge and technology in the field of semantic interoperability with a work addressed to the practical implementation and deployment of a so-far experimental approach to actual document structure conversion and contents translation. More precisely,

- Much care is taken to minimise and simplify the manual activity of annotating the proprietary or standard data models of the involved legacy systems with the concepts of the reference ontology. In particular, the difficulties arising from representation in research-related languages - such as OWL - are overcome by introducing easy drag & drop user interfaces.

- The run-time document transformation function is completely automatic and transparent to the sender (receiver) user who will maintain its own legacy ERP system and be relieved by the need to know the data model and language of the receiver (sender) user. In other words, the effort spent in ontology construction and manual annotation is widely recovered at run-time.

2.4 The FInES roadmap

In developing the research work towards autonomic management of networked small-medium factories special attention is spent to follow the indications reported in the Future Internet Enterprise Systems (FInES) Research Roadmap (FInES, 2010). First, the concept of enterprise Quality of Being is introduced as an extension of the current notion of enterprise quality. Six distinct FInES Grand Objectives are identified to characterise the enterprise Quality of Being, which are taken as reference by our research activity in term of exemplary quality of future enterprises:

- Inventive enterprise. Flexibility and distributed autonomy are the main features addressed by the research. Very important, the target factories have a partially organised structure with strong delegation to lower operational levels. In addition, the research aims at supporting a distributed control functionality based on feedback mechanisms and autonomous reactive behaviours.
- Cloud enterprise. The autonomic approach directly targets distributed organisations where raw materials and intermediate products are supplied by different organisations (often located in different places or regions) that are in or out depending on the conditions, but with a coordination level that is based on clearly defined agreements about supply time and quality of the supplied goods and services.
- Cognisant enterprise. The research aims at pushing companies towards this quality in two ways. First, it adds further knowledge including formalised process models and decision-support rules. Second, it keeps and elaborates data on performed activities and their effects thus enabling the networked factory to learn from experience and adapt consequently its behaviour.
- Community-oriented enterprise. The intended virtual factory is characterised by transparency and accountability. Transparency is assured by the definition and application of behavioural rules at the company and the network levels, accountability comes from objective measures of performance in the perspective of revising or improving relations with peers and customers.
- Green enterprise. The research explicitly addresses the identification of optimal distributed planning solutions including minimisation of transports between the virtual factory nodes. This will have an immediate positive effect on freight traffic, especially at the regional scale where over 80% of transports occur at a distance of less that 100 km, and the resulting reduction of $CO2$ emission.
- Glocal enterprise. The real-life cases taken as reference by the research are all run by virtual factories acting in the global market and however deeply rooted in their territory and culture. In particular, the proposed approach facilitates the target companies in looking ahead to get immediate advantages for future improvements or, at least, to reduce future detriments.

Going deeper into the FInES Research Roadmap the approach we propose is consistent to a large extent with the six operational and two (final) strategic research challenges defined in the document:

- RC1 - Federated open application platform. This is the architectural view of the proposed technology assuring the construction of a federated collaboration environment open to the addition of further value-added services. At the same time the autonomic service platform should be normally provided in Software-as-a-Service (SaaS) mode with no restraint on the place where data are stored and functions are executed.
- RC2 - Awareness and intelligence platform. The ability of an enterprise to understand its status with respect to the market, identify innovation needs and grasp collaboration opportunities requires the adoption of proper modelling methods and tools. The formalisation of this knowledge is a condition to make it available for defining a medium-term development perspective and business process improvement.
- RC3 - Innovation-oriented continuous (re)design environment. Modelling methods and tools in combination with autonomic network configuration and work optimisation tools enable companies to simulate hypothetical organisational changes and estimate their effects, including strategic indications on how to empower the network composition with involvement of further members (or exclusion of inefficient members).
- RC4 - Implementation recasting platform. The proposed solution facilitates the dynamic constitution of manufacturing networks meeting at best, time by time, the current requirements set by customers. This implies configuring the single network instance by selecting the best fitting candidate SMEs, establishing the needed relations between them and making them cooperate and exchange data and documents.
- RC5 - Meta-knowledge infrastructure. The intended services use an underlying meta-knowledge made of two main parts, reference ontology and history of past behaviour. The former provides the network nodes with the necessary semantics to interoperate with no cultural restraints, the latter supports their daily decisions with a clear view of the accessible resources, their features and performances.
- RC6 - Interoperability & cooperation infrastructure. Distributed processes regulate cooperation of distributed entities within a production network. To this purpose the autonomic platform supports document exchange assured by an automatic transformations and translation engine executing the directives resulting from annotation of legacy data models by the concepts of the reference ontology.
- RC7 - The FInES constituent. According to the roadmap definitions our collaboration environment includes: (a) enterprises of different complexity and variously linked and organised, (b) people establishing the rules of play and performing the planned processes, (c) intangibles like knowledge and services dynamically produced and consumed, (d) tangibles represented by the products constituting the final process outcome, and (e) public bodies as possible promoters and multipliers of the new model.
- RC8 - The FInES science base. The devised work on collaborative networks aim at providing sound scientific foundations to the technical work addressed to introduce more efficient collaboration practices and their support functions. This makes the project achievement suited to be adapted to a wide spectrum of practical cases and sufficiently technology-independent to cover the operational needs of the target companies and networks over a long period of time.

3. Research methodology and components

In this section the defined objectives are examined in more detail to show the directions of our research and technological development effort. In particular the research approach and methodology are presented while decomposing the research project into its components, namely the various aspects and challenges that are taken into consideration in order to satisfy the wide spectrum of networked small-medium factory requirements.

The first critical aspect to investigate is the new collaboration practices to make the networked factory and its member companies gain competitiveness in the global (electronic) market. This leads to the definition of a collaboration environment that each networked factory will interpret and adapt to its specific contextual conditions.

The adoption of the new collaboration model calls in turn for developing a suite of software services to be made available to each of the target companies. More precisely, the autonomic tools and their ancillary services are conceived for being provided as functions of a web-based platform published at every networked factory, or possibly at public service providers, and then accessible to user companies in SaaS (Software-as-a-Service) mode.

3.1 Collaboration environment

3.1.1 Collaboration habits, needs and exchanged documents

Even though SMEs are already running production projects together, the collaboration potential of companies belonging to a networked factory is often not completely exploited because of its low organisational efficiency. Other variable factors are industrial sector, company nature and size, local norms and regulations. Therefore a specific investigation was carried out to assess in depth the actual situations represented by a sample of networked factories. This knowledge is the first measurable result of our research, consisting of business and technical specifications to be taken as basis for the definition of the collaboration environment and for the design of the software platform and its functions.

While studying the target collaboration scenarios it was also possible to find out and classify the data and documents the involved actors are used to exchange. This is the condition to assure that the collaboration environment provides the needed communication channels and safeguards, at the same time, the investments in legacy ERP systems of the member companies. Analysing the exchanged documents means documenting their structure and extracting the terms used to express the relevant concepts, both normally coming from the available information systems. The construction of a comprehensive ontology to serve as semantic reference for automatic document transformation was obtained by integrating this information from the sample of small-medium factories.

The characteristics of these first research outcomes are briefly summarised in the following points:

- The first critical aspect of the collaboration model is identifying the decision steps in the frame of the distributed processes, and formalise them so as to enable the network leader and every node to customise their respective behaviours according to habits and preferences. This is of paramount importance to reach the needed level of confidence in an automatic system to which the network nodes will delegate the future operational decisions.

- Concerning the business documents exchanged between the network leader and its nodes much attention was found around requests for quotations and task assignments. Both document trigger an intense communication from leader to suppliers proposing the execution of a certain task, from supplier to leader replying with estimated lead times and costs, and finally from leader to supplier confirming or cancelling the task.
- Finally, the ontology representing the semantics used in the network for communication between parties accounts for an order of magnitude of 300-400 concepts and 700-800 terms, independently to some extent from the industrial sector. For instance, concepts are "company", "payment mode" or "quotation" while terms are the concept names themselves plus enumerations like "cash", "direct debit" or "credit transfer" for payment mode.

3.1.2 Process modelling and mapping

A necessary condition for autonomic collaboration in the networked small-medium factory is the formalisation of the knowledge on distributed processes that is normally hidden in the experience of skilled persons and daily operational practices. This requires a preliminary (design-time) effort establishing the correspondence between the distributed process activities as seen at the leader tier and the internal operations generated by those activities at the candidate suppliers tier.

In general the distributed process P(X) for manufacturing the product X can be represented as a typical workflow diagram, like that sketched in Figure 2, including a number of activities linked to each other by sequences, alternatives (e.g. A2 and A5 branches) and parallelisms (e.g. A3 and A4 branches). Moreover, every activity has associated the candidate supplier, or the list of candidate suppliers in case of networks with competing nodes, that could perform it. In Figure 2 the list of candidate suppliers for activity A7 includes the leader (Self) to represent a typical make-or-buy choice.

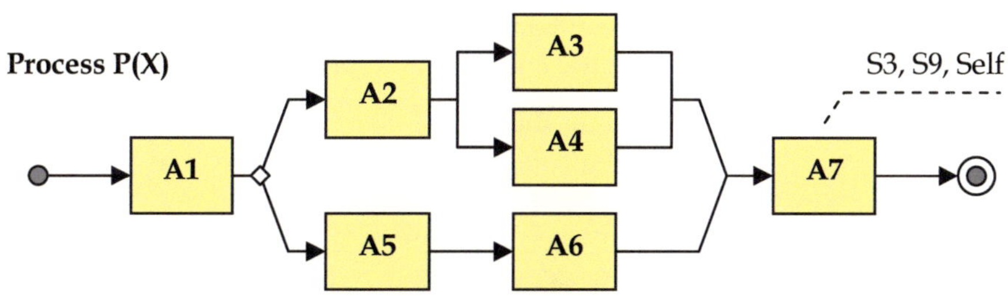

Fig. 2. Example of distributed manufacturing process.

Once defined the distributed processes for all the products offered by the network, and known the candidate suppliers for every activity, two more operations are required to complete the information needed by autonomic services, namely:

- Mapping activities with supplier internal processes. An activity as seen by the leader is normally characterised by a name and a set of parameters expressing in detail its execution conditions. For instance the activity "painting" will specify paint material, painting technique, colour, thickness and other possible parameters. In turn the candidate supplier presents the painting service in its offer, likely represented with a different name and different parameters. These data must be mapper at design-time to establish a clear match for the autonomic algorithms.

- Modelling the supplier internal processes. Every supplier must finally document the offered products and services with the internal operations required by their manufacturing. If leader in turn of a networked factory it will associate each product or service with the relative distributed process. If leaf in the networking structure it will represent the shop-floor operations in terms of routings, phases and resources, that is, the typical information required by a scheduling algorithm as well as by the autonomic Self-Optimisation service.

3.1.3 Key performance indicators

Appropriate key performance indicators (KPIs) are the enabler for decision making activities in our approach. The major decision areas that are based on KPIs include the timely selection of the right supply chain configuration, the set-up of priority rules for resource allocation at the shop-floor tier, as well as triggering events for replanning and rescheduling tasks. These decision areas are not limited to the single factory and are also relevant for the network as a whole. Besides their role for decision making processes, KPIs are used to evaluate the success in terms of factory and network efficiency as well as customer order satisfaction and supply and demand matching accuracy and flexibility.

Therefore a specific analysis of the objectives and requirements of KPIs is inevitable. The analysis consists of technical specifications on functionalities of the proposed approach and the deduction of their KPI requirements. This calls for a differentiation of the KPIs between those that have to be delivered for decision making and those that measure the performance of the autonomic processes, the involved factories and the total network. It is also necessary to find out and classify the quality criteria that KPIs have to meet for being successfully used in the autonomic environment. These criteria include the measurability of KPIs, their real-time availability for decision making purposes and their hierarchical and operational data structure. In terms of measurability the broad application focus of our research in different companies and industries calls for a higher customisable specification.

Therefore KPIs must be classified into those that are critical for the core autonomic functions, those leading to an optional increase in planning accuracy and transparency and those that are only substitutes for other KPIs. The conceptual aggregation of these analysing steps leads to a performance measurement system supporting requirement analysis for KPI definition, deduction of the application areas, quality criteria for KPI definition, process of data collection and interpretation, overall structure of the KPI system and information flow for KPI creation and usage.

3.1.4 Contracts, reliability and dispute resolution

Reaching agreements and contracting are important elements in the process of creating and operating dynamic goal-oriented networks. Research on this issue focuses on indentifying how concerns on conflict-related risks avoidance can be supported by negotiation and contracting. The dynamics of the negotiation process and the necessary support functionalities are influenced by factors such as nature and characters of the involved organisations, their expectations regarding the collaboration opportunity, affective aspects, the governance principles adopted in the networked factory, as well as the historic traces of past collaborations.

Since decision-making as well as the individual and joint behaviours in a collaborative network depend on and are reflected by the underlying value system of network participants, it is important to properly model value systems and devise methods for value systems alignment analysis. Complementarily, the roles of the various participants in the network need to be characterised and taken into account in the negotiation and contracting processes as a basic condition for trust building.

Together with value systems alignment, collaboration readiness is another relevant aspect in partner selection, which is relevant for anticipating potential conflicts. The needed work on collaboration readiness focuses on understanding, reasoning, and measuring how ready an actor is for collaborating with others, and to estimate how well an organisation is likely to perform in a partnership.

3.1.5 Identification and traceability issues

Planning processes are only as good as they can be realised in supply chain operations. By that, matching supply chain planning and supply chain execution is a key factor for the overall efficiency and flexibility of supply networks. This calls for ubiquitous information that forms the link between planning decisions and their operational consequences on the shop floor. Such an objective is influenced by ubiquitous information in three different directions:

- First, ubiquitous information leads to an information effect, what means that operational data needed for KPI formulation and overall transparency can be measured more quickly, accurately and efficient by the use of identification technologies.
- Second, ubiquitous information can be used for gaining automation effects in operational processes. While planning processes are automated by the autonomic approach itself, the automation of their execution can be managed by identification technologies. That can reduce costly media discontinuity, when automated planning results trigger manual execution processes.
- Third, ubiquitous information is often linked with process transformation effects enabling new business models and process configurations in supply chains. When implementing the autonomic system, networks have to meet physical and organisational process requirements including a high level of flexibility and traceability in supply chain operations needed for the execution of dynamic supply chain configurations, relations and processes.

Dealing with these objectives of ubiquitous information in supply chain execution the cause-and-effect relations between the autonomic approach on the one hand and the operational processes in factories and networks on the other has to be analysed. Based on this analysis the requirements on information technologies in terms of information, automation and transformation effects can be defined and afterwards being linked to those technologies like RFID or satellite and indoor telematics, best capable to realise the needed effects.

Knowing about the potential benefits of information technologies for the devised co-manufacturing environment leads to the question of process redesign and technology implementation. By using the insights of the industry partners, reference processes and implementation guides can be formulated for establishing the operational requirements for successful autonomic planning processes.

3.1.6 Risk management and security issues

It is widely accepted that a networked factory is as secure as its weakest element, since an attack against that element can lead to the collapse of the entire supply chain. This makes a holistic approach of risk and security management in a network environment such as the autonomic approach even more important. Security can be achieved in the entire network only if it is borne in mind at an early stage when planning its design. Furthermore, security should not be forgotten in the company everyday life, since even small security gaps may lead to major risk issues. Supply chain security therefore needs to simultaneously address both the entire network and its constitutive elements.

Therefore in the first instance a risk and security management framework is needed that arranges the elements and relationships of the autonomic approach in a high level of abstraction using a selected structuring of reference modelling. This reference model can be arranged into different levels, every level representing a security-relevant view of a networked factory:

- Security strategy. At first, a security strategy for the entire network has to be selected. As the autonomic supply chain configuration should be dynamic, the network precise design is not defined in this level, but in the subordinate ones.
- Network topology. A network topology is derived from the security strategy by defining the number, type, and location of linkages and nodes of the networked factory. By that, the basic policies for automated partner selection, customer acceptance and supply chain configuration functions are determined.
- Linkage. Each linkage within the network is further managed. Security measures for each route are selected that trigger the planning and scheduling routines in the system.
- Node. Each node within the network is further managed, bequeathing the security requirements of the network topology. Security measures for each network member that reflects its individual risk profile and secures the entire network has to be selected.

For the definition and design of each of these security levels the relevant requirements and risk triggers must be analysed in cooperation with the network leader. Afterwards appropriate management approaches and security technologies can be selected to be implemented in the software platform and translated into policies and practical guides for each networked small-medium factory member.

3.2 The software platform

3.2.1 Self-configuration service

Imagine the leader company in the virtual factory is receiving a customer order. This company must activate its supply chain and to this purpose it will carry out a planning activity to decide which tasks are conveniently performed internally, which are better assigned to partners and which are the preferred partners in those specific cases (the same holds to much extent in case of request for quotation). The planning activity requires that the leader company submits to the candidate suppliers the proposals for the tasks it could assign to them, and every supplier replies with its estimation of execution time and cost. Those suppliers leading in turn a supply chain will carry out a similar planning activity with the involvement of further companies to fulfil the proposed tasks (see Figure 3).

Fig. 3. Order (task assignment) propagation in the networked factory.

According to the received answers and the rules of its policy the leader company will select the best configuration of the working supply chain for that customer order, and will confirm the assigned task to each of the involved partners and a fast and detailed order acceptance to the customer. The configuration will be static if partners are predefined for the tasks to assign, it will be dynamic if partners are selected within constellations of competing companies. Should instead that configuration be not profitable to the networked factory, this will be established in a very short time and on the basis of objective data, and in that case the network leader will cancel the proposed tasks (with propagation to the lower-level networks) and reject the customer order.

It is worth noting that with the autonomic approach no resource has to be diverted from the company core business to manage real-time interactions with partners. The communication flows down from the function triggered by the customer order to the homologous functions of the lower-level nodes that lead in turn a supply chain, and the responses flow back inversely. The final response to the customer and the confirmation of the assigned tasks to the selected partners are fast and accurate when all of them are provided with the same, or equivalent, technology. No interference from other orders is expected to occur in such a brief planning time, therefore the plan is perfectly compliant with the current conditions of the involved companies.

This distributed planning process and its propagation across the ecosystem mirrors the present time-wasting human interactions but it will lead to the devised solution very efficiently. Moreover it can take into account often neglected variables, such as transport times between manufacturing steps, and compare alternative solutions to measure their actual profitability.

Details on the Self-Configuration algorithm are reported in the next section.

3.2.2 Self-optimisation service

During the planning process every supplier company is asked time by time to express its decision about acceptance and conditions for the execution of the proposed task. This decision is related to task profitability and is taken through the estimation of execution time and cost coming from the shop-floor. This requires the availability of a specific tool which, basically, is an internal resource scheduler. Every task is represented as a routing made of a number of operational phases, that is, atomic operations to be carried out by a certain resource type. The resource type is normally composed by a combination of primary and secondary atomic resources (persons, machines, equipment, etc.) for each of which the company has available one or more instances.

Then, scheduling the execution of a task means allocating the internal resource instances that are able to perform the phases of the relative routing. To this purpose the work calendars of the proper resource instances are examined and one or more combinations are identified under the given constraints. The choice of the best combination will be based on the policy established by that company. Further improvements are obtained by periodic revisions of the current schedule aimed at squeezing the resource work calendars to remove possible idle times generated by odd durations or cancellations or modification of the scheduled tasks. Note that this approach results in an optimal solution for the single company and, since combined with the Self-Configuration process, assures the local optimisation of network processes.

If a node in the network is already provided with its own resource scheduler the distributed planning function (Self-Configuration service) will obviously interface it. This will decrease the efficiency and responsiveness of the whole systems unless it is possible operating the legacy scheduler in autonomic mode. Otherwise that company can choose to adopt the Self-Optimisation service of the software platform being thus sure to obtain a full autonomic behaviour.

Details on the Self-Optimisation algorithm are reported in the next section.

3.2.3 Self-healing service

Later, during order execution, the distributed plan will be put in action and the tasks will be progressively carried out by the selected companies employing the scheduled resources. Should a task be affected by problems (e.g. on delivery date or product quantity), these will be taken as exceptions requiring a rapid replanning and rescheduling effort for damping down the perturbation. With respect to the viewpoint of the network leader company the exception could come from two different directions: from the customer in terms of order reduction/extension or changed due date, or from a supplier in terms of renounce, delay or loss of materials. In turn, with respect to the viewpoint of a supplier node the exception could come from two directions: from the network leader behaving as a customer, or from its own shop-floor in terms of missing resource, delay or loss of materials.

The company, dependently on its role, will try to solve the exception and damp down its effects by replanning the supply chain (if leader) and/or rescheduling its internal resources (if supplier). If this Self-Healing effort will be successful there will be no impact on the rest of the network, otherwise the residual perturbation will be properly propagated to the partners that will use the same tools in turn.

Because of the strong need to solve fast those kinds of problems the Self-Healing service should be autonomic even more that others. In fact if managed manually they are very stressing and can seldom take into account all the possible alternatives since their immediate practical objective is finding in any case a solution although not the best one. Instead, autonomic functions can do it as native behaviour. Of course, the autonomic Self-Healing service will rely on the same planning and scheduling algorithms supporting the Self-Configuration and Self-Optimisation phases. Every small-medium factory will be put in condition to customise them further to take into account the additional policies used in critical conditions which, in principles, could differ from those adopted for normal conditions.

Details on the Self-Healing algorithm are reported in the next section.

3.2.4 Self-protection service

The proposed model addresses the construction of self-regulating networked factories where the good practices are not imposed by a global optimisation functions but, on the contrary, they come from the ability of every actor to organise at best itself and the surrounding partners. This ability calls for the computation of proper key performance indexes to feed the partner selection mechanism. Partner selection occurs at both design-time and run-time. Design-time partner selection corresponds to the modelling stage when every company associates the activities of its distributed processes with the suppliers that could perform them. Run-time partner selection occurs during the Self-Configuration stage as consequence of the application of the customised policies autonomously defined by every actor.

In addition to criteria like profitability, execution time and cost, performance indicators play a relevant role in partner selection since they can represent the actual quality of candidates. For instance, under equivalent conditions a reliable supplier is preferred to others; moreover, the selection policy could introduce additional times or costs in presence of underestimating partners. Key performance indicators are conveniently computed from the past network member behaviour that, to this purpose, must be adequately traced. Besides contributing, as actually objective data, to implement a suited trust building mechanism, their values are considered in the policy rules and used by the planning tools. Their periodic update is ensured by the Self-Protection service.

Finally, the knowledge about partner quality, as expressed by the performance variables, must flow across the network, in at least two ways: (a) whenever considering a company as possible partner its quality must be immediately known, and (b) time and cost estimations coming from possible suppliers must be accompanied by the integrated quality measure of the underlying supply chains, if any.

3.2.5 Document transformation service

Business collaboration implies an intense data exchange in terms of business documents. The software platform is provided with a technology realising an easy communication channel taking into account that customers and partners are normally operating different information systems (meaning, with different data models) and possible using different languages.

This requires the construction of a Reference Ontology - a data model and multilingual vocabulary possibly specialised for the specific application domain. The data model is used to "annotate" (or "map") the concepts of every interesting legacy systems while the vocabulary will cross-reference the terms from the origin language to English as "lingua franca" up to the destination language. The experience gained with the already mentioned SEAMLESS project showed that the reference ontology is conveniently built up as a controlled user-defined semantic representation based on the analysis of the exchanged business documents. Once constructed, the reference ontology is passed to an ontology Editor to be stored, managed and updated off-line by the appointed experts. Then the ontology is moved to an ontology Mapper where the data model and vocabulary of every interesting legacy system is put in relation, by a graphic user interface, with the homologous concepts of the reference ontology.

The outcome of the Mapping function is an executable fragment of code that, if applied to an incoming document, assures its structure conversion and content translation from the origin data model (and language) to an internal representation in English and, vice versa, from this internal representation to the destination data model (and language). The Transformation service is applied in two important cases:

- Document import/export. Every document moved from the legacy system of a user company to the autonomic platform, and vice versa, will undergo an automatic transformation process from the origin data model to the destination data model.
- Document exchange. Every document exchanged through the software platform with a customer or a partner will undergo a similar transformations process plus the automatic translation of contents between the origin and destination languages, if different.

While the Editor and Mapping functions are off-line, and used manually once for all at design-time, the Transformation function is automatically executed at run-time by the import/export and document exchange services so as to provide the user with a very easy communication tool hiding the intrinsic complexity of the transformation process.

4. Details on autonomic functions

This section provides some details on the algorithms of the run-time autonomic functions, namely Self-Configuration, Self-Optimisation and Self-Healing, and of the relations between them in the construction of a full autonomic network management environment. Although schematic and neglecting minor aspects, they should be sufficient to show the basic computation and interaction mechanisms that have been developed and are presently tested at networked small-medium factories working in different industrial sectors.

4.1 Self-configuration algorithm

Figure 4 shows the general schema, in form of activity diagram, of the Self-Configuration algorithm processing a customer order (or a request for quotation) received by the network leader. Let us assume that the customer order refers to a single product, that is, is made of just one order line (and the request for quotation as well). Double-line boxes correspond to invocations to other autonomic services.

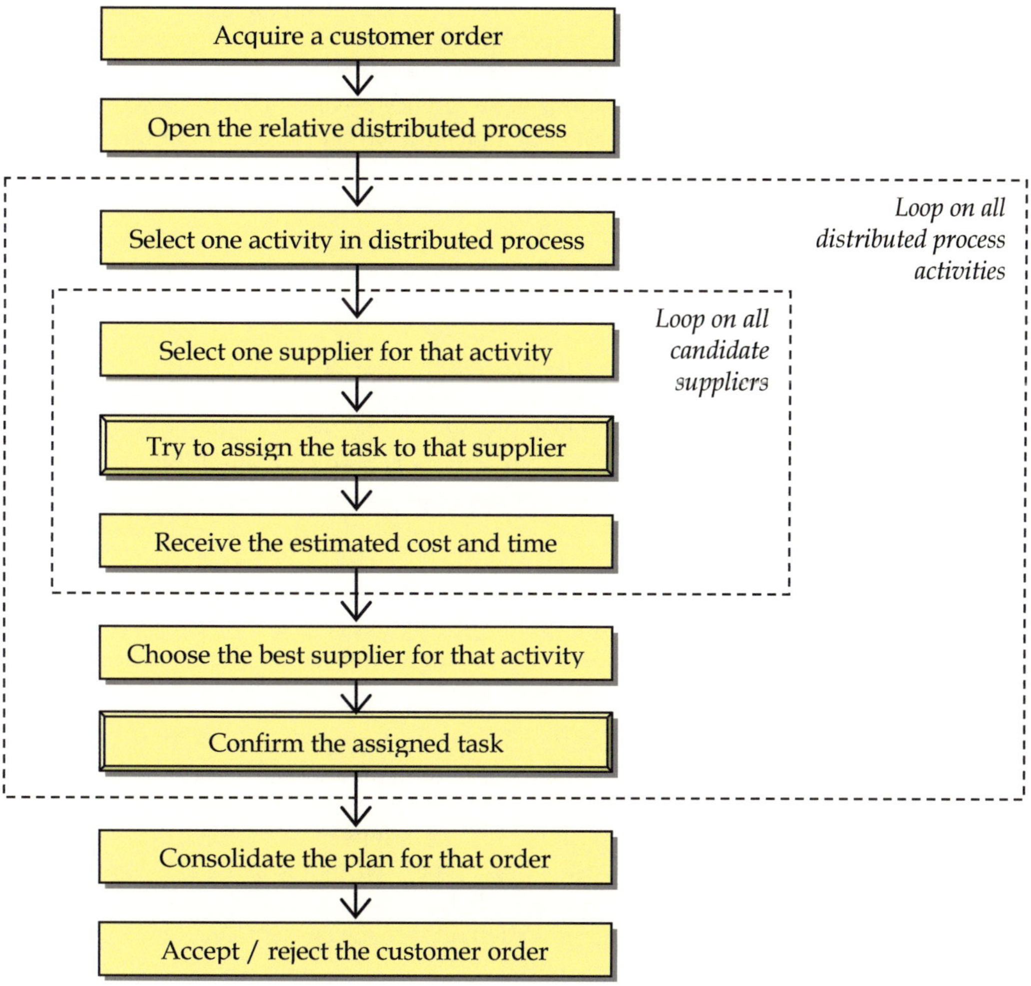

Fig. 4. General schema of the Self-Configuration algorithm.

Some comments and explanations:

- This algorithm holds for both customer order and request for quotation. In the latter case the function "Confirm the assigned task" is removed since there is nothing to confirm until the customer will issue an order. Moreover, the "Accept / reject customer order" is replaced by the simple communication of the estimated lead time and cost.
- The distributed process is scanned one activity at a time. In fact only after an activity has been assigned to the best supplier its end time is known to become the start time for the next activity. In case of alternative branches the algorithm follows the branch corresponding to the given parameters. In case of parallelism its branches are scanned in parallel to determine the overall execution time (the time computed on the slowest branch) to become again the start time for the next activity.
- More precisely, not always the end time of an activity becomes the start time of the next one. In fact, whenever materials must be moved from one company to the other it is necessary to plan the transport and then estimate its duration and cost. The duration is added to the end time of the previous activity to determine the real start time of the next one, the cost is added to the overall process cost for profitability analysis.
- The invocation of all the suppliers under consideration is indeed the moment showing the benefits of the autonomic mechanism. While planning successive customer orders a certain supplier associated to different activities will be subject to many invocations whose only objective is estimating its execution times and costs. Much better if this occurs in a way that is transparent to the company personnel, provided that the supplier autonomic service is certainly applying the established customised rules.
- The choice of the best supplier for a given activity is based on a multiplicity of criteria including cost and lead time but also supplier quality, reliability and correctness. This is in fact one of the decision points that is worth customising so as to mirror the usual human decision mechanisms.
- Once the best supplier is chosen for a given activity its autonomic service is invoked again with the only objective to confirm its booking to perform that assigned task. This occurs within the same transition of the previous invocation therefore the renewed invocation will confirm the same results as before.

It is worth underlying that this process is completely automatic, hence it achieves the devised solution in a very short time having considered a number of alternatives that are completely out of the reach of any human operator. When invoking the autonomic service of a candidate supplier two different situations may occur: (a) the supplier is in turn the leader of a networked factory, then the assigned task will be processed by the same Self-Configuration algorithm; or (b) the supplier directly operates with its shop-floor, then it schedules its internal resources with the Self-Optimisation algorithm described in the next sub-section.

4.2 Self-optimisation algorithm

Figure 5 shows the general schema, in form of activity diagram, of the Self-Optimisation algorithm scheduling a task tentatively assigned to that supplier by the network leader. The Self-Optimisation algorithm has different effects depending on whether it is invoked just to estimate lead time and cost or to book definitely the internal resources. In the latter case the algorithm outcome updates the work calendars of the allocated resources while no change results in the former case.

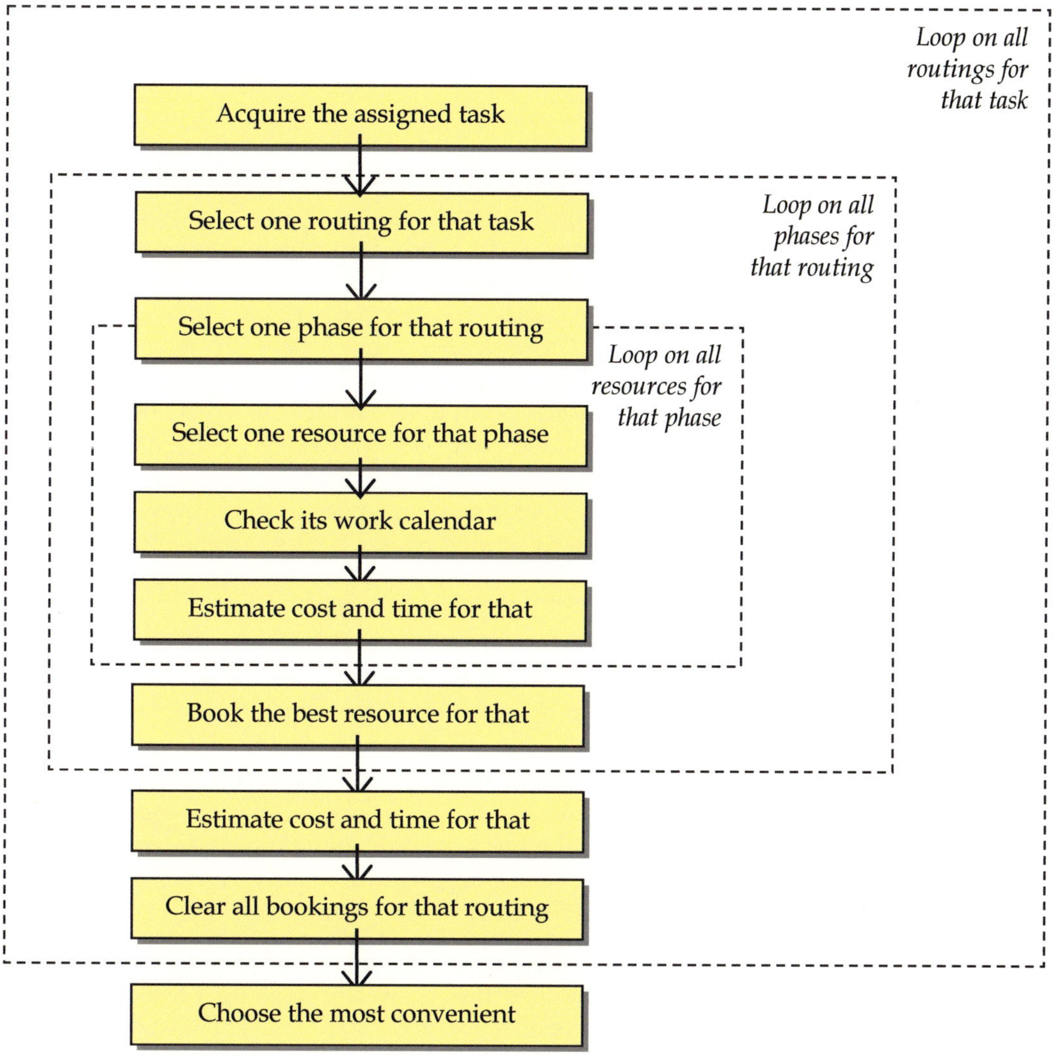

Fig. 5. General schema of the Self-Optimisation algorithm.

Some comments and explanations:

- We assume that, as it often happens, a supplier can perform a certain task according to different routings normally defined for the employment of alternative resources. Therefore the first scheduling loop is intended to verify which is the routing assuring the most convenient solution in the specific case, according to the policy of that company.
- We also assume that a routing is represented as a rule by a sequence of phases that are scheduled following their execution order. Once scheduled one phase its end time is known, which becomes the earliest start time for the next phase. Delays on the latter phase can result from late availability of the required resources or possible lack of the needed materials.
- Then, scheduling a phase means identifying the internal resource instances that are able to execute it, simulate the employment of each of them by checking its work calendar, and eventually choose the best resource instance for that phase. Even in this case the final choice is (automatically) performed according to the company policy and decision criteria coded at the system customisation stage.
- Resource is a generic term. Sometimes it indicates a specific machine whose availability is the only constraint for allocation to a given phase. In other cases it is a combination of primary (finite capacity) and secondary (infinite capacity) resource instances including persons, machines and possibly consumables. In the latter case the scheduling algorithm must ensure that all these components are simultaneously available for that phase.
- Once found out the best resource instance for a certain phase its allocation must result in a booking on its work calendar, so as to mark it as busy in the corresponding time period. In fact when scheduling the next routing phases it might happen that the same resource instance is considered again, therefore its availability must exclude the booked periods to avoid phase overlapping.
- Instead, once completed the scheduling of one routing and produced the expected estimation of execution time and cost, all the bookings of the involved resources shall be cancelled. The reason is that the scheduling of the next routing, for the same assigned task, must start from the same initial conditions found by the previous routings: this is a strict condition to allow a correct comparison of their results.
- At the end of this effort the best schedule on the most convenient routing is finally known, and its data are returned to the invoking Self-Configuration algorithm. If that schedule is afterwards confirmed the scheduling algorithm will be called again, within the same transaction, to just book the chosen resource instances.

This algorithm is not so different from other scheduling functions found on the market. However, it presents two main peculiarities that is worth recalling here: (a) a deep integration with the upper-level Self-Configuration algorithm assuring intense recursive interactions, fast decisions and an overall efficient behaviour of the autonomic system, and (b) the real possibility to be executed with no human intervention while assuring the full application of the customised rule for the benefits of the single user company.

4.3 Self-healing algorithm

Self-healing means enabling every actor in the supply chain to manage exceptions and contribute, according to its role, to damp down the perturbation across the network. This is done by triggering the execution of the proper autonomic services with the purpose of replanning the affected distributed processes and rescheduling the affected tasks.

Exceptions come, for a leader, from a customer or from a selected supplier; in turn they come, for a supplier, from its leader or from its shop-floor. Moreover exceptions are of many types since they include order or task cancellation, advance or delay on the due date, increased or decreased product quantity, and the like. Therefore it is necessary to consider a wide spectrum of cases and the actions that shall be undertaken to face each of them with the maximum level of efficiency.

Let us consider one of the many possible situations of exception (see Figure 6): supplier S4 that had assigned activity A2 of process P(X) informs the leader of a 4-days delay to complete that activity. The leader activates the Self-Healing service that performs the following automatic activities:

- The plan generated by the Self-Configuration service for process P(X) is scanned to find out the activities depending on the A2 results or, however, not started yet (the shaded area in Figure 6) since they are the only activities whose planning can be revised.
- The autonomic services of suppliers S1 and S9, respectively assignees of tasks A3 and A4, are invoked for new estimations of time and cost according to the changed conditions. Once known the new latest delivery time of those two parallel activities, that time is taken as new start time for replanning A7, and the autonomic service of the Self supplier is invoked in turn.
- An alternative, more aggressive, replanning approach consists of two steps: (a) the autonomic functions of suppliers S9, S1 and Self are called to immediately revoke the respective tasks assigned; then (b) a radical replanning of activities A3, A4 and A7 is carried out by means of the Self-Configuration function with the involvement of all the candidate suppliers.
- In either cases the conclusion is a new overall lead time and cost, hopefully compliant with the indications of the customer order. If not, the change shall be communicated to the customer.
- Note that in this situation the perturbation is transmitted to suppliers S9, S1 and Self which could in turn propagate it to the respective sub-networks, if any. Otherwise they shall solve the problem at the shop-floor level by cancelling the scheduled tasks and rescheduling them according to the new conditions.

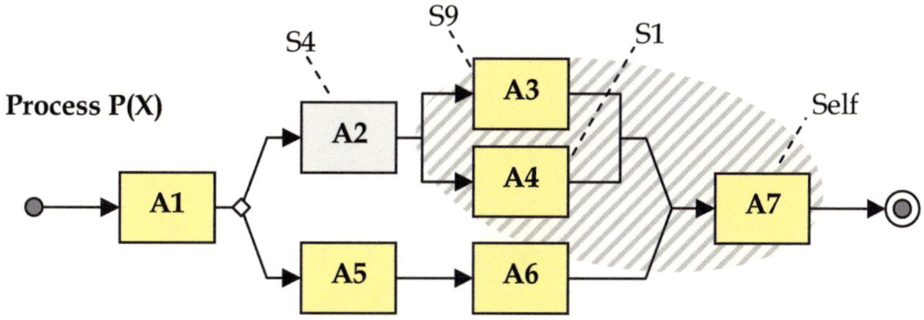

Fig. 6. Partial replanning of a distributed manufacturing process.

Although representing one of the possible cases of exception, this example shows one of the most important benefits brought to the networked small-medium factory by the autonomic approach. Indeed exceptions occur very often and much time is wasted in phone or fax communications between the involved actors to solve the problem. The same or likely a better result is achieved by the autonomic system in a very short time with no human intervention and considering all the available options.

5. Conclusions

To sum up, the vision is for a manufacturing environment where each small-medium factory is provided with the devised software functions to relate with customers and suppliers, and possibly with its shop-floor, in such a way to accelerate its reactivity to orders and perturbations and increase its control over order profitability without the need of diverting resources from the core business:

- The solution is individualised. Even though the autonomic system assures full benefits only when all the network members adopt it, the underlying model is itself an extraordinary opportunity for every company to formalise processes, roles and decision criteria thus increasing its understanding of network management aspects.
- The solution is customised. Each company understands that the software platform reacts automatically, that is, with no human intervention, to customer requests, orders and exceptions. In order to maintain its independence and decision power the company is put in condition to parameterise the autonomic services during the set-up phase to correctly represent its policy.
- The solution is collaborative. Once provided with its own customised software services each company can use them to manage its supply chain and to participate in one or more networks. The adaptation to requirements of different customers is quite simple, and mirrors the present ability of the sales office to establish profitable relations.
- The solution is easy to adopt. Each company can understand the behaviour of the autonomic services, needs a limited support to initialise and customise them, and safeguards its previous investments in ICT having assured the possibility to import/export documents from/to its legacy ERP system (if any).
- The solution is reversible. Should a user company decide to leave the new network coordination environment, it can do it at any time with no restraints or loss of operational data. Rather, it may happen that when leaving the autonomic tools system it will hardly find other ICT solutions able to host the knowledge generated by its use.

The proposed autonomic system is in an advanced development stage. A preliminary version of the devised autonomic and additional services has been developed and is presently under thorough test by some networked factories acting in the fashion, mechanical and global service industrial sectors.

For information completeness sake we can specify that the software platform is a web application developed in open-source technology using Java as programming language, Apache Tomcat as web container, Java Server Faces for UI specification and Ice Faces for its implementation, Hibernate as object-relational mapper, PostgreSQL as relational DBMS, Apache CXF for web service development, Eclipse Rich Client for desktop applications and Spring as configuration and integration platform.

The first on-field experiences highlight most of the foreseen benefits and then represent an encouraging boost to this research project. At the same time they show some problems that are to be considered as precious indications to improve the quality of the autonomic solution and its ability to meet at best the practical need of the target companies. The main problems concern the effort required for platform initialisation and the resistance to delegate important decisions to a software application.

As discussed in Section 3 of this Chapter the initialisation stage must lead to formalise some critical aspects of the interactions within the network. Let us recall them:

- For the leader company: represent products or services produces through the network, associate each product or service with the corresponding distributed process, establish precedence, alternatives and parallelisms between process activities, characterise each process activity with proper parameters and options, couple each process activity with the list of candidate suppliers and, very important, code the network management policy in form of decision criteria and rules to be applied by the Self-Configuration and Self-Healing functions.
- For every supplier company in the network: represent its products and services offered to the network, associate each product or service with the relative routings, split the routings into sequences of operational phases, couple each phase with the resource instances that can perform it, provide each resource instance with its work calendar and, again very important, code the shop-floor management policy in form of decision criteria and rules to be applied by the Self-Optimisation and Self-Healing functions.

This formalisation effort takes time and needs often to be supported by an external advice, thus constituting an objective obstacle to innovation. To this purpose a further research work is presently devoted to minimise the cost of the initialisation stage by offering a library of distributed process templates, pre-coded decision rules and other defined elements among which it will be easy to find out the desired solution.

Concerning the resistance to delegate important decisions to the autonomic platform, experience shows that in most cases there is no explanation or estimation of economic benefits that can overcome it. In order to reach the needed level of confidence in and acceptance of the proposed model there are two complementary initiatives that can positively contribute:

- Document successful cases from other networks possibly operating in the same sector and region. As a matter of fact innovation at SMEs is normally introduced by imposition from main customers or by imitation of neighbouring partners and competitors.
- Introduce for each company in the network the possibility to swap from the automatic to the manual mode until the correct application of its individual policy and, ultimately, the brought benefits will be actually proved by a systemic evaluation of actual system performances. This will slow down the distributed decision process but it could be very appreciated, at least during the initial stage, to overcome the expected distrust in such a new technology-supported collaboration model.

6. References

Arana, J., Berasategi, L., & Aranburu, I. (2007). Collaborative Innovation Networks Management in the Elevation Sector. eChallenges Conference, The Hague, Netherlands.

Bonfatti, F., & Monari, P. D. (2007). Putting semantic interoperability between small and micro companies into practice, eChallenges Conference, The Hague, Netherlands.

Bonfatti, F., Martinelli, L., & Monari, P. D. (2010). Autonomic approach to planning and scheduling in networked small factories, Proc. of PRO-VE '10, 11th IFIP Working Conf. on Virtual Enterprises, St. Etienne (France).

Bonfatti, F., Monari, P. D., & Martinelli, L. (2010). Proposal for a logistic domain ontology, IST Africa 2010 Conf. Proceedings, ISBN 978-1-905824-15-1 P. Cunningham and M. Cunningham Eds., IIMC Publisher.

Brock, D. L., Schuster, E. W., Allen, S. J., & Kar, P. (2005). An introduction to semantic modelling for logistical systems. *Journal of Business Logistics*.

Camarinha-Matos, L. M., & Afsarmanesh, H. (2008). Collaborative Networks: Reference Modeling, Springer, New York.

Camarinha-Matos, L. M., Afsarmanesh, H., & Ollus, M. (Eds.). (2008). Methods and Tools for Collaborative Networked Organizations. Springer.

Dekkers, R. (2010). A co-evolutionary perspective on distributed manufacturing. In: Distributed Manufacturing, Springer.

Deussen, P. (2007). Model based reactive planning and prediction for autonomic systems, Proc. of First Int. Conf. on Autonomic Computing and Communication Systems (AUTONOMICS 2007).

Di Ferdinando, A., Rosi, A., Zambonelli, F., Lent, R., & Gelenbe, E. (2008). A platform for pervasive combinatorial trading with opportunistic self-aggregation, Proc. of 2nd IEEE WoWMoM Workshop on Autonomic and Opportunistic Communication.

FInES Task Force. (2010). http://cordis.europa.eu/fp7/ict/enet/documents/task-forces/research-roadmap /fines-researchroadmap.pdf.

Fujii, K., & Suda, T. (2006). Semantics-based Dynamic Web Service Composition, International Journal of Cooperative Information Systems.
http://www-01.ibm.com/software/tivoli/autonomic/ (visited in March 2010).

Hülsmann, M., & Grapp, J., 2006. Why Do Service Logistics Need Self-Organization? - A Competence-Based Analysis of Global Supply Chains in the Movie Production Industry, in Baltacioglu, T. (ed.), Proc. of the 4th International Logistics and Supply Chain Congress.

IBM Autonomic Computing Initiative. (2010).

Lima, C., Bonfatti, F., Sancho, S., & Yurchyshyna, A. (2006). Towards an ontology-enabled approach helping SMEs to access the Single European Electronic Market in the enlarged Europe, CE2006 conference.

Msanjila, S. S., & Afsarmanesh, H. (2008). Trust Analysis and Assessment in Virtual Organizations Breeding Environments. In the Int. Journal of Production Research, Taylor & Francis publisher.

Ommeren, E., Duivestein, S., deVadoss, J., Reijnen, C., & Gunvaldson, E. (2009). Collaboration in the Cloud. Microsoft and Sogeti. ISBN 978 90 75414 24 0.

Pouly, M., & Huber, C. (2009). Collaborative networks: human aspects and corresponding IT support, Proceedings of CSCWD'09: the 13th IEEE International Conference on Computer Supported Cooperative Work in Design.

Quitadamo, R., Zambonelli, F., & Cabri, G. (2007). The Service Ecosystem: Dynamic Self-Aggregation of Pervasive Communication Services, Workshop on Software Engineering of Pervasive Computing Applications, Systems and Environments.

Romero, D., Galeano, N., & Molina, A.. (2009). Mechanisms for Assessing and Enhancing Organisations' Readiness Collaboration in Collaborative Networks. International Journal of Production Research, Taylor & Francis.

Tidd, J. (2006). A review of innovation models. Imperial Collage London.

2

Land Cover/Use Dynamics and Vegetation Characteristics in the Rural District of Simiri (Tillabery Region, Niger)

Karim Saley[1,2], M. Zaman Allah[1], Boubé Morou[2], A. Mahamane[1,2] and M. Saadou[1,2]
[1]*Université de Maradi BP Maradi*
[2]*Département de Biologie, Faculté des Sciences, Université Abdou Moumouni de Niamey, BP, Niamey Niger*

1. Introduction

Niger is a landlocked country with an economy largely dependent on the rural sector (Anonymous, 2004). In recent decades this country faced natural resources degradation especially in the Sahelian agro-ecological zones. Climatic and anthropogenic factors were the main causes of this degradation. The western part of the country, including the regions of Dosso and Tillabery (departments of Ouallam, Tillabery and Filingué) belongs largely to the Sahelian zone characterized by difficult climatic conditions (low rainfall, high temperatures, etc.). In this area, overexploitation of cultivated land and other natural resources resulted in serious environmental issues including loss of biodiversity. The pressure on agricultural soils and marginal lands was increasing due to high population growth rate. The remaining areas covered by trees or shrubs were improperly used for agricultural purposes or for domestic energy or for building shelters.

In this area the observatory network ROSELT (ROSELT / OSS, 2004) has set up an observatory called "Tahoua North Tillabéry Observatory" which included the rural commune of Simiri (Department of Ouallam). This Commune is characterized by recurrent droughts episodes with serious consequences on natural resources, the main sources of livelihood for the local population. Indeed, agriculture and animal husbandry are the main activities of the population, often leading to increased wood cut and overgrazing, adding pressure on the limited available natural resources. Measures to reverse trends are therefore needed for environmental protection and restoration as well as to ensure an ecological balance for a sustainable development in that region. However these measures require reliable information on the resources available in the commune. This study was devoted to analysis land cover/use dynamics in the commune of Simiri between 1975 and 2006 and characterize the vegetation to help for decision making on the use of these resources.

2. Material and methods

2.1 Experimental site

The rural municipality of Simiri is located in the Department of Ouallam (Tillaberi). The county town of the Commune is Simiri (14 ° 08'N and 02 ° 08E), located 70 km north of Niamey, Niger's capital. The Municipality covers an area of 2233 km ^{2}. It is bounded on the north by the urban municipality of Ouallam, south by the rural municipalities of Hamdallaye, Karma, on the east by the municipalities of Dingazi, Tondi Kandia and, on the west by the rural town of Kurthèye.

The climate is arid (Mahamane et al., 2005) with annual rainfall below 500 mm. The town is located in the western Iullimeden Basin that consists mainly of formations of tertiary Continental Terminal (Ct). Ct is a set of plateaus and hills occasionally separated by valleys. In some places, the valleys host permanent ponds. The altitude varies between 200 and 350 m on average. The south and east of the Commune were covered by dunes and constitute a serious threat for the moist valleys. The soils are of leached tropical ferruginous type.

In terms of phytogeography, Simiri is located in the South Western Sahel compartment (Saadou, 1990), characterized by:

- A substrate formed by sediments of the Continental Terminal (CT3) on lateritic plateaus, fixed sand dunes and sandy terraces occupying the bottom of dry valleys;
- South-Sahelian climate with a rainfall index ranging between 400 and 600 mm, a relative humidity between 20% (February) and 73.5% (August), a mean minimum temperature of 24 ° 35 (January) and mean maximum temperature of 33 ° 64 (April) and a thermal amplitude of 9° 29;
- A vegetation consisting of *Combretum* thickets on lateritic plateaus and steppes on sandy terraces in the dry valleys and on the fixed dunes.

The flora was composed of the following species: *Guiera senegalensis* G. F. Gmel., *Commiphora africana* (A.Rich.) Engl., *Combretum micranthum* G. Don, *Acacia macrostachya* Reich. ex Benth., *Lannea acida* A. Rich., *Croton gratissimus* Buch., *Acacia ataxacantha* DC., *Combretum nigricans* var elliotii (Engl. Ex Diels) Aubrev., *Boscia senegalensis* (Pers.) Lam. Ex Poir., *Boscia angustifolia* A. Rich. on lateritic plateaus and *Hyphaene thebaica* (L.) Mart., *Bauhinia rufescens* Lam., *Annona senegalensis* Pers., *Combretum glutinosum* Perr. ex DC. et *Acacia albida* Del. in the dry valleys. Millet, sorghum and cowpeas were the food crops used in that area.

The municipality of Simiri had 95,727 inhabitants (2005/2006) living in 74 villages. The average density was 43 inhabitants / km^2. The growth rate of this population was about 3.2% leading to high pressure on natural resources because of an increasing demand for agriculture.

2.2 Mapping methods

The map was developed based on two satellite images: Landsat MSS (Multi Spectral Scanner) 1975 and Landsat TM (Thematic Mapper) 2006 of the studied area and from the IGNN database. UTM coordinates were recorded using a GPS. Data processing was done using softwares for image processing Arcview and Env. 4.2 and Excel.

High resolution Landsat images (30 m x 30 m) covering an area of 500 km^2 were used to extract the study area which covered 2233 km^2 using the satellite imagery processing software ENVI 4.2.

The geographic coordinates of selected points, chosen in the different defined units, were loaded onto a CX Garmin GPS receiver to facilitate their identification. At each of these points, observations were recorded using forms established for this purpose. The coordinates were then used on the satellite images to generate thematic maps including that of land use with the software ArcView 3.2 ArcView GIS 3.2 (Environmental System Research Institute, Inc. 1992-1999).

A table of attributes associated with each class of land cover unit was generated with Arcview, using a stratification of the different land cover units. Assisted classification was used for the digitization of images which were divided into homogeneous zones on the basis of the classification of land cover commonly used as reference for land use mapping. This was followed by a check using field data to make corrections wherever needed on the map.

2.3 Data collection

Data were collected in three units of land occupation: tiger bush, degraded fallow and cultivated lands (ROSELT, Saadou, 2005). Floristic surveys were conducted in plots of 1000 m^2 (20 m * 50 m). In each plot, a complete census of the species was made using a survey form on which was indicated the presence of each species found in the plot. The collected data were then gathered in an absence-presence table of species.

These data were used to determine:

- Systematic diversity including the number of species (or species richness), genus and families;
- Biological types identified according to the classification of Raunkiaer (1916), adapted to Trochain (1966, *In* Saadou, 1984) that was used by several authors in tropical Africa (Saadou, 1984 , 1990; Mahamane, 2005; Oumorou, 2000).
- Phytogeographical types
- Shannon-Weaver diversity index (α) that provides information on the diversity of species in a community and gives an indication on the relationships between species within the community. It is calculated using species richness of plant groups:

$$H' = \sum_{i=1}^{S} pi \log 2pi$$

With Pi= ni/S; ni: frequency of the species i in the group and S: total number of species

- Equitability index of Piélou:

$$E = H' / \log(S)$$

- Diversity β based on floristic communities: index (or coefficient) of Sorensen:

$$K = 2\,C/(A + B)$$

with A = number of species on list a;
B = number of species on list b;
C = number of species common to list a and b.

The collected data were analyzed using Detrended Correspondence Analysis (P. Legendre and Legendre L., 1999) with PC-Ord 5 (McCune and Mefford, 2006) software.

3. Results

3.1 Dynamics of land cover/use

Four maps were developed; two on land cover/use in 1975 and 2006 and two on the status of the tiger bush in 1975 and 2006.

Land cover units in 1975

The distribution of tiger bush, cultivated areas, riparian cords and ponds are shown in Fig. 1a. The tiger bush consisted of a regular tiger bush and a degraded tiger bush. The regular tiger bush is a well-structured unit consisting of wooded and denuded strips (Ambouta, 1997). It covered an area of 90.373 hectares or 34.06% of the total area of the Commune. The degraded tiger bush derived from the regular tiger bush, with degraded structure as a consequence of anthropogenic and climatic effects. It covered an area of 68.738 hectares or 25.91% of the total area of the municipality.

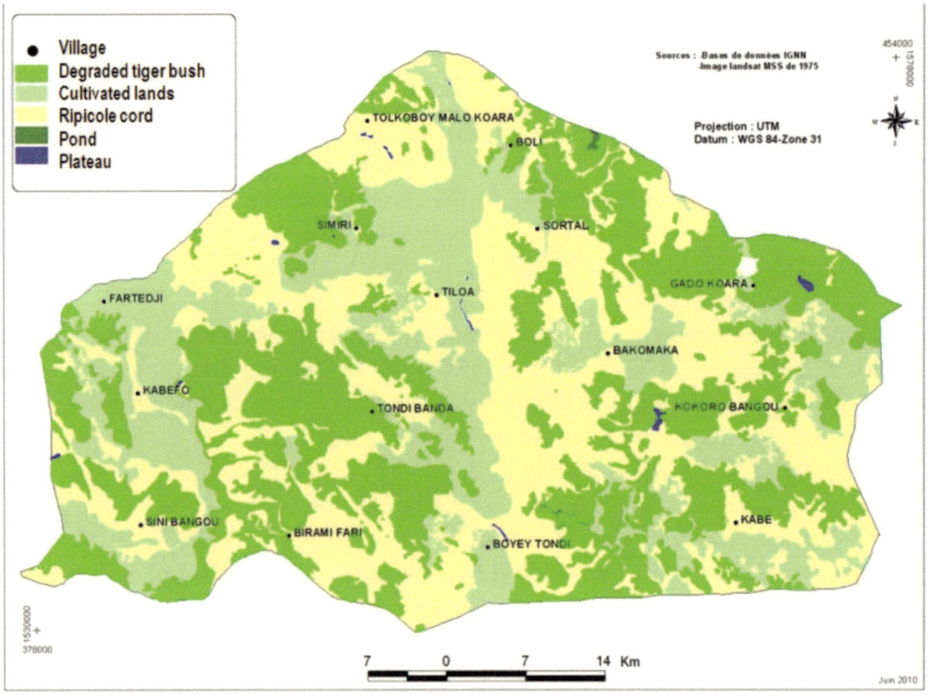

(a)

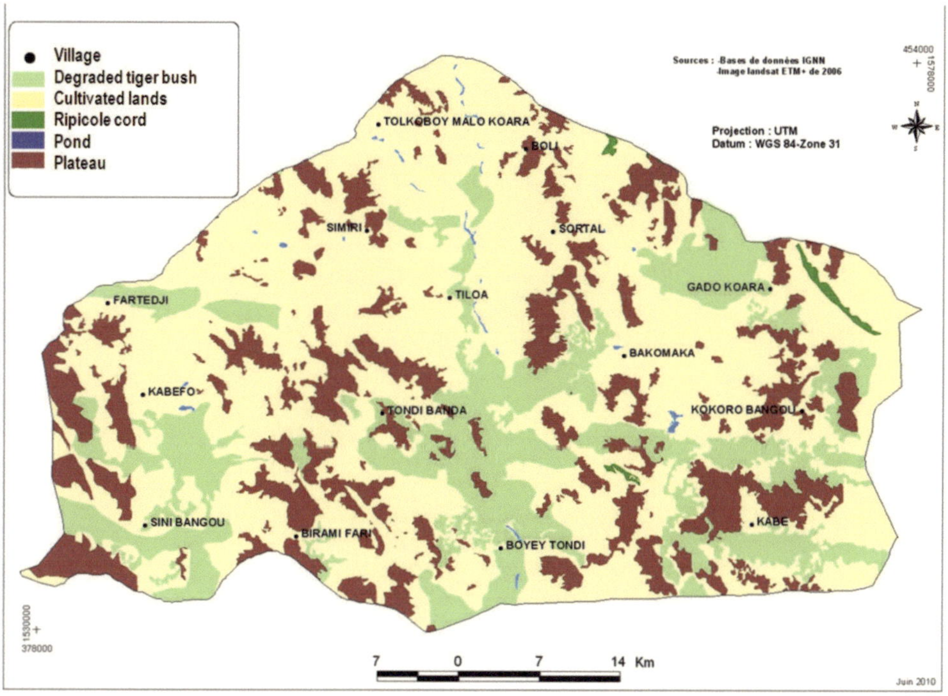

(b)

Fig. 1. Land cover in (a) 1975 and (b) 2006.

Regarding rainfed cultivated areas they were located on dunes, sandy skirts and lowlands and covered 105.700 hectares or 39.84% of the area of the municipality. Riparian Cords and ponds represented respectively 0.06 and 0.11% of the Commune area.

Land cover units in 2006

In 2006, the land cover units were the degraded tiger bush, rainfed cultivated lands, ripicole cords and ponds Fig. 1b. The regular tiger bush disappeared completely. The degraded tiger bush covered 55.691 ha or 20.99% of the total land cover units. Compared to its surface in 1975, there was a reduction of 13.07%. Rainfed cultivated lands accounted for 59.64% of the total area of the municipality. This corresponded to an increase of 19.79% compared to 1975. Riparian cords covered 807 ha representing 0.30% of the total area of land cover units with a slight increase of 0.24% compared to 1975. By contrast, the surface covered by Ponds did not show any significant change between 1975 and 2006.

On the other hand, the plateau appeared in 2006 and consisted of bare soil. This unit covered 18.97% of the Commune. It was located on the lateritic plateau where the tiger bush was replaced by bare soil due to vegetation loss and soil erosion.

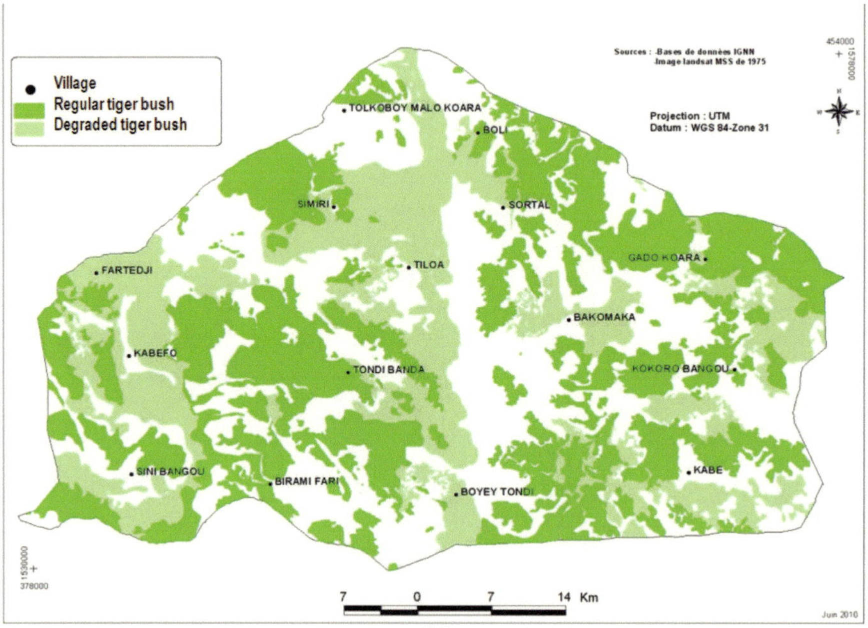

(a)

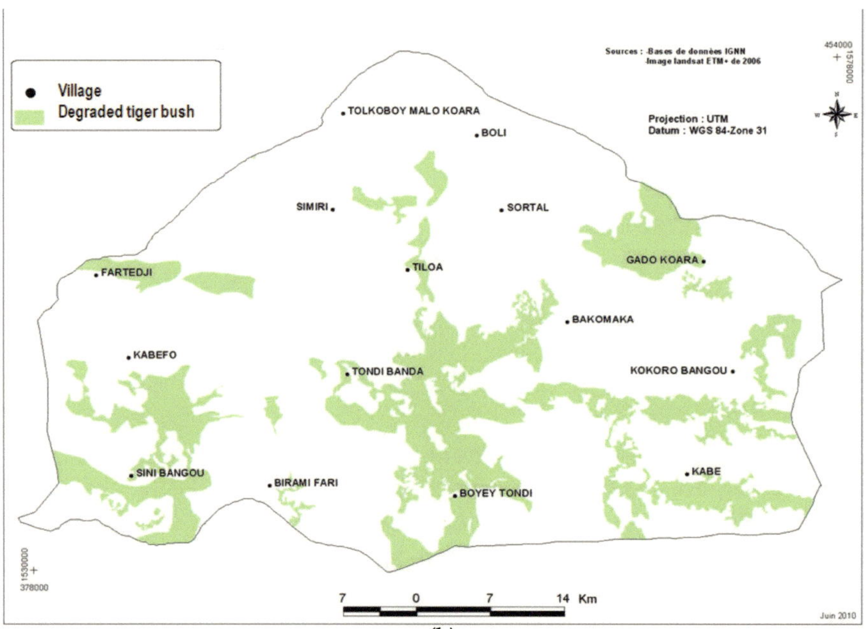

(b)

Fig. 2. Tiger bush status in (a) 1975 and (b) in 2006.

3.2 Dynamics of land cover between 1975 and 2006

The diachronic analysis of land cover units between 1975 and 2006 showed spatial changes in the unit types (Fig. 2 & 3). The regular tiger bush and the degraded tiger bush were both present in 1975. But due to human pressure and drought events (1973 and 1984), the regular tiger bush was completely transformed into a degraded tiger bush Fig. 2. In addition, the area covered by the degraded tiger bush was significantly reduced and in some places replaced by denuded soil.

Cultivated lands were extended as a consequence of both the increase of population and soil degradation. The surface of riparian cords was increased while that of ponds decreased.

Riparian cords covered a larger surface in 2006 than 1975, probably due to important runoff following the destruction of mainly the vegetation of plateaus where drainage into the lowland by several "koris" occured.

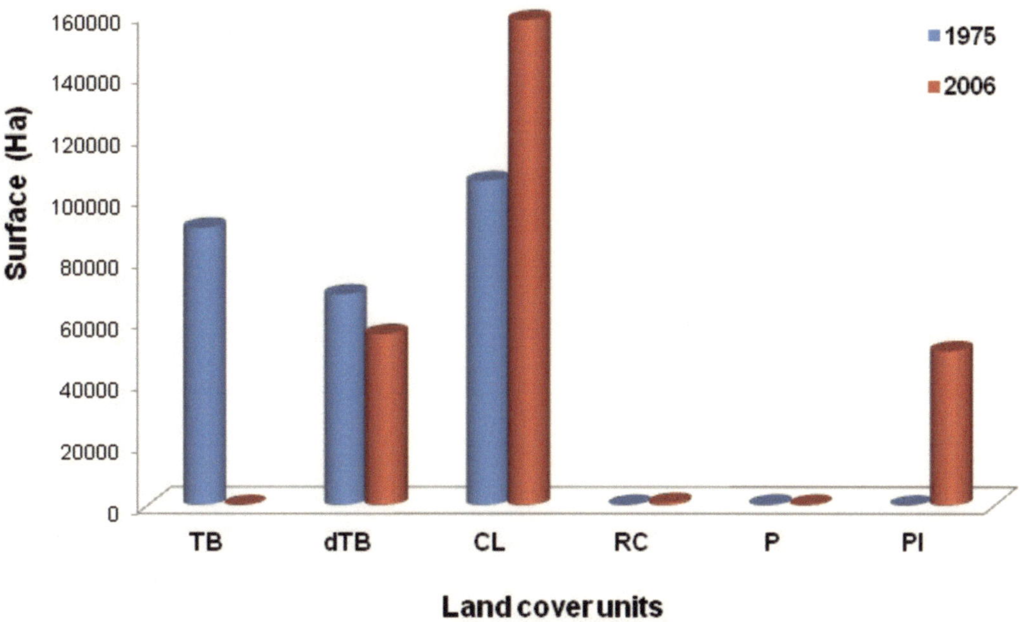

Fig. 3. Evolution of land cover units between 1975 and 2006.
TB: tiger bush; dTB: degraded tiger bush; RP: riparian cord; CL: cultivated lands; P: pond, Pl: denuded plateaus.

3.3 Characteristics of vegetation

Data ordination

A matrix of 38 records associated with 103 species was subjected to detrended correspondence analysis, which pointed out four groups (Fig. 4). Axis 1 contrasted group GI (consisting mainly of records associated with the rainfed agricultural lands) and GII (consisting of records relating to fallows). Axis 1 was therefore a gradient of land use or anthropogenic impact. From a geomorphological point of view these records concerned the sandy skirt, one of the three geomorphological units in Western Niger (Courault et al., 1996).

The axis 2 contrasted groups GIII and GIV. GIII was composed of records located on the sandy skirt while GIV was composed of those on lateritic plateau, an area of the tiger bush (Ambouta, 1984). Axis 2 was therefore a topographic gradient.

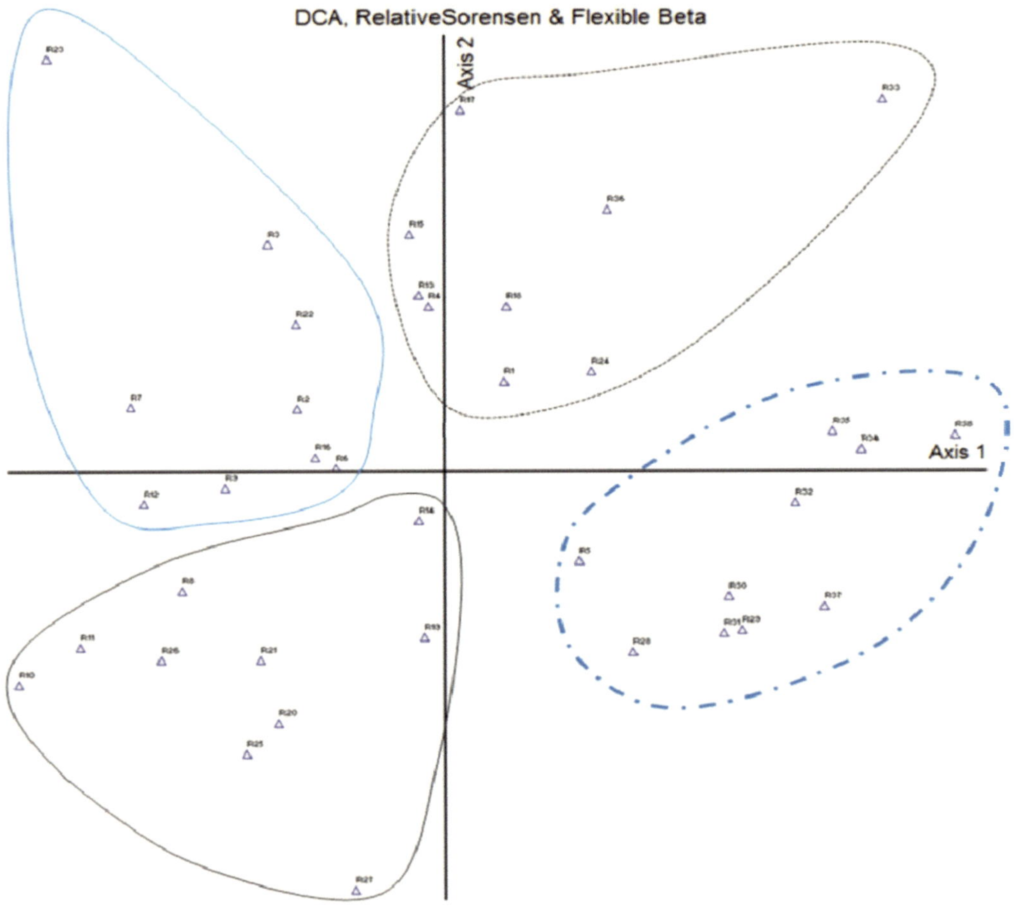

Fig. 4. Detrended Correspondence Analysis (DCA) diagram (axis 1 and 2).

Systematic diversity

In total 113 species were recorded. They were divided into 73 gender and 28 families. The most represented family was Poaceae with 24 species (21.24%) followed by Fabaceae with 18 species (15.93%). The least represented family had a single species (0.88%) (Table 1).

Family	**Nb of species**	%
Poaceae	24	21,24
Fabaceae	18	15,93
Convolulaceae	8	7,08
Combretaceae	6	5,31
Mimosaceae	7	6,19
Amaranthaceae	5	4,42
Cucurbitaceae	4	3,54
Rubiaceae	4	3,54
Acanthaceae	3	2,65
Caesalpiniaceae	3	2,65
Capparaceae	3	2,65
Commelinaceae	1	0,88
Caryophyllaceae	3	2,65
Cyperaceae	3	2,65
Euphorbiaceae	3	2,65
Malvaceae	3	2,65
Pedaliaceae	2	1,77
Sterculiaceae	2	1,77
Tiliaceae	2	1,77
Annonaceae	1	0,88
Araceae	1	0,88
Asclepiadaceae	1	0,88
Asteraceae	1	0,88
Balanitaceae	1	0,88
Lamiaceae	1	0,88
Molluginacae	1	0,88
Rhamnaceae	1	0,88
Schrophulariaceae	1	0,88
Total	113	100,00

Table 1. Repartition of species per family.

Biological types

The flora of all the studied stations was dominated by Therophytes. The dominance ranged from 52.94% (GII) to 70.15% (GI). They were followed by microphanerophytes regardless of station. The latter had the highest rate in fallow (20%) and lowest in the cultivated areas (11.94%).

Phytogeographical types

The distribution of plant species or phytogeography is presented in Table 2. Species of Guineo-Congolian-Sudano-Zambezian (29.17% to 35.82%) and the Sudano-Zambezian

(26.47 to 37.80%) distribution were dominant regardless of the station. However, those with Sudano-Zambezian distribution were absent on plateaus. These two phytogeographical types were followed by Guineo-Congolian-Sudano-Zambezian-Sudano-Zambezian-Saharo-Sindian and Sudano-Zambezian-Saharo-Sindian types in the first three stations. By contrast, on plateaus introduced species were dominant (31.25%). This was due to restoration activities carried out in that area.

	Biological types (%)									
Stations	CH	H	**Hé**	**Hy**	**LCH**	**Lmp**	LT	mp	**np**	T
GI (Cultivated lands)	2,99	2,99	-	-	-	-	7,46	11,94	4,48	70,15
GII (Fallows)	2,94	4,41	1,47	2,94	1,47	1,47	8,82	20,59	2,94	52,94
GIII (Sandy skirt)	2,44	3,66	1,22	-	2,44	-	9,76	12,20	3,66	64,63
GIV (Plateaus)	4,17	6,25	-	-	-	-	8,33	14,58	6,25	60,42

CH: Chamephytes, H : Hemicryptophytes, Hy : Hydrophytes, LCH : Liana chamephytes, Lmp : Liana microphanerophytes, LT : Liana Therophytes, mp : Microphanerophytes, np : Nanophanerophytes , T: Therophytes.

Table 2. Distribution of biological types (%).

	Phytogeographical types (%)				
Stations	GC-SZ	GC-SZ-Sah.S	i	SZ	SZ-Sah.S
GI (Cultivated lands)	35.82	11.94	1.49	35.82	14.93
GII (Fallows)	32.35	20.59	1.47	26.47	19.12
GIII (Sandy skirt)	30.49	17.07	-	37.80	14.63
GIV (plateaus)	29.17	20.83	31.25	-	18.75

GC-SZ: Guineo-congolese-soudano-zambesian ; GC-SZ-Sah.S : Guineo-congolese-soudano-zambezian-saharo-sindian; i : introduced species ; SZ : Soudano-zambezian; SZ-Sah.S : Soudano zambezian-saharo-sindian.

Table 3. Repartition of phytogeographical types (%).

Alpha diversity, Pielou equitability and specific richness:

Results are presented in Table 4. The diversity index was high regardless of the station. This indicated that all stations were diverse. Indeed, the diversity index varied from 5.17 (plateaus) to 6.02 bits (sandy skirt).

The equitability index ranged from 0.70 to 0.75, confirming the floristic richness of the groups, in other words, there is no dominance of one or few species over others. This index was identical between fallow (0.75) and skirts sandy (0.75) and similar between cultivated lands (0.71) and plateaus (0.70). Unlike the Shannon-Weaver index, equitability index was not dependent on the floristic richness of the station.

The floristic richness (S) was identical between the first two groups (68 species). It was weak on plateaus (48 species) and high in the GIII group. This was explained by the presence of lowland record in the GIII group, where the ground water conditions were more favorable to the development of various species.

Stations	H'	E	S
GI (Cultivated lands)	5.68	0.71	68
GII (Fallows)	5.69	0.75	68
GIII (Sandy skirt)	6.02	0.75	83
GIV (Plateaus)	5.17	0.70	48

Table 4. Diversity index (H '), equitability index (E) and floristic richness (S) per station.

Bêta diversity

β diversity was analyzed by calculating the coefficient of Similarity of Sörensen. This coefficient was relatively high between GI and GII (0.54) and between GIII and GIV (0.51). This indicated a similarity between the cultivated lands and fallows and between the sandy skirts and plateaus. This index was however weak between fallows and plateaus, indicating that these stations do not share the same floristic characteristics.

Stations	GI (Cultivated lands)	GII (Fallows)	GIII (Sandy skirt)	GIV (Plateaus)
GI (Cultivated lands)	**0**			
GII (Fallows)	**0,54**	**0**		
GIII (Sandy skirt)	**0,43**	**0,38**	**0**	
GIV (Plateaus)	**0,49**	**0,31**	**0,51**	**0**

Table 5. Bêta diversity indices.

4. Discussion

Drivers of land cover dynamics

The factors governing landscape dynamics were anthropogenic and climatic conditions. Indeed, the population of the municipality has one of the highest population growth rates in the world: 3.2% (RGP / H, 2001). This population growth resulted in new needs like more agricultural lands and wood with subsequent negative impact on the natural resources of the region. This pressure was coupled with the effects of climate change that resulted in recurrent droughts and heavy runoff leading to ravine, sandy skirts units and the silting of lowlands and pools. As consequence, soils were degraded and the income of the population in the commune was significantly decreased.

Determinants of plant diversity

The flora of the Commune was dominated by therophytes (annual species) and nanophanerophytes. According to Mahamane et al. (2007), annual species are sensitive to changes in rainfall. The nanophanerophytes were found in fallow, tiger bush and lateritic plateaus. Despite the various constraints including climatic conditions in that area, the flora seemed to be significantly diverse as indicated by diversity and equitability indices. This is an indication of the adaptation of local of species to climate change and human pressures. Indeed, the population has no other business than exploiting natural resources (energy, timber, medicines, sale of dry grass to the city of Niamey). The proximity of Niamey as big market is one reason why trees were often cut green, resulting in destruction of natural vegetation of plateaus and fallows.

5. Conclusion

This study showed that the dynamics of land cover in the rural area of Simiri was governed by both anthropic activities and climate change. A fragmentation or even absence of some cover units like regular tiger bush was recorded. However a well adapted flora to both anthropic and climate change pressures appear to remain in the various studied biotopes. Restoration activities with introduced species showed significant positive impact; however, the use of local species better adapted to local conditions would provide better results.

6. References

Anonyme, 2004. *Stratégie de Développement Rural* (SDR). Cabinet du Premier Ministre, Secrétariat Permanent de la SRP. 75 p.

AberlinJ-P., Daget P., 2003 Etablir et comparer les spectres biologiques de plusieurs groupements végétaux. *Rev Elev. Med. Vet. Pays trop.* 56 (1-2) : 57-61.

Ambouta J.M.K., 1997. Définition et caractérisation des structures de végétation contractée au Sahel : cas de la brousse tigrée de l'ouest nigérien. *In* D'Herbes J.-M., Ambouta J.-M.-K. et Peltier R., 1997. Fonctionnement et gestion des écosystèmes contractés sahéliens. Ed. John Lybbey Eurotext, Montrouge, France, 274 p.

Courault D., D'Herbes J.-M. et Valentin C., 1996. *Le bassin versant de Samadey*, premières observations pédologiques et phytoécologiques. HAPEX-SAHEL, 31p. *de l'Ouest de la République du Niger, de la longitude de Dogondoutchi au fleuve Niger.*

Legendre P., Legendre L., 1999. *Lexique anglais-français d'écologie numérique et de statistique.* Montréal ULaval.

Mahamane A., 2005. *Etudes floristique, phytosociologique et phytogéographique de la végétation du parc régional du W Niger.* Thèse présentée á l'Université Libre de Bruxelles pour obtenir le titre de Docteur, 516 p.

Mahamane A., Ichaou A., Ambouta J.M. K., Saadou M., Morou B., Amani I., Hango M., D'Herbès J. M., Gineste P., Wata I., Abassa I., 2007. Indicateurs écologiques de la période optimale de remise en culture de jachères au Niger. *Sécheresse,* 18 (4) : 289-95.

Mahamane L., Saadou M., Nonguierma A., 2005. Détermination du degré d'aridité bioclimatique de sept localités du département de Tillabéri (sud-ouest du Niger) : classement en zones bioclimatiques. *Sécheresse* 16 (2) 107-114.

McCune B., Mefford M.J., 2006. PC-ORD. *Multivariete Analysis of Ecological Data.* Version 5, MJM software Gleneden Beach, Oregon.

Oumorou M., 2000. *Etudes écologique, floristique, phytogéographique et phytosociologique des inselbergs du Bénin.* Thèse de doctorat, Université Libre de Bruxelles, Faculté des Sciences, Laboratoire de Botanique Systématique, 210p.

ROSELT/OSS, 2004. Conception, organisation et mise en œuvre de ROSELT/OSS. Collection ROSELT/OSS, document scientifique n°1, 112 p.

Saadou M., 1984. *Contribution à l'étude de la flore et de la végétation des milieux drainés de l'ouest de la République du Niger, de la longitude de Dogondoutchi au fleuve Niger.* Thèse de 3e cycle, Univ. de Bordeaux II, 177 p.

Saadou M., 1990. *La végétation des milieux drainés nigériens à l'Est du fleuve Niger.* Thèse de Docteur ès Sciences Naturelles, Université de Niamey, Niger, 1990 ; 395 p. + annexes.

3

Organizational Ecosystems: Interaction and Alignment Towards Innovation

Adalberto Mantovani Martiniano de Azevedo, Ana Karina da Silva Bueno, Antonio José Balloni, Moacir de Souza Dias Neto and Marco Antonio Silveira
Center for Information Technology Renato Archer
Brazil

1. Introduction

The decrease in the time required for the introduction of innovations in the marketplace, the increasing technological complexity of these innovations, the requirement of diverse competencies for its development and the high costs implied in the introduction of new products and processes in the marketplace have made unviable the single development of innovations in any public or private organization individually, especially those small and medium-sized (De Pellegrin *et al* 2007). Because of this, multiorganizational arrangements have often had the role of joining competences and resources of heterogeneous organizations aiming to develop conjointly costly and complex innovations. In most cases, the organization of these arrangements are integrated to sectoral development policies headed off by State entities, who consider the introduction of innovations as a safe path to competitive insertion of regional and national industries in the markets of high added value products and with the potential of generating high-level jobs.

However, the constitution of these arrangements is not a magic formula that assures the success of innovative efforts. Firstly, in order to meet the goal of joining the different skills required for developing complex innovations, the multiorganizational arrangements must be composed of heterogeneous institutions. These heterogeneous kinds of arrangements generate very frequently communication and integration difficulties,[1] whose overcoming depends on effective mechanisms for coordination. Secondly, often such arrangements are limited in terms of participating institutions, fact that limits the possibilities of induction of a complete technological transfer project, generator of innovation, for which the participation of a wide range of organizations that fulfill specific functions in the process

---

[1]A very commonly cited example of this problem is the integration between universities and private companies, made difficult by factors such as the definition of intellectual property rights, lacks of communication, inadequacy of human resources, funding inconstancy, sociocultural conflicts and differences in acceptable deadlines in project execution (Rapini, 2007). In Brazil, such difficulties reflect in a science and technology system with respectable performance in scientific production, however generating few innovations in private companies (Velho, 2004; Lotufo, 2009).

is essential (such as technological, regulatory and marketing functions). In most of the cases,[2] to complete an innovative process, the participation of private sector organizations, research institutes and government agencies are required.These institutions as a whole are the constituting elements of organizational ecosystems related to an industry or a technology. In view of this problem, this chapters goal is to describe the Brazilian National Institute of Science and Technology in Micro and Nanoelectonic Systems (INCT/NAMITEC), showing its main indicators and forms of network coordination, focusing on the activities of the Coordination of Knowledge Transfer to the Productive Sector of INCT/NAMITEC (Coordination A.7.). Hereby are also presented the main advances and problems in terms of multiorganizational cooperation and transference of technological knowledge in the area of microelectronics to the productive sector and to society.

The remainder of this chapter is organized as follows. Section 2 presents selected approaches regarding concepts of multiorganizational arrangements towards innovation: techno-economic networks, innovation systems, triple helices and organizational ecosystems. Section 3 presents INCT/NAMITEC, emphasizing its main management instruments, results and limitations towards technological transfer promotion. Section 3 ends with a proposal of a multiorganizational arrangement management model based on the concept of organizational ecosystems developed within the activities of Coordination A.7. Section 4 enriches the previous discussion with the participatory approach to organizational ecosystems. And finally, section 5 presents the findings of the chapter: an efficient management of multiorganizational arrangements towards innovation must promote the integration of several organizations that compose the organizational ecosystem and that carry out essential functions in the innovative process.

2. Multiorganizational arrangements to technological and innovation transfer: Techno-economic networks, innovation systems, triple helices and organizational ecosystems

Several denominations are given to multiorganizational arrangements directed to technological transfer for innovation. Amongst the most utilized ones, here will be described the concepts of techno-economic networks (Callon, 1992), national, local and sectorial systems of innovation (Nelson, 1993; Cassiolato e Lastres, 2000; Malerba, 2002; Hekkert *et al*, 2008) and Triple Helix (Leydesdorf and Etzkowitz, 1996; Etkowitz e Leydesdorf, 2000). These concepts define arrangements deliberately organized to join the efforts of the public and private organizations, aiming towards the development of one or more innovations. They frequently inspire innovation policies based on the organization and coordination of multiorganizational arrangements:

"In several countries, the technological policies have emphasized cooperation programs between the public and private sectors to stimulate and support the company's efforts, reduce risks and maximize the results of the scientific training built locally. These efforts, besides

---

[2]The nature of the institutions involved in the innovative process (ecosystem) depends on the particular characteristics of the artifacts and institutions involved, such as sector, localization, technologic complexity, ethical and legal aspects, among others.

encouraging partnerships between universities, research institutes and companies, are also oriented towards a larger interaction between the companies themselves, either as research "cooperation networks", shared centers, common infrastructure, or by mean of explicit politics of support in arrangements and local systems of innovation." (Lotufo, 2009: 42).

The network concept is utilized in various fields of science to describe complex systems constituted of diverse components (Börzel, 1997). In social sciences, stand out the studies about industrial networks, business management networks and public policies networks. In the field of science, technology and innovation studies, Börzel (1997) considers seminal Michael Callon's article, *The Sociology of an Actor Network; the Case of the Electric Vehicle*, published in the book *Mapping the Dynamics of Science and Technology: Sociology of Science in the Real World.*

In a posterior article, Callon (1992) introduces the concept of techno-economic network, describing the components of the technological and innovation transfer networks as being formed by several heterogeneous actors belonging to three poles: a scientific pole, producer of knowledge; a technological pole, oriented towards application of knowledge; a market pole (companies and customers), where the innovation spreading is accomplished . Between these poles are intermediaries that mediate the interactions between the actors, such as scientific and technological documents (articles, patents), competencies and capabilities (that circulate, for example, through courses and professional mobility), financial resources (funding, sales), as well as artifacts (scientific and technological equipment). Callon (1995) proposes a typification that differentiates convergent and divergent networks: a convergent network is one where the actors present a consensus on what actions should take place (which technology to adopt, for example); on the other hand, in a divergent network there is no established consensus towards what actions should take place, coexisting different opportunities of action, fact that makes it harder to align the actors. In general terms, the convergent networks are built around technologies with well stable and consolidated trajectories; divergent networks are typically those around new technologies, technologies upon which rest great uncertainties regarding the trajectory to follow, thus requiring a more intense negotiation process between the actors, based on the exchange of intermediaries, therefore turning the technological transfer process into a complex and uncertain one.

Corallo and Protopapa (2007) refer the concept of innovation networks to the studies of Manuel Castells, organizer of the book *The Rise of the Network Society*, of 1996. Castells considers the private company the central *locus* of the transfer of technology on innovation networks. These networks are nothing more than a form of organization adequate to environments of high degrees of uncertainty, where it is necessary to change from a vertical governance structure (concentrated on individual companies) to a horizontal structure (dispersed amongst diverse organizations). This new form of governance, facilitated by the technologies of digital communications, has as its main characteristics the exploration of complementarities and collaboration between organizations, pointing organizations towards processes of co-evolution nurtured by the exchange of knowledge and initiated by a common characteristic (geographical, institutional, ideological or technological) that approaches organizations around common interests.

De Pellegrin *et al* (2007) proposes a network management model named Rede de Inovação Horizontal Induzida (Horizontally Induced Innovation Network) (RIHI). In a RIHI, the Government and/or a group of companies[3] develops action plans to enhance or develop the cooperation between organizations in a sector, aiming their convergence in the innovation process. To reach this objective, it is necessary to create an organization whose objectives and structure are decided by the participating actors, including a coordination center that harmonizes the network's member's different objectives, keeping in sight the companies objectives, considered by the actors the *locus* of innovation. It is up to this coordination center to recruit the members of the network, articulate the cooperation and technological transfer relations (utilizing, for example, calls for cooperative research projects between universities and companies), organizing information of interest (such as market prospection), acting on environmental factors (promoting actions focused on specific markets), creating learning mechanisms (like courses and trainings) and acting along with supporting institutions (facilitating technological service provision, such as certification/accreditation, for example). This way it is up to the network coordination center to reduce the risk of companies engaging in a collaborative project of technological innovation through prospection of opportunities and promotion of collaboration between companies, as well as between companies and other institutions.

The concept of Innovation Systems,[4] according to Edquist (2001), was introduced by three authors: Christopher Freeman in 1987, in the book *Technology Policy and Economic Performance: Lessons from Japan*; Bengt-Aake Lundvall, who in 1992 organized the book *National Systems of Innovation: Towards a Theory of Innovation and Interactive Learning*; and Richard Nelson, who organized in 1993 the book *National Innovation Systems: a Comparative Study*.

A Sistem of Innovation (SI) is composed by all the economic, social, political and organizational elements that have an influence on the development, diffusion, and utilization of innovation, generated through the learning process that occurs in interactions between institutions that exchange knowledge and technology (Edquist, 2001). The main components of a SI are the organizations and institutions. The organizations are the components consciously created by the actors, with explicit objectives (companies, educational and research institutes and public promoting agencies, for example). The institutions comprehend a set of routines and rules that regulate the interactions (marketwise or not) between the organizations (for example: rules of intellectual property, ethics codes, commercial laws, technological transfer contracts, etc).

The innovation induced by the creation of interinstitutional collaborative arrangements that manage the virtuous cycle of learning and technological transfer can be facilitated by the formation of sectorial innovation systems (Malerba, 2002). These Si's include diverse components with specific functions (Hekkert *et al.*, 2007): development and diffusion of

---

[3]The author also considers that induction can be achieved by an anchor firm (*Top down* network). In the example cited in the article (Petro-RS Network), the anchor company is Petrobras, who coordinates a network of suppliers and laboratories.

[4]A system in formed by components and by the relationship between the components. It is always a simplification of reality, due to the practical necessity in defining the components and system limits in an arbitrary and idealized manner (Edquist, 2001).

knowledge; influence on the direction of new technology searches; promotion of experimental ventures; market formation; human and financial resource mobilization; legitimating of new technologies and generation of positive externalities. Thus, the definition of functions of a SI is a management instrument that helps to define the attributions of their components, as well as controlling the execution of these functions.

The concept of Triple Helix is an explanation model of the multiorganizational innovation process presented in 1995 by Henry Etzkowitz and Loet Leydesdorf in the article *The Triple Helix of University-Industry-Government Relations: A Laboratory for Knowledge Based Economic Development*[5] (Leydesdorf and Etzkowitz, 1996). The model points out the role of universities as the central agent of the innovative process in knowledge-intensive economies, role induced by institutional innovations created by the government that pursues approaching universities to companies[6] (Etzkowitz and Leydesdorf, 2000).

Therefore are identified the three helices of the model: universities, companies and government. The relations between the helices can undertake three forms (Etzkowitz and Leydesdorf, 2000): the triple helix I, characterized by government control over universities and companies; triple helix II, in which the institutions are sharply separated and where relationships are circumscribed to eventual contracts (liberal model, counterpoised to triple helix I); and triple helix III, represented by the juxtaposition of the three spheres, which means that each one of the spheres assume roles traditionally attributed to the others: for example, the university leading multiorganizational arrangements, the government producing or consuming goods and companies generating scientific and technological knowledge in their field of interest. Simplifying, the triple helix model searches to understand the relationships between these three actors, that combine and recombine themselves constantly in forms that adapt to the conditions of human resources, financial constraints and technology transfer.

The concept of organizational ecosystems refers to the tradition of economists of several schools of thought that stresses the similarities between biological evolution and economic development (Corallo and Protopapa, 2007).[7] The authors attribute the first utilization of the term "business ecosystem" to the article *Predators and Prey: A New Ecology of Competition*, published by James F. Moore in 1993 in the *Harvard Business Review*. According to Moore, a firm is not only a member of a specific industry, but part of a business ecosystem that involves several industries, where the capabilities evolve conjointly around a set of technologies produced by these firms.

---

[5]The article was published in the *European Association for the Study of Science and Technology Review*, v.14, n.1, 1995, p. 11-19.

[6]In the authors points of view, the inclusion of universities as agents of economic development (the "third mission" of the academy) represents a academic revolution comparable to that occurred in late XIX century, when research ("second mission") was introduced as a complementary academic mission to education ("first mission").

[7]Corollaro and Protopapa (2007) observe two problems in the use of biological analogies in economics: firstly, the biological evolution does not have goals as well defined as economic development; secondly, complex species do not combing through crossbreed; in the other hand, economic systems very often combine artifacts and competencies. Even though these limitations, the authors consider that this does not invalidate biological ecosystems as a metaphor to the comprehension of multiorganizational networks.

Kay *et al* (1999) define organizational ecosystems as complex systems (non explainable by linear relations of causality), ecosystems called by the authors as Self-Organizing Holarchic Open Systems (SOHO).[8] This systems are characterized by permanent interaction between their components, by flexible hierarchies and by constant reconfiguration of one organizational state to the other. This instability requires and promotes a constant and adaptative learning, fundamental in rapidly changing environments composed by heterogeneous institutions, such as innovative organizational ecosystems.

The concept of SOHO organizational ecosystems considers the idea of managers as omniscient coordinating agents to be a negative factor for the sustainability of organizational ecosystems, since it creates a inertia harmful to dynamic learning and doesn't take in regard the need for adaptive learning, monitoring and constant adjustments of the structures of governance. In this manner, the efficient management of organizational ecosystems requires the constant identification of changes in the environment and available resources, planning interventions that facilitate the system's self-organization. Depending on the particular characteristics of each actor, these interventions must be performed taking into consideration the forces that give cohesion to the group. These are called on by the actors as attractors,[9] and they are nothing more than the objectives and aspirations of the components of an ecosystem that are responsible for its maintenance in a specific domain of activities. Therefore, the creation or maintenance of attractors is an instrument to maintain or change the state of an ecosystem (attractors to stimulate scientists focused on academic production to dedicate resources and time to entrepreneurial activities, for instance).

The SOHO organizational ecosystems have as an inherent characteristic the uncertainty of the motivations and behavior of the actors, thus resulting in the impossibility to accomplish any "anticipated" management system, capable of predicting the decision's consequences. The solution presented by Kay *et al* (1999) is to utilize management schemes that take under consideration the possibilities of complex systems, characterized by the following elements: 1. The actors and their context; 2. The hierarchical characteristics of the system; 3. The attractors that delimit the "orbit" of the system, and how actors respond to the attractors; 4. the entries and exits of information and resources that organize the ecosystem around the attractors, concerning attractive forces as well as forces of repulsion.

With this concept in mind, the authors propose a manual to plan and manage organizational ecosystems, composed by stages of action divided into subtasks (table 1).

The concepts here presented are quite functional as conceptual and methodological guides in implementing multiorganizational arrangement management models for technological transfers towards innovation, and coherent with the approach of research-action that motivated the elaboration of this chapter. The next item will present an organizational ecosystem management model that will absorb the contributions of this approach, illustrating the proposition with the experience of Coordination A.7.: Knowledge Transfer to the Public Sector within INCT/NAMITEC.

---

[8]*Self-Organizing Holarchic Open Systems.* The term holarchic designates systems where the components are, simultaneously, a part and the whole system, in continuous interaction.

[9]The term attractor is an analogy to the forces of gravity, which maintain the planets circumscribed to their orbits.

A. Characterization of the ecosystem
1. Definition of the analysis perspective (economical, management, cultural) and scope (limits of description).
2. Definition of the processes that define the ecosystem.
3. Definition of the actors and of their vertical and horizontal relationships.
B. Description of the ecosystem as a self-organizing entity.
1. Characterization of the attractors associated to the existent organizational states.
2. Assessment of development trends of the ecosystems
3. Evaluation of the reactions of members of the ecosystems to different attractors.
4. Identification of possible changes in the attractors.
5. Characterization of resource and information flows (entry and exit).
6. Identification of synergic relations and characterization of resource swaps.
C. Evaluation of the ecosystem's sustainability as a function of the proposed goals and objectives.
1. Identification of acceptable states of the ecosystem as a function of the proposed objectives.
2. Identification of the economical, technical, political, etc. processes necessary to fulfill the proposed objectives.
3. Identification of inacceptable attractors, desirable actors and possible tradeoffs involved in the choosing of the actors.
D. Maintenance of the integrity (sustainability) of the ecosystems.
1. Identification of actions to mitigate threats to the integrity of the ecosystems.
2. Identification of actions to promote positive attractors.
3. Definition of ecosystem monitoring actions to detect changes.
E. Actions to deal with the complexity of the ecosystems.
1. Elaboration of anticipative management schemes that allow adaptation to changes based on organizational learning.

Source: Adapted from Kay *et al* (1999)

Table 1. Step-by-step for planning/management of SOHO organizational ecosystems.

3. INCT/NAMITEC's management model: Proposal of a knowledge transfer model to the productive sector

The Ministry of Science, Technology and Innovation (MCTI) in Brazil has been using the scientific-technological network model frequently in the application of resources towards fomenting Science &Technology activities. Between 2001 and 2009, eleven public calls for research projects of MCTI were facing specifically the promotion of this type of arrangements (SIGCTI, 2011). Aligned with this model, the MCTI Ordinance n° 429/2008, that established the National Institutes of Science and Technology Program (INCTs), clearly shows that the management model adopted by the INCTs is based on the concept of science and technology network:

Art. 2nd. The National Institutes will be formed by a host institution, characterized by the excellence of its scientific and/or technological production, high qualification in the formation of human resources and with the capacity of leveraging resources from other sources, and by a set of laboratories or associated groups to other institutions, articulated **in**

the form of scientific-technologic networks. (MCT Ordinance n° 429/2008, author highlight).

An example that illustrates the possibilities of multiorganizational cooperation in microelectronics is the explanation given by Saxenian (1990) for the resumption of competitiveness of the United States Silicon Valley companies in the 1980's. In the author's view, it was neither the companies separately, neither the Government alone the promoters of the region's recovery, but the emergence of collaborative networks between specialized producers and a collective learning process inside these networks.

"The dynamics of Silicon Valley´s revitalization are reflected in this new wave of semiconductors start-ups. These firms, together with hundreds of neighboring technology firms, are forging a flexible model of production in the region. By building on the social networks and industrial infrastructure which were created and then abandoned by the established semiconductor firms, these small and medium-sized enterprises are pioneering a new Silicon Valley- one which fosters collaboration and reciprocal innovation among networks of specialist producers" (Saxenian, 1990: (89-90)".

In Brazil, the microelectronic industry is still an incipient one, and has been weakening since the end of the 80's, when there were 23 active companies in the semiconductor sector in Brazil. Since then, both R&D and industrial activities suffered a sharp reduction. This weakening becomes evident in the sector's deficit trade balance: in 2009, the country has imported an amount equivalent to US$ 3,2 billions in semiconductor components; on the other hand, in the same year the exports reached only US$ 57 million (Swart, 2010).

This fragility, in addition to the microelectronic industry's strategic and economic relevance, guarantied its inclusion in the public policy agenda (Swart, 2010). Table 2 resumes these initiatives, aimed towards the formation of human resources, creating integrated circuit project companies (*design houses*) and setting the regulatory framework in order to induce private investments in this industry.

Year	Action
2002	Launch of the National Microelectronics Program.
2004	Inclusion of semiconductors amongst the priorities of the Industrial, Technological and Foreign Trade Policy (PICTE).
2005	Creation of the CI-Brasil Program, targeted to human resources formation.
2007	Inclusion of the area of electronic *displays* amongst PICTE's priorities.
	Creation of the Semiconductor Industry Technological Development Support Program (PADIS).
2008	Inclusion of the microelectronic field in the Science, Technology and Innovation Action Plan.
	Creation of the National Center of Advanced Electronic Technology (Ceitec).
	Creation of INCT-NAMITEC.

Source: Swart (2010), adapted

Table 2. Incentives to Brazil's microelectronic industry.

Within a group of joint actions proposed to stimulate the microelectronic industry's development, the creation of INCT/NAMITEC in 2008 continued NAMITEC's project of Institutes of the Millennium[10], of 2005. NAMITEC counts on several educational and research institutes in the fields of physics, chemistry, computer science and electric/electronic engineering. Nowadays NAMITEC is constituted of 137 researchers of 27 departments in 23 institutions in 13 Brazilian states (INCT/NAMITEC, 2011).

NAMITEC has as its host and coordination institute the Center for Information Technology Renato Archer (CTI). It is managed by a committee composed by five members of different institutions. Its research activities are organized in eight coordination areas, being five technological and three administrative ones. (Table 3).

Field	Specific Area
Technological Development	A.1. Wireless sensor networks.
	A.2. Integrated circuit projects and library of intellectual property.
	A.3. Automatic integrated circuits projects.
	A.4. Semiconductor's material.
Administrative	A.6. Human resource formation.
	A.7. Transfer to the productive sector.
	A.8. Transfer to society.

Source: Adapted from INCT/NAMITEC (2010)

Table 3. Research areas of INCT/NAMITEC.

The network was conceived having as an integrator axis the technologies of wireless sensors networks, correspondent to the first area of technological development (A.1.). The other areas were conceived to subsidize the production of autonomous electronic systems (intelligent sensor network, embedded systems and self-adjustable systems), and contemplate all the necessary knowledge in research to develop wireless sensor network, ranging from materials and fabrication techniques to integrated circuits projects and intellectual properties library. The coordination and interaction mechanisms between NAMITEC's participants are briefly described in Table 4.

Coordination	Area Coordinators receive quadrimestral reports from project coordinators in order to evaluate and elaborate each areas reports
Interaction	Resources for exchange between NAMITEC's members and other INCTs. Support the participation in events are granted only for papers with inter-institutional co-authors. Post-doctoral scholarships are restricted to collaborative projects

Source: Adapted from INCT/NAMITEC (2010)

Table 4. NAMITEC's coordination and interaction mechanisms.

NAMITEC's network gives an important contribution to the formation and of human resources in the field of microelectronics, which is the scope of action of coordination A6,

[10]The Program Institutos do Milênio was transformed in the Progama de Institutos Nacionais de Ciência e Tecnologia by the ordinance MCT n°429/2008.

and of diffusion of the microelectronic area to society, object of coordination A.8. The indicators of academic production and human resource formation in postgraduate levels are significant and synthesized on Table 5.

Indicators	Number
Books	14
Chapters in Books	34
Articles in national journals	22
Articles in international journals	175
National conferences	270
International conferences	274
National summaries	46
International summaries	30
Softwares	1
Product patents	7
Processes patents	1
Technical bulletins	1
Concluded scientific initiations for undergraduate students	59
Concluded Master degrees	103
Concluded PhDs degrees	30
Concluded Postdocs	12
Scientific initiations for undergraduate students in progress	102
Masters in progress	138
PhDs in progress	123
Postdocs in progress	22
Concluded oriented graduations	35
Scientific Conferences organized	12
NAMITEC Colloquiums	6
Short term courses	8
Courses ministered in events	8
News in the open media	8
Lectures/round tables	71
Participation in fairs and workshops	11

Source: INCT/NAMITEC (2011)

Table 5. NAMITEC: Indicators of HR formation and diffusion.

The coordination A.7. (Knowledge Transfer to the Productive Sector) has as its objective to develop strategies and actions aiming to enhance the transfer of technology developed in NAMITEC to the productive sector. In order to do so, several strategies are applied, including direct contact with the companies, arrangement of meetings with business associations and participation in events. As a result, cooperation arrangements have been firmed with the twenty eight companies listed on Table 6.

Company	NAMITEC Institution
Pronatus Amazônia	Center for Science, Technology and Innovation of Manaus Industry Center (CT-PIM)
Datacheck	Center for Information TechnologyRenato Archer (CTI)
Curitiba International Center for Software Development	CTI
Vale do Rio Doce Company University (Univale)	CTI
Transpetro/Petrobras	Rio Grande do Sul Federal University/Engineering School (UFRGS/EE)
ARM	UFRGS/EE
Texas Instruments (USA)	UFRGS/EE
Amplivox	Santa Catarina Federal University (UFSC)
Potychip	UFSC
Tydex (Russia)	Mackenzie University
INO (Canada)	Mackenzie University
Embú Scientific	(Integrated Systems Laboratory / São Paulo University (LSI/USP)
Dixital Technology	LSI/USP
KBA	LSI/USP
High Comm	LIS/USP
LG Electronics	LIS/USP
Novus Electronics	LIS/USP
Treetech Digital Systems	LIS/USP
Digicrom Analyses	LIS/USP
Brasília Technology	Brasília University (UNB)
Z Technology	Integrated Circuits and Devices Laboratory (LDCI)/UNB
Wise Informatics	LDCI/UNB
Tipo D engineering services	LDCI/UNB
Digital Technical Systems	LDCI/UNB
DFChip	LDCI/UNB
Hewlett-Packard Laboratories	LDCI/UNB
São Francisco hydropower company	Electrical Engineer Department/ Campina Grande Federal University(DEE/UFCG)
Tocantins power company	DEE/UFCG

Source: Adapted from INCT/NAMITEC (2010)

Table 6. NAMITEC´s cooperative arrangements with private companies.

The interactions with the productive sector occur by establishing agreements between NAMITEC participant institutions and companies, aiming towards technology and knowledge transfer. These interactions can be classified accordingly to Table 7, which shows the distribution of the types of interactions in the 41 company cooperative projects.

Types of Interactions	**N° of cases**
Know-how transfers on product fabrication	19
Process and equipment transfers to the productive sector	15
Analogical, digital and radio frequency Intellectual Property Libraries (IP)	1
Workshops	6

Source: Adapted from INCT/NAMITEC (2010,2011)

Table 7. Types of interactions: NAMITEC Institutions with the productive sector.

The interactions with the companies may be considered peripheral, since these are not institutions that directly integrate the NAMITEC network. Despite the fact that the five technical areas of INCT/NAMITEC are integrated logically in a productive chain, each institution develops interactions with the companies in an isolated manner, thus not involving any of the remaining institutions within NAMITEC.

The network formed by NAMITEC is essentially an academic network, since the majority of the participating institutions are public institutions of education and research, promoting an important scientific cooperation in the formation of human resources as well as in scientific and technological production (articles, patents and *softwares*) (Table 5).

These characteristics indicate that in NAMITEC, as conceptualized by Callon (1992), the emphasis is on the scientific pole (production of knowledge) and the technological pole (application of knowledge). It lacks, however, an active participation of the market pole, including the companies and users that materialize the innovation. In the INCT/NAMITEC network, the majority of exchanged intermediaries between the poles (that mediate the interactions of the actors) are academic documents, lectures and courses objectifying human resource formation and bound to events directed to scientific production.

In order for NAMITEC to promote innovative activities, it is necessary that the private companies assume a bigger role in this network, as proposed by the network concept by Castells (1996, *apud* Corallo e Protopapa, 2007). Furthermore, joint and coordinated action by the Government and the Education and Research Institutions are required so that the several organizations that are a part of the Brazilian microelectronic sector's organizational ecosystem[11] acquire a beneficent convergence to the generation of useful innovations for the society as a whole. Be it in the form of a Horizontal Network of Induced Innovation (Pellegrin *et al*, 2007) or a Triple Helix (Leydesdorf e Etzkowitz, 1996), a broader interaction between the actors is necessary.

In Brazilian's microelectronic industry, the challenges are even greater, given the fragilities of the companies. Nowadays there are in Brazil seven Design Houses and two designer training centers created within the CI-Brasil Program. In the other segments in the semiconductor's production chain, and specially the participant companies in the sector, Brazil still in industry infancy:

> "However, there isn't a single semiconductor manufacturer that has benefited from the Informatics Law. This finding, combined with the fact that today the semiconductor segment counts on only one SDRAM memory encapsulation plant, two discrete semiconductors (isolated components, not integrated circuits, such as diodes, transistors, etc, for instance, for the fabrication of power supplies) and one integrated circuit design company, belonging to a multinational company, clearly shows the amount of effort that the country needed and needs to **create and develop a microelectronic ecosystem** in Brazil" (Swart, 2010, 276; authors' emphasis)

In order to overcome the difficulties of creating a microelectronic ecosystem in Brazil, NAMITEC's Coordination A.7. established a set of objectives, which unfold into three macro-functions with the goal of beaconing management activities of technology transfer in the NAMITEC network (Table 8). In essence, these macro functions have as a final purpose to act on the ecosystem that the public and private institutions of NAMITEC are inserted, creating attractors that stimulate the cooperation towards innovation and mechanisms that allow the planning and efficient management of the organizational ecosystem delimited by the integrating institutions of INCT/NAMITEC (Kay *et al*, 1999).

With these efforts, the idea is to induce the diverse actors in the ecosystem to the state desired by Coordination A.7., stimulating collective and interactive learning in the participating institutions (academies and firms) in order to circulate the knowledge that has the potential of generating innovations that strengthen the Brazilian microelectronic industry. The organizational ecosystems concept adopted in the management model proposed by Coordination A.7. does not consider the technology transfer activity a unilateral relation, in which knowledge flows from the Educational and Research Institution to the productive sector. Coordination A.7. sees the NAMITEC network as a self-organizing ecosystem (Kay *et al*, 1999), where changing to a desirable state depends on collective learning processes, highlighting the continuous organizational learning of all the components of the ecosystem: Network managers, researchers, demanding companies of NAMITEC's technologies and government institutions.

---

[11]The challenges in integrating multiple actors towards an innovative action in Brazil, mainly in the private sector, are a part of the characteristics of our Sistema Nacional de Inovação (National Innovation System). The integration between universities/research institutes and companies are especially problematic (Velho, 2004; Rapini, 2007; Lotufo, 2009).

Macro Function	Actions
Enhance NAMITEC's cooperation and intern alignment.	Map purposive and supportive knowledge/technology. Identify actual and potential relationships between researchers and technologies. Stimulate cooperation and new projects of common interest. Identify the technological areas with greatest integration potential. Comprehend the potential synergies existent amongst groups that may facilitate the diffusion of knowledge and organizational learning. Develop means to stimulate greater levels of cooperation between research groups. Enhance comprehension on the complex relationships between organizational learning, technology development and innovation in INCTs. Identify mechanisms to enhance diffusion of knowledge.
Enhance alignment of NAMITEC's technologies with company demand.	Initial survey on the potential applications of existent technologies. Verify technological areas with greater potential of application/transfer to companies. Provide conditions so that NAMITEC's technologies can leverage companies. Identify company demands that mobilize new research in NAMITEC. Identify NAMITEC's most demanded technologies (purposive and supportive).
Develop means to potentiate the assimilation process of NAMITEC's technologies by the companies.	Map company demand. Enhance the comprehension on organizational learning relationships, technological development and innovation in the cooperation processes with the companies. Develop mechanisms to enhance the organizational learning processes within NAMITEC and in the technologic transfer activities. Make means available in order to integrate organizational learning processes to labor activities in the target organizations (of the productive sector). Publicize NAMITEC technologies. Identify mechanisms to enhance diffusion of knowledge outside NAMITEC. Stimulate university-company cooperation. Contribute to a greater level of cooperation amongst research groups (suppliers) and the productive sector (demand). Foment complementary projects. Study more effective transfer mechanisms. Study protection of intellectual property mechanisms.

Source: Coordination A.7./NAMITEC (2009)

Table 8. Macro functions of NAMITEC's Coordination A.7.

4. Organizational ecosystems: The participatory approach

The concept of organizational ecosystems can be a very effective management instrument. The management of people is necessary to fulfill the required functions of generation and dissemination of innovation within Namitec, in a way that it allows the integrated management of several important institutions within this process. So, to give practical substance to the case study presented in this chapter, it is presented a brief view of what has been called by Balloni et al as "the six characteristics of the participatory organizacional ecosystem" (Balloni, 2011). These six characteristics are suggested to be adopted as a model for organizational ecosystem, such as the INCT/NAMITEC:

1. Open and lateral dissemination of know-how.
   To facilitate teamwork, every participant must be encouraged to engage in yokoten (a short for yokoni tenkaisuru), which literally means "unfold or open out sideways."This approach encourages everyone to share their individual know-how and expertise openly with others.
2. Freedom to voice contrary opinions.
   The organization (private and/or public managers) should also be open to criticism and contradiction for the nerve system (which we call here as "Organizational Ecosystem") to function properly. This means everyone has to feel free to voice contrary opinions, even to top management and headquarters.
3. Frequent face-to-face interaction.
   Any managers to reach senior positions have a must to acquire and embrace the skill of listening thoroughly and intently to what employees have to say and continually questioning and probing to find a better way.
4. Making tacit knowledge explicit: Organizational Ecosystem.
   Tacit knowledge is converted to explicit knowledge every time someone verbalizes or writes down the knowledge he or she has.
5. Formal and informal organizational support mechanisms.
   Formal and informal support mechanisms have been established in the organization to contribute to the effective functioning of the nerve system.
6. The current global economy, driven by accelerated growth strategy, is unsustainable since it leads to the population bomb ("more customers" is good for business), ecological bomb, and eventually to the depletion of strategic resources bomb. Today civilization faces - the *gene* versus *mind* evolutions and *globalization* versus *sustainability*. Hence, the following principles-strategies of the wise civilization should be applied to steer long and short-term planning and execution of national goals, objectives, and targets (Balloni, 2011). Where among equals, the ecosystem is more equal; Cognizing (education) **and** Ecoism is a new world business system, which would provide the preference to the ecosystem, not to the capital (capitalism) or social prosperity (socialism). The nation-state concept should be sustained in order to prevent diversity of the world society and secure cultural heritage of nations. The Knowledge-Wise Society, which should promote the education, cognition, and knowledgeable and wise decision making and wise use of e-Global Village, which should support the *mind* evolution and also vice versa, is supported by the *mind* evolution to promote the sustainability of any Organizational Ecosystem.

These 6 key principles-strategies above, if adopted as a model for Organizational Ecosystems such as INCT/NAMITEC, should lead to the development of a wise ecosystem. These principles-strategies should be incorporated into all level of NAMITEC decision-making. In short, these key principles-strategies should lead to the development of a sustainable Organizational Ecosystem. The accountability is now in the management system of an Organizational Ecosystem instead of in the hands of the manager (peoples).

5. Conclusion

The inherent complexity of innovation, the diversity of required competences to develop them and the high costs of contemporaneous innovative processes have led public and private institutions to adopt multiorganizational cooperation as a tool to make viable the development of innovations. Such arrangements, however, do not guarantee the achievement of such objectives, due to the difficulties of interaction and alignment between heterogeneous institutions.

Several studies and concepts deal with the virtues and difficulties of multiorganizational arrangements. The concepts examined in the present chapter are a small part of these studies. However they help to clarify some of the difficulties in the multiorganizational arrangement chosen as a case study: the INCT/NAMITEC. It may stated the concept of organizational ecosystems can be a very effective management instrument, managing the people necessary to fulfill the required functions of generation and dissemination of innovation in a way that it allows the integrated management of several important institutions within this process, as pointed out by NAMITEC case.

The multiorganizational arrangement created by NAMITEC has a fundamental part in the training of human resources and in the generation of scientific and technological knowledge, since it is composed by the main universities and research institutes in Brazil. However, Callon's (1992) concept of techno-economic networks allows the pinpointing of a weakness in this network: the lack of a market pole, where innovation becomes effective. In Castells network concept (1996, *apud* Corallo and Protopapa, 2007) this weakness is in the peripheral position that private companies have in NAMITEC. Speaking in terms of Innovation Systems (Edquist, 2001; Malerba, 2002), the weakness resides in the Brazilian Sector System of Innovation in microelectronics: the fragility of the national industry in this field (Swart,2010).[12] And finally, the concept of Triple Helix (Leydesdorf and Etkowitz, 1996; Etkowitz and Leydesdorf, 2000) reveals that the multiorganizational arrangement created by NAMITEC refers to the type I triple helix, where the companies are seen as users of the technologies generated in the Educational and Research Institutions, without a active participation in the innovative process, fact that makes the appropriation of these technologies by the productive sector, during the transfer process, quite difficult.

In short, this chapter shows ways to improve NAMITEC's management. Pellegrin's *et al* (2007) propositions indicate that the presence of one or more inducting agents (anchor) would be desirable to stimulate cooperation between agents, thus acting in a convergent manner (Callon, 1992) in generating innovations. In the words of Hekkert *et al,* (2007), to create stimulus for the components (actors) of Brazil´s sector system of innovation in microelectronics in fulfilling the necessary functions to the completeness of the innovative process.

The theories presented in the chapter suggest actions aiming to improve the interaction and alignment between the diverse acting components in the multiorganizational arrangement created by NAMITEC. These suggestions align themselves to the concept of organizational ecosystems (Kay *et al,* 1999), complex systems composed by heterogeneous actors, where the change to a desirable state dispenses coordination actions. However, the concept of organizational ecosystems goes beyond, characterizing such systems as self-organizing, where adaptive learning of the involved allow them to deal with constant change, be it change in the components or the environment in which the ecosystem is inserted. Such knowledge represents a change from an "anticipatory" management scheme to an "adaptive" management scheme.

---

[12]Actually, the Brazilian industry of microelectronics assembles "national" equipment composed by imported components (Gutierrez and Alexandre, 2003).

Considering the participatory approach (see section 4 of this chapter) two weakenesses can be pointed out in NAMITEC'S network: firstly, the lack of open and lateral dissemination of know-how, due to the poor communication mechanisms between researchers and interested companies. Secondly, face to face interaction is not as frequent as it should be: NAMITEC's researchers only discuss collectively network issues in Namitec workshops, that occur twice a year.

With this vision of organizational ecosystems, coordination A.7. generated a proposal aiming to promote interaction and alignment between actors in the technology transfer process that occur within INCT/NAMITEC. The proposal involves a set of activities in three macro-functions: enhance cooperation and alignment between participating institutions in NAMITEC; enhance alignment of NAMITEC generated technologies with demands of the productive sector; create mechanisms that allow assimilation by the productive sector of the technologies generated by NAMITEC network.

6. References

Balloni, A.J., Resende, J.P. and Targowski, A., Brazil of the Future: Strategizing with the Socio-Technical Management Approach. To be published in the proceedings of 8th Annual International Conference on Computer Science and Information Systems to be held in Athens, Greece on 21-24 May 2012.

Börzel, T.A. What's so special about Policy Networks? An Exploration of the Concept and its Usefulness in Studying European Governance. European Integration online papers (EIoP), v. 1, n.16, 1997. Available at: http://citeseerx.ist.psu.edu/viewdoc/download?doi=10.1.1.122.6885&rep=rep1&type=pdf Access in: 12/05/2011

Callon, M. The dynamic of techno-economic networks. In: R. Coombs *et al.* Technological Change and Company Strategies, Academic Press, London, 1992, p. 72 - 102.

Callon, M. Technological conceptions and adoption network: lessons for the CTA practitioner. In: RIP, A.; MISA, T. J.; SCHOT, J. (eds) Managing Technology in Society: The Approach of Constructive Technology Assessment. Pinter Publishers: London, 1995. p. 307-330

Cassiolato, J. E. e Lastres, M. H. M. Sistemas de inovação: políticas e perspectivas. Parcerias Estratégicas, n. 8, 2000, p. 237-255.

Corallo, A. and Protopapa, S. Business networks and ecosystems: rethinking the biological metaphor. In: Nachira *et al* (Eds.). Digital Business Ecosystems. Luxembourg: Office for official publications of the European Communities, 2007. Available at: <http://www.digital-ecosystems.org/book/2006-4156_PROOF-DCS.pdf> Access in: 09/05/2011

Conselho Nacional De Desenvolvimento Científico e Tecnológico (CNPq). Programa Institutos Nacionais de C&T: documento de orientação aprovado pelo comitê de coordenação. CNPq/MCT, 2008. 12 p. Available at: <http://www.cnpq.br/editais/ct/2008/docs/015_anexo.pdf> Access in: 10/05/2011

Coordenação A.7./NAMITEC. Transferência de conhecimento para o setor produtivo-Relatório de Atividades, Abril a Julho/2009

Coordenação A.7./NAMITEC Proposta de plano de trabalho. (Apresentação). s/d.

De Pellegrin, I. *et al.* Redes de inovação: construção e gestão da cooperação próinovação. Revista de Administração da USP [online]. 2007, vol.42, n.3, p. 313-325. Available at: <http://www.revistasusp.sibi.usp.br/pdf/rausp/v42n3/v42n3a5.pdf> Access in: 06/05/2011

Edquist, C. The systems of innovation approach and innovation policy: an account of the state of the art. Draft presented at the DRUID Conference, Aalborg, 2001. Available at: <http://folk.uio.no/ivai/ESST/Outline%20V05/edquist02.pdf> Access in: 11/05/2011

Etzkowitz, H., and Leydesdorff, L. The dynamics of innovation: from national systems and "mode 2" to a triple helix of university-industry-government relations. Research Policy, v. 29, n.2, 2000, p.411-424.

Gutierrez, R.M.V.; Alexandre, P.V.M. Complexo eletrônico brasileiro e competitividade. BNDES Setorial, n.18, Brasília, 2003. p. 165-191.

Hekkert, M.P. *et al.* Functions of innovation systems: a new approach for analyzing technological change. Technological Forecasting and Social Change, v.74, n.4, p.413- 432, 2007.

INCT/NAMITEC. Annual activity report (Apr. 2009-Mar 2010)- year one. INCT/NAMITEC, 2010. 74 p. Available at: <http://NAMITEC.cti.gov.br/index.php?option=com_galeria&tmpl=component&no_html=1&task=getfile&id=553&name=relatorio-abril2010_anexos-13mai.pdf> Access in: 13/05/2011

INCT/NAMITEC (internet page). Available at: <http://NAMITEC.cti.gov.br/index.php?option=com_page&Itemid=2> Access in: 07/06/2011

Kay, J.J *et al.* An ecosystem approach for sustainability: addressing the challenge of complexity. Futures, v. 31, n. 7, September 1999, p. 721-774

Leydesdorff, L. and Etzkowitz, H. Emergence of a triple helix of universityindustry-government relations. Science and Public Policy, n.23, 1996, p.279-86.

Lotufo, R.A. A institucionalização de núcleos de inovação tecnológica e a experiência da Inova Unicamp. In: Santos, M.E.R.; Toledo, P.T.M.; e Lotufo, R.A. Transferência de tecnologia: estratégias para a estruturação e gestão de núcleos de inovação tecnológica, Campinas: Komedi, 2009, p. 41-108.

Malerba, F. Sectoral systems of innovation and production. Research Policy, v.31, n.2, p.247-264, 2002.

Nachira et al (Eds.). Digital business ecosystems. Luxembourg: Office for Official Publications of the European Communities, 2007. Available at: http://www.digitalecosystems.org/book/2006-4156_PROOF-DCS.pdf Access in: 09/05/2011

Portaria MCT N° 429, DE 17.07.2008. Available at: <http://www.mct.gov.br/index.php/content/view/73306.html> Access in: 10/05/2011

Rapini, M. S. Interação universidade-empresa no Brasil: evidências do Diretório dos Grupos de Pesquisa do CNPq. Estudos Econômicos Cedeplar [online]. 2007, vol.37, n.1, p. 211-233. Available at: <http://www.scielo.br/pdf/ee/v37n1/08.pdf> Access in: 06/05/2011

Saxenian, A. Regional networks and the resurgence of Silicon Valley. California Management Review, 1990; v.33, n.1, p. 89-112.

Sistema Integrado de Gestão de Ciência, Tecnologia e Inovação (SIGCTI) MCT (Internet Page). Available at: <http://sigcti.mct.gov.br/fundos/rel/ctl/ctl.php?act=demanda.buscar_form> Access in: 12/05/2011.

Swart, J. Relatório da sessão "Microeletrônica". Parcerias Estratégicas. Edição Especial. v.15, n. 31, jul.-dez 2010, p.267-280.

Velho, L.; Velho, P. e Saenz, T. W. P&D nos setores público e privado no Brasil: complementares ou substitutos? Parcerias Estratégicas, n.19, Dezembro de 2004, p.87-129

4

The Effects of the N-Fixing Tree *Pentaclethra macroloba* on the Above and Below Ground Communities Within a Primary Forest in the Northern Zone of Costa Rica

W.D. Eaton, D. Shebitz, K. Niemiera and C. Looby
Kean University, School of Environmental and Life Science,
New Jersey
USA

1. Introduction

The soil microbial community is not only critical to nutrient recycling and mineralization of organic material, but also plays a fundamental role in influencing plant community composition (Kent & Triplett 2002, Buckley & Schmidt 2003, Leininger *et al.* 2006, Wardle *et al.* 2004, Ibekwe *et al.* 2007, Fierer *et al.* 2007, Litton & Giardina 2008). Conversely, differences in vegetation community also affect the soil fungal and bacterial composition, which then drives the decomposition and nutrient cycle processes. For example, a complex relationship exists between vegetation-derived carbohydrates, lignin, soil inorganic N, increased soil complexity and microbial community development that affects the soil organic C levels and productivity (Guggenberger et al. 1994, Zech and Kögel-Knaber 1994, Brookes 1995, Anderson 2003, He et al. 2003, Moscatelli et al. 2005, Bradford et al. 2008). Furthermore, increases in soil inorganic N stimulate increased production of more labile root-derived carbohydrates by plants which are used by the bacterial community, while preliminary plant decomposition selects for fungi which degrade the lignin, cellulose, hemicelluloses, and other complex materials (Guggenberger et al. 1994, 1995; Padmanabhan et al. 2003; de Boer et al. 2005; Fierer et al. 2007; Bradford et al. 2008; Talbot et al. 2008). Thus, it is the fungi that decompose more complex organic substrates more efficiently than bacteria leaving behind more recalcitrant residues and enhancing the organic carbon matter in the soil, and are more important as decomposers in older or restored soils (Holland and Coleman, 1987, Bardgett et al. 1993, Cambardella and Elliot 1994, Guggenberger et al. 1994, 1995, 1999; Beare 1997, Bardgett and McAlister 1999; Stahl et al. 1999, Griffith and Bardgett 2000, Zeller et al. 2001; Bailey et al. 2002, Talbot et al 2008). The soil bacteria, on the other hand, are more critical in decomposition and N nutrient cycling in managed, young, or recovering ecosystems (Moore and de Ruiter 1991, Lovell et al. 1995).

Many scientists have recently recognized the value in studying the linkages between above and below-ground communities in order to better understand nutrient resource availability and the dynamic processes associated with nutrient cycling and terrestrial ecosystem function and condition (Wardle et al 2004, Bardgett et al 2005, Wardle 2006, Fierer et al. 2010, Kardol and Wardle 2010). It is becoming clear that the simultaneous assessment of the soil biota and above ground vegetation communities, and the associated nutrient components can shed light on many questions and issues associated with terrestrial ecosystem condition, the nutrient cycle processes, drivers of forest succession, and ecosystem remediation following disturbance. The below-ground indicators of organic matter decomposition and utilization, the N cycle processes, soil carbon (C) and nitrogen-(N) biomass and rates of development, and the relative contribution of fungi and bacteria are of special interest in understanding ecosystem dynamics especially when examined in concert with above-ground vegetation metrics (Anderson 2003, Buckley & Schmidt 2003, He *et al.* 2003, Carney *et al.* 2004, Moscatelli *et al.* 2005, Pendall *et al.* 2008). However, more information is needed to connect the soil and vegetation parameters together and also to connect them with ecosystem functioning (Anderson 2003, Carney & Matson 2005, Wardle 2006, Pendall *et al.* 2008).

Most of the scientific efforts associated with these findings have occurred in either temperate forests or agricultural systems, with far too little work occurring in tropical forest ecosystems. The paucity of research in the tropics is remarkable when one considers that tropical forests contain some of the greatest biodiversity and biomass on the planet, 20% of the planet's C within the first 3 m of their soil, and are considered one of the most critical biogeographic zones for global nutrient cycling and C and N sequestration (Jobbágy and Jackson, 2000). As such, understanding how vegetation and soil fungal and bacterial communities interact to drive the decomposition and nutrient cycle processes in the tropics is especially critical now as deforestation of these unique and critical habitats has resulted in deleterious impacts at local, regional and global scales, including an atmospheric rise in greenhouse gas levels (Keller et al 1993, Cochrane and Laurance 2002, Laurance and Peres 2006), shifts in biodiversity (Ehrlich and Wilson 1991), and negative impacts on nutrient cycles (Reiners et al 1994, Nüsslein and Tiedje, 1999, Cleveland et al 2003, Decaëns et al 2006), thus, representing a great threat to the global nutrient cycles and C and N sequestration (e.g.,Laurance et al 1997, Wolters et al 2000, Campo et al 2001, Carney et al 2004, Feldpausch et al 2004, Waldrop and Firestone 2006, Clark 2007, Ewing et al 2007, Sahrawat 2008). Yet, little is known of the important factors associated with establishing, maintaining, and/or changing the level and rates of activity of the biogeochemical nutrient cycles within tropical forest communities, ecosystems, and landscapes. These factors may be biotic (e.g., adundance of nitrogen-fixing microbes in soil, fungal dominance in soil, plant species and percent cover) or abiotic (e.g., fire, moisture, temperature, soil texture). Whatever their nature, these factors must be critical in shaping and maintaining ecosystem conditions, yet there is a large gap in the knowledge about these drivers of nutrient processes within tropical ecosystems.

One such set of biotic factors presumed to be important in ecosystem development and condition in tropical forests are the symbiotic and free-living bacteria and fungi associated with the so-called "N-fixing" trees and understory vegetation. However, there is little information regarding the effect of this group of plants and the associated microbial community on above and below ground biotic community structure and function, and on nutrient cycle processes and dynamics in tropical forest ecosystems around the world. This lack of information is despite the knowledge that plants with N-fixing symbiotic root nodules, bacteria and archaea associated with nitrification, and both saprotrophic and ecotmycorrhizal fungi represent a biotic consortium that play critical roles in forest ecosystem function, which can be affected by changes in these biotic communities.

Pentaclethra macroloba is a member of the Fabaceae, and considered a dominant N-fixing tree in Central American hardwood forests. As such, this tree is thought to be an early succesional tree that is important in N and C cycle dynamics and biomass enhancement, and ecosystem development in these forests (Hartshorn & Hammel 1994, Wang and Qui 2006, Pons *et al.* 2007). Despite the critical roles this species is thought to play, the above and below ground community structure and function associated with it is poorly characterized at best, and generally under-studied. Eaton et al (2012) provided the first preliminary look at the composition of the soil community associated with these trees in the tropical lowlands of Costa Rica. They found that *Frankia, Rhizobium,* Archaea, and Type II methanotrophs were present and likely involved in recuperating the soil N and enhancing the microbial biomass C via more efficient use of organic C. To our knowledge, there have been no studies that assess the role of *P. macroloba* in development of both the understory vegetation and the associated bacterial, fungal, and archaeal communities in these primary forests; how the structure, abundance and biomass of *P. macroloba* may drive the structure of the understory plant and microbial communities; and how these collectively may drive the C and N cycle dynamics in these soils. These concepts were the focus of our project reported here, which represents the first attempt to identify the role of *P. macroloba* on the nearby above and below ground ecosystem in tropical primary forests.

This work was conducted in two distinct habitats that are located in the San Juan-La Selva Biological Corridor, which is in the Northern Zone forests of Costa Rica: a primary forest dominated by *P. macroloba* and a primary forest with very little of these trees. To determine the role of *P. macroloba,* as the dominant N-fixing tree in these tropical forests, we compared the following components within the 2 different habitats:

- the composition of the understory vegetation (N-fixing and non-N-fixing);
- the DNA-based abundance of the fungal and bacterial DNA, the ammonium oxidizing bacteria, methyltroph, and Archaea DNA;
- the C and N cycle dynamics, rates of production of NH_4 and NO_3;
- the soil dissolved organic C and total N levels, and the C:N ratios;
- The amounts of Phosphorus (P) and the N:P ratios;
- the amount and rates of development of production of C and N biomass;
- the density and percent cover of *P. macroloba* in the two stands;
- the richness and evenness of distribution of the understory vegetation; and
- the richness and evenness of the N-fixing understory vegetation.

2. Methods

2.1 Field locations

In 2001, the Costa Rican Ministry of Environment and Energy helped establish the San Juan-La Selva Biological Corridor (SJLSBC) to help protect the Northern Zone ecosystems (http://www.lapaverde.org.cr) from further damage due to 3-4 decades of extraction-based land management practices (Monge *et al.* 2002). The core conservation unit of the SJLSBC is the Maquenque National Wildlife Refuge (MNWLR; Figure 1), which is located in Northeast region of Costa Rica, about 15 km south of the Nicaraguan border, which conserves the highest percentage of forest cover, and contains the most valuable habitat for biodiversity within the corridor (Chassot & Monge, 2006). Our study area was a primary forest within the MNWLR (10°27'05.7''N, 84°16'24.32W.) and located within the private lands of Laguna del Lagarto Lodge that has not been harvested in at least 80 years, if at all, and contains regions of the forest that are dominated by P. *macroloba* (PM-D) and regions with few to none to these trees (PM-L).

Fig. 1. Location of the Maquenque National Wildlife Refuge (MNWLR) in Northern Costa Rica (map is from Dr. Olivier Chassot, Scientific Director, Centro Cientifico Tropical, E-mail: ochassot@cct.or.cr).

During July of 2006 and 2007, as part of a Tropical Ecology research course, a preliminary study of these forests was conducted in which microbial activity-linked measurements were taken in ten 5 m x 5 m plots in regions of these forests that were visibly dominated by *P. macroloba,* with either tree trunks and/or canopy coverage from these trees being within the plots, and ten plots in regions that had no trunks or canopy coverage within the plots—thus, these plots were considered to be not dominated by *P. macroloba*. In this unpublished preliminary work, we found that soil respiration (808±372 vs. 554± 323 µg CO_2/g/24h), soil total mineral N (2.92±0.61 vs.1.73± 0.27mg mineral N/g), the percent of the mineral N as nitrate (31% ±22% vs. 19% ± 9%), were all greater, and the qCO_2 (8±6 vs. 4±3) as an indicator of C use efficiency (lower value = greater efficiency)/g/24h) was less in the *P. macroloba*-dominant forests. However, the soil C biomass levels were greater in the forest regions not dominated by *P. macroloba* (137±48 vs.101± 41 μg C/g soil). These results suggested that *P. macroloba* may be playing an interesting role in the C and N dynamics in these soils, warranting more work.

Based on this work, studies were conducted in the summers (June and July) from 2008- 2011 in these forested areas. Four 15 x 20 m plots were established in *P. macroloba*-dominant (PM-D) and limited (PM-L) regions, using the definitions above. during the summer of 2008, 2009 and 2010 (n = 12), these plots were assessed for the amounts of soil dissolved organic C (DOC), total mineral N (TMN), the rates of ammonium oxidation, Total N, soil C biomass (Cmic), soil N biomass (Nmic), and the abundance of various microbial groups were determined. In 2010 (n = 4), we also determined the rates of Cmic and Nmic development, soil laccase activity, the structural characteristics of *P. macroloba* and the understory vegetation community composition in these forests.

2.2 Soil collection and carbon, nitrogen, phosphorus, and laccase analysis

Each year, using sterile technique, we collected 25 randomly located 2 cm wide x 15 cm deep soil cores within each plot, over two consecutive days, composited the soil by plot, and sieved it at field moist conditions through 8 mm mesh. Percent saturation, pH, and bulk density were determined at 10 randomly located sites within each plot. All nutrient and microbial activity data presented have been adjusted for dry weight and bulk density of the soil.

The amount of total mineral N (TMN) as the total amount of ammonium (NH_4-N) and nitrate (NO_3-N) were determined following 2M KCl extraction of 10 g of soil (Alef and Nannapieri, 1995) using the ammonium salicylate and cadmium reduction spectrophotometric methods using the HACH DR 2700 system (Hach Company, Loveland, Colorado, 80539-0389; HACH methods 8155 and 8192 respectively). Nitrification rates were measured as the difference in nitrate levels of unincubated samples and samples incubated for 4 days. The microbial biomass N (Nmic) was determined as the difference in Total N in chloroform fumigated vs. unfumigated soils, using the potassium thiosulfate oxidation methods of Jiménez *et* al (2008); and the rates of Nmic development were determined as the difference in Nmic from unincubated samples and samples incubated for 4 days.

Several indicators of decomposition and potential C-sequestration activities, and organic C-use efficiency were examined. The soil dissolved organic C (DOC) was determined by the Walkley-Black rapid dichromate procedure, modified by (Nelson and Sommers, 1996). The soil biomass C (Cmic) determined as the difference between the DOC levels in chloroform-fumigated and unfumigated (total DOC) soil samples using the methods of Anderson and Ingram (1993). The rates of Cmic development were determined as the difference in Cmic from unincubated samples and those incubated over 4 days. The C use efficiency (Cmic/DOC) was calculated by the methods of Moscatelli *et al.* (2005) to suggest the efficiency at which the soil community uses organic C and incorporates it into the biomass. Phosphorous (P) content was measured following Bray 1 extraction from 2 g of soil using the molybdenate reduction method (Method # 8048) and the HACH DR 2700 system. From these data the C to N and N to P ratios were calculated. The phenol oxidase assay (Saiya-

Cork et al. 2002) was also determined to measure the rate of laccase activity over time as an indicator of the fungal-associated degradation of lignin (Sinsabaugh 2010).

2.3 Abundance of soil microbial groups

Soil microbial community DNA was extracted from three 0.3-g replicate samples of pooled soil using the Power Soil DNA Isolation Kit (MO BIO Laboratories, Inc., Carlsbad, CA, Catalog #: 12888), and the DNA extracts from each replicate then pooled. The soil DNA concentration was determined by agarose gel electrophoresis, using the BioRad Precision Molecular Marker Mass Standard and GeneTools software. The percent relative abundance (%RA) of a variety of microbial groups important in the C and N cycles were estimated by qPCR analysis. The %RA was determined for bacterial 16s rRNA, fungal 18s rRNA, fungal ITS regions, Types 1 and 2 methanotroph 16s rRNA, Proteobacteria AOB, and Archaeal 16s rRNA using the published PCR primers and reaction conditions of Martin-Laurent et al. (2001), Smit et al. (1999), Gardes et al. (1993), Chen et al. (2007), Webster et al. (2002), Kemnitz et al. (2005), respectively, and a MJ Research Opticon 1 Real Time Thermal Cycler. Each PCR product was assessed by agarose electrophoresis to confirm the presence of the correct size DNA bands. For the qPCR analyses, the fluorescence values were determined for sample DNA and for known concentrations of cloned control target DNA (7 to 30 ng/µL of cloned target gene DNA with sequences confirmed in GenBank). These values were used to compare the threshold cycle (C_t) for sample DNA to the C_t of the positive control DNA and then to calculate the abundance of the different target gene DNA concentrations in relation to the total abundance for all target genes to result in the %RA calculations.

2.4 Vegetation assessment

During the summer of 2010, an assessment of the vegetation community was conducted. Within each of the plots, five 1m^2 quadrats were randomly placed to sample understory vegetation. Within each quadrat, percent cover and density were calculated for all plant species occupying over 1% of the area and the total percent cover of the understory vegetation and overstory was estimated for the 1m^2. Aboveground herbaceous and shrub biomass was calculated by clipping all above-ground vegetation in the quadrats down to the soil level, bringing it back to the lab, drying in an oven and weighing the material. Species of plants occurring in the quadrats were identified in the field if possible, or pressed and brought back to be identified by local experts.

The diversity of the understory vegetation was categorized also into N-fixing and non-N-fixing vegetation groups. The richness and abundance of N-fixing species were calculated in each of the five quadrats in each plot and summed to get these measurements per 5m^2. We used the number of observed species per 5m^2 as an index of species richness and stem density as a measure of abundance and to suggest the diversity of N-fixing plants between the sites.

2.5 Statistical analyses

For data analysis, forest type (PM-D stands vs. PM-L stands) was the independent variable. Dependent variables that were compared between forests using SPSS (α=0.1) on the plot scale included all nutrient, biomass, and microbial metrics; the number of *P. macrloba* seedlings, *P. macroloba* seedling height (in classes of <1m and >1m), *P. macroloba* tree diameter at breast height and heights of the trees). Dependent variables from the quadrat-scales included the percent cover of total overstory, total understory, percent cover for each species, and density of each woody species in the quadrats. Additional analyses were performed comparing the sum of the 5 quadrats in each plot for biomass, the total density of legumes (including *P. macroloba*) rooted in the quadrats, the species richness of legumes, and the total percent cover of legumes in each quadrat.

A weight of evidence statistical approach was used to compare differences in the mean values of all metrics determined from the two habitats. The mean and standard deviations were determined for each metric, and the percent differences (PD), T-test p values, and the Hedge's d effect size values were used to suggest biologically meaningful differences between means, consistent with the recommendations for analysis of small sample sizes by Di Stefano *et al.* (2005). We used a combined approach of T-test p values ≤ 0.1, PD $\geq$ 20%, and Hedge's d values ≥ 0.7 (>0.7 is considered a large effect size difference) as a weight of evidence to define biologically important differences in the mean values for this project.

3. Results

3.1 Soil nutrient, biomass and microbial analyses

There were many biologically important differences (defined as T-test p values ≤ 0.1, PD $\geq$ 20%, and Hedge's d values ≥ 0.7) found in the soils between the *Pentaclethra macroloba*-dominant (PM-D) and *P. macroloba*-limited (PM-L) stands (Table 1). The PM-D stand soils had greater amounts of inorganic nutrients in general. Specifically, these soils had greater levels of phosphate, TMN, and percent of inorganic N as NO_3, nitrification rates, and N-biomass than found in the PM-L soils, which had a greater rate of N-biomass development and more total N. Consistent with this was the finding that there was a greater amount of bacterial DNA, in general, and AOB DNA specifically in the PM-D stand soils. In addition, there was a much lower N:P ratio found in the PM-D than PM-L soils.

There was a greater amount of DOC and soil C-biomass found in the PM-L stands, which also had greater levels of the indicator of C-use efficiency (Cmic/DOC), and a greater C to N ratio (Table 1). The rate of C-biomass development was somewhat greater in the PM-L stand soils, but did not reach our defined critical level of biological importance (p = 0.112, d = 0.68). However it does suggest a trend toward a more rapid C-biomass development rate. Consistent with these data was the greater amount of fungal rRNA gene and fungal ITS DNA. The laccase activity for lignin degradation, an indicator of more complex organic C decomposition (Guggenberger et al. 1994, 1995, Guggenberger and Zech 1999, de Boer et al. 2005, Bradford et al. 2008), was also greater in the PM-L stand soils (Table 1). The amount of Archaea and Methylotroph Type 2 DNA was about equally high in both soil types, but the Methylotroph Type 1 was greater in the PM-D soils. This suggests that the Archaea and Type 2 Methylotrophs are playing some kind of important role in the C and N cycle activities in both soils types.

	P. macroloba- Limited		*P. macroloba-* Dominant					
Metric	mean	SD	mean	SD	% Difference	T-Test *p* Value	Hedge's *d* value	Standard Error of E.S.
Nitrification (mg/CC/24h)	0.194	0.034	0.253	0.027	30.40	0.00200	1.53	0.57
Total N(mg/cc)	1.93	0.41	1.54	0.42	20.10	0.08200	0.75	0.52
TMN (mg/cc)	2.21	0.62	6.51	0.66	194.57	<0.0001	5.36	1.07
% as NO3	16.73	9.33	33.48	18.41	100.12	0.03767	0.92	0.53
Phosphorus(µg/cc)	9.95	5.64	29.24	14.03	193.90	0.00300	1.43	0.56
N:P	0.23	0.14	0.04	0.01	82.6	0.0018	1.53	0.57
Nmic (mg/cc)	0.19	0.06	0.7	0.29	268.42	0.00025	1.94	0.61
Nmic(mg/cc/24h)	3.22	2.47	1.24	0.96	61.49	0.05301	0.84	0.52
DOC (µg/cc)	671	141	523	108	22.10	0.03400	1.95	0.61
Cmic (µg/cc)	107	23	53	17	50.47	0.00010	2.13	0.63
Cmic(µg/cc/24h)	53	21	36	19	32.08	0.11200	0.68	0.51
Cmic:DOC	0.16	0.042	0.126	0.034	21.25	0.09700	0.63	0.51
C:N	0.39	0.09	0.29	0.12	25.64	0.08000	0.75	0.52
Laccase (OD/h)	0.0049	0.002	0.0013	0.001	73.47	0.00450	1.82	0.59
% RA UB rRNA	10.9	1.2	19.9	2.12	82.57	0.00000	4.17	0.89
%RA UF rRNA	18.35	9.83	7.21	3.12	60.71	0.00856	1.22	0.54
%RA Fungal ITS	16.1	6.41	9.01	4.12	44.04	0.01972	1.05	0.53
%RA Arch	16.57	7	16.59	5.12	0.12	0.99489	0.00	0.50
%RA Meth1	9.49	8.21	16.76	6.12	76.61	0.06433	0.80	0.52
%RA Meth2	26.8	9.01	27.22	7.12	1.57	0.91908	0.04	0.50
%RA AOB	1.79	0.84	3.45	0.48	92.74	0.00026	1.94	0.61

Table 1. A comparison of the mean levels (± standard deviation) of carbon, nitrogen, biomass, and microbial group-related metrics from soils within a primary forest in regions dominated by *Pentaclethra macroloba* (PM-D) and in regions with little to no *P. macroloba*: Presented are differences between the PM-L and PM-D regions in the rates of Nitrification (mg/CC/24h), the amount of Total N(mg/cc), the total mineral nitrogen levels, or TMN (mg/cc), the percent of the nitrogen that is present as nitrate (% as NO3), the levels of Phosphorus (µg/cc), the ratio of N:P,the amount of soil biomass nitrogen or Nmic (mg/cc), the rate at which the Nmic develops (mg/cc/24h), the amount of dissolved organic carbon, or DOC (µg/cc), the amount of soil biomass carbon or Cmic (µg/cc), the rate at which Cmic develops (µg/cc/24h), the ratio of Cmic:DOC, the ratio of C:N, and the amount of laccase activity (OD/h). Also presented are the percent mean relative abundance (% RA) values (± standard deviation) of DNA using qPCR targeted the total bacterial 16S rRNA (%RA UB rRNA), total fungal 18S rRNA (%RA UF rRNA), the fungal internal transcribed space region (%RA Fungal ITS), Types 1 and 2 Methylotrophs 16S rRNA (%RA Meth 1 and 2), Archaea 16S rRNA (%RA Archaea), and ammonium oxidizing bacteria 16S rRNA (%RA AOB). Data were analyzed to determine biologically important differences between the means, which was defined as mean differences having T-test p values ≤ 0.1, % difference levels $\geq 20\%$, and Hedge's d values ≥ 0.7 (>0.7 is considered a large effect size difference) as a weight of evidence.

3.2 Vegetation analyses

There were also differences in mean values of the metrics describing the vegetation structure between the two habitats that met our definition of being biologically important (Table 2). The *P. macroloba* density was greater in the PM-D stands (p=0.014), as was the percent cover of the overstory due to *P. macroloba* (p=0.005). However, the plots seemed to be similar in overall vegetation structure as this study did not find strong evidence of a difference in biomass (p=0.208), the percent cover of shrubs in the quadrats (p=0.846) or the total overstory cover, of all tree species combined (p=0.306) between the PM-D and PM-L plots. Thus, although there are similarities in structure, it appears that different species are fulfilling these roles. As well, there were a total of 51 plant species were identified in the PM-D forest sampling quadrats and 39 were identified in the PM-L quadrats. Therefore, the PM-D stand overall plant richness was greater than in the PM-L (p=0.060). There was a total of nine N-fixing plant species identified in the PM-D stands, and six such species in the PM-L stands, resulting in a greater richness index of N-fixing plants in the PM-D stands (p=0.094). There was also a greater percent of *P. macroloba* seedlings in the PM-D stands (p=0.074).

	P. macroloba-Limited		*P. macroloba*-Dominant					
Metric	mean	SD	mean	SD	% Difference	T-Test p Value	Hedge's d value	Standard Error of E.S.
Total Understory S	18.25	2.29	25.75	2.29	29.13	0.06000	2.61	0.96
N-Fixing Veg S	1.75	0.48	3	0.41	41.67	0.00745	2.23	0.90
biomass	442	93	830	542	46.75	0.20790	0.80	0.73
PM density	0.25	0.5	5.25	2.9	95.24	0.01453	1.92	0.85
% PM Overstory	16.25	10.9	80.5	27.8	79.81	0.00507	2.43	0.93
Total Overstory	96.2	6.8	100	0	3.80	0.30646	0.63	0.72
%Shrub Cover	68.6	22.7	65.4	21.8	-4.89	0.84558	0.11	0.71
%PM as Seedlings	1.5	1.9	19.8	16.8	92.42	0.07359	1.22	0.77

Table 2. A comparison of the vegetation characteristics in tropical lowland forested regions dominated by *Pentaclethra macroloba* (PM-D) and in regions with little to no *P. macroloba* within the Maquenque National Wildlife Refuge in the Northern Zone of Costa Rica: Presented are the differences between the PM-L and PM-D regions in the mean levels (± standard deviation) of the total richness of the understory vegetation (Total Understory S), the richness of the nitrogen-fixing vegetation (N-Fixing Veg S), the biomass of the vegetation (biomass), the *P. macroloba* density (PM density), the percent of the overstory that is from *P. macroloba* (% PM Overstory), the total amount of the overstory vegetation (Total Overstory), the percent of the understory that is shrub cover (%Shrub Cover), and the percent of the seedlings that are *P. macroloba* (%PM as Seedlings). Data were analyzed to determine biologically important differences between the means, which was defined as mean differences having T-test p values ≤ 0.1, % difference levels ≥ 20%, and Hedge's d values ≥ 0.7 (>0.7 is considered a large effect size difference) as a weight of evidence.

4. Discussion

The data from this work suggest that PM-D stands are very important in recuperating and maintaining the N levels in these soils. They also appear to be facilitating the growth of more N-fixing understory vegetation, more young growth of plants, and an overall greater vegetation richness in the PM-D stands, suggesting a vegetation structure that is less homogenous (and perhaps less stable) than in the PM-L stands. Both the N-fixing understory and the *P. macroloba* trees in the PM-D stands are providing the needed inorganic N to stimulate development of a more complex organic C and N composition somewhat distance from these trees. This is likely a dynamic situation that requires a balance between N cycle bacterial-associated stimulation and fungal stimulation and inhibition in order to generate what appears to be a greater potential for C and N sequestration in the PM-L soils. It appears that the structure of the PM-L stands is more similar to an older, more established forest with a more stable and somewhat less diverse vegetation, yet with high levels of biomass, but these parts of the forest are clearly dependent on the PM-D stands and the nutrients they provide.

There is a complex relationship between increases in plant-derived carbohydrates, lignin, celluloses and other more recalcitrant organic compounds, soil inorganic N, increased soil complexity (Guggenberger et al. 1994, Zech and Kögel-Knaber 1994; Bradford et al. 2008) and soil biomass development (Anderson and Domsch 1989; Brookes 1995; Anderson 2003; He et al. 2003; Moscatelli et al. 2005), and the clear differences in vegetation structure between the PM-D and PM-L stands appear to be playing a role in the dynamics of this relationship in these forests. Increased levels of soil inorganic N stimulate rhizodeposition and increased production of more labile root-derived carbohydrates used to enhance the bacterial community, although increases in inorganic N also inhibit fungi (Bittman *et al.* 2005, de Vries *et al.* 2007). Plant lignins and other more recalcitrant organic compunds select for Basidiomycota and other fungi that degrade the lignin, celluloses, etc (Sinsabaugh 2010). Both of these types of microbial-directed processes enhance the soil organic matter complexity, stimulating complex microbial activities associated with decomposition, soil respiration, and mineralization of organic matter, followed by an increase in organic compounds and soil biomass (Andrews and Harris 1986; Powlson et al. 1987; Guggenberger et al. 1994, 1995; Zhang and Zak 1995; Arunachalam et al. 1997; Eaton 2001; Padmanabhan et al. 2003; de Boer et al. 2005; Schwendenmann and Veldkamp 2006; Fierer et al. 2007; Bradford et al. 2008), a more efficient use of the soil organic matter, and more organic C and N available to the foodweb (Anderson and Domsch 1989; Brookes 1995; Anderson 2003; He et al. 2003; Moscatelli et al. 2005).

The N-cycle data from the current study suggests that in the PM-D stands, there was a greater amount of microbial activity involved in the development of inorganic nutrients, typically associated more with the bacterial than fungal communities (e.g., Moore and de Ruiter 1991; Lovell et al. 1995). This is likely due to the greater amount of N-fixing vegetation in the PM-D stands, including the greater amount of *P.macroloba*. This would explain the greater amount of N cycle activity, greater amount of inorganic N and Nmic, and bacterial and AOB DNA in the PM-D soils, suggesting a greater amount of N-fixation and ammonium oxidation was occurring there. Fungi, and especially some groups like the Basidiomycota, are known to harvest organic C and N, moving it into the biomass and making it available to the other parts of the forest soil food web (e.g. Talbot et al 2008).

However, increases of inorganic N in soils have been shown to suppress fungal biomass development (Bittman *et al.* 2005, de Vries *et al.* 2007) and the decomposition of lignin and lignaceous material. The latter occurs through inhibition of lignolytic enzyme (i.e., laccase) synthesis (Worrall *et al.* 1997, Carreiro *et al.* 2000, Sinsabaugh *et al.* 2002 and 2004, Knorr 2005, Waldrop and Zak 2006) and/or by reacting with lignin degradation products to form more recalcitrant compounds (Dijkstra *et al.* 2004, Hobbie 2008). This is what appears to be happening in the PM-D soils. The PM-L stands have fewer N-fixing plants and *P. macroloba,* but still and equally strong overall vegetation community, which is associated with the lower levels of inorganic N, and greater amounts of organic N (as part of the total N measurement), N-biomass, and rates of development of the N-biomass in the PM-L stands, along with the greater amounts of fungal DNA and laccase activity. This all suggests that the PM-L stands are facilitating development of a more fungal-dominant soil microbial community, and the PM-D stands, due to the inorganic N, are facilitating the development of a more bacterial-dominant soil microbial community. The fungal dominant community in the PM-L stand soils would result in the production of more N and C biomass and stable forms of organic C and N material in these soils than in the PM-D soils.

The C cycle data from this study showed there was a greater amount of soil DOC, amount and rate of development of C biomass and laccase activity, and more fungal DNA in the PM-L soils. The greater amount of laccase activity, that is associated with lignin degradation, is an indicator of an enhanced Basidiomycota fungal population (see de Boer *et al.* 2005 for a review) and a more complex organic C decomposition (Guggenberger et al. 1994, 1995, Guggenberger and Zech 1999, de Boer et al. 2005, Bradford et al. 2008). Again, this is associated with a more homogenous vegetation community in the PM-L stands that have less N-fixing vegetation. The increased efficiency of C use in the PM-L stand soils, as measured by the ratio of C biomass to the DOC, is also an indicator of an enhanced soil fungal population. An increase in this metric suggests there is an increase in the amount of organic C being made available for the microbial community and for transfer up the food web (Anderson & Domsch 1989, Brookes 1995, Anderson 2003, Moscatelli *et al.* 2005), and usually is associated with a shift towards fungal dominance (Anderson 2003). An increase in this ratio often occurs as a result of an increasing fungal biomass in comparison to a relatively static bacterial biomass (Ohtonen *et al.* 1999, Van der Wal *et al.* 2006). As well, due to the inhibition of fungi by inorganic N and the increase in organic C often associated with an increase in fungal biomass, an increase in the C to N ratios is also an indicator of a fungal-dominant soil biota (Anderson, 2003).

As stated in the Introduction, fungi are thought to decompose organic substrates more efficiently than bacteria (Holland and Coleman, 1987; Griffith and Bardgett, 2000), leaving behind more complex fungal biomass organic residues than bacterial residues, thus increasing the amount of recalcitrant organic matter and an enhanced DOC (e.g., Malik and Haider, 1982; Guggenberger et al., 1999; Sinsabaugh 2010). The saprotrophic fungi play major roles in the decomposition of dead plant material, particularly cellulose and lignin in the litter and organic soil horizons (Luis et al., 2004; O'Brien et al., 2005), and are thought to be very diverse in lowland tropical soils (Bills and Polishook, 1994). Mycorrhizal fungi are mutuallistic symbionts of plants (Malloch et al., 1980), and are especially abundant in the O and A soil horizons (Luis et al., 2004; O'Brien et al., 2005) and are now being shown to play important roles in both providing materials to plants and in organic C and N harvesting from surrounding soil regions (Talbot et al 2008). The observations from the soil analyses in our study were consistent with a more fungal-dominant soil ecosystem in the PM-L soils. The greater amount of N-fixing vegetation enhances the bacterial community in the soils associated with the inorganic N-cycle processes, resulting in greater amounts of inorganic N which would somewhat inhibit the Basidiomycota and/or other laccase producing fungi. In addition, it is also possible that the vegetation in the PM-L stands are releasing materials that are more conducive to establishing a more dominant fungal community. This should be examined in the future. Nonetheless, it is evident that the N-fixing vegetation in the PM-D stands are both inhibiting the soil fungi, and providing what is needed by them for use at some distance to the trees. This is a critical dynamic that must be balanced in order to enhance the fungal community in these older primary forests to allow them to function properly.

This study also showed that Archaea and Methylotrophs may be important in both the PM-D and PM-L stands in enhancing the organic C and N. The methylotrophs play an important role in soils by using methane as their sole energy source, converting it into forms of organic carbon (C) that can be utilized up the food web and are indicators of a complex soil system (Bastyiken et al 2003, Bull et al 2000, Murase and Frenzel 2007, Mancinelli 1995, Hanson and Hanson 1996, Whalen et al 1990), and that some are capable of N fixation (Chu and Alvarez-Cohen 1999, Auman et al. 2001). The role of the archaeal community in forest soils is much less clear, although some of their various functions are known or have been proposed. The terrestrial representatives of the Crenarchaeote Group 1.1b (Chaben et al 2006) have been found in a wide variety of soil types (Ochsenreiter et al 2003, Sliwinski and Goodman 2004). Many Crenarchaeotes have been found associated with a wide variety of plant species within the rhizosphere (Simon et al 2000 and 2005; Chelius and Triplett 2001; Sliwinski and Goodman 2004), and are thought to be aerobic heterotrophs (Rutz and Kieft 2004). More recently, ammonia-oxidizing archaea (AOA) within the *Crenarchaeota* group have also been found to play this important role in soils (Hallam *et al.* 2006; Schleper *et al.* 2005; Treusch *et al.* 2005). In some cases, the AOA have been shown to be more numerically dominant over the AOB group in terrestrial soils (Adair and Schwartz 2008; Chen *et al.* 2008; He *et al.* 2007), although this is still controversial as it has been pointed out that abundance may not correlate with contribution (i.e., Wessen et al 2010). Given both the wide range of plant species the Crenarchaeota have been associated with, and that they are known AO microbes, it is not a surprise that they are fairly equally distributed in these two habitats. Although they are clearly important in developing appropriate amounts and composition of C and N materials in soils, there is a significant gap in the knowledge about how these microbes interact or compete with plants and other soil members of the microbial communities for C and N, and how they change during plant growth or changing plant species during succession in tropical forest ecosystems.

The rate and amount of soil N-fixation activity is associated with the development of the organic C and N composition in soils, and is also known to be regulated by a feedback inhibition mechanism that is activated in the presence of higher N:P ratios or lower concentrations of P (Eisele *et al.* 1989, Smith 1992, Israel 1993, Almeida *et al.* 2000, Schulze 2004, Pons *et al.* 2007, Reed *et al.* 2007). This is consistent with the results of our work in which there were somewhat greater concentrations of total N and much lower concentrations of P in the PM-L soils, resulting in higher N:P ratios (i.e., P limited), greater soil C biomass, greater rates of both C and N biomass production, and a more fungal-dominant microbial community in the PM-L soils.

Changes in N:P ratios have been associated with increases in microbial activity and biomass, N and P cycle processes, and microbial community structure (Leahy and Colwell 1990, Smith 1992, Smith *et al.* 1998, Cleveland *et al.* 2004, Cleveland and Townsend 2006, Allison *et al* 2007, Cruz *et al.* 2008). It has also been shown that well-established tropical forest soils tend towards more P than nitrogen limitation (Vitousek & Farrington 1997, Sollins 1998, Hedin *et al.* 2003), resulting in higher N:P ratios, in part due to the extensive amount of nitrogen fixation that occurs (Vitousek & Howarth 1991, Cleveland *et al.* 1999, Galloway *et al.* 2004), and as a result of the decomposition of greater amounts of forest litter and vegetation tissues (see McGroddy *et al.* 2004 for a review). Our results lead us to pose that the differences in the above-ground vegetation, especially the N-fixing species, in the PM-D and PM-L stands, and the subsequent differences in stimulation of either a more bacterial or fungal-dominant biota, are altering the ratios of N to P and are, thus, helping to drive the microbial community function in these soils. This results from changes in the N-fixation and ammonium oxidation, and the abundance of the associated bacterial groups responsible for these processes. As well, lower concentrations of inorganic N facilitate development of fungal populations, in particular, fungi such as Basidiomycota and others associated with an increase in laccase activity—due to the reduction in inorganic N-associated inhibition of these groups.

The increased fungal dominance and laccase activity observed in the current study occurs along with an increase in DOC, total N, rates of incorporation of N into the biomass, rates of C biomass development, and a more efficient use of organic C and N suggest that the PM-L soils have a more fungal-dominant biota. These differences in patterns created by the structure of the *P. macroloba* forest stands have very important implications for P, N and C cycle dynamics, soil organic matter development, lignin degradation and other types of more complex organic matter decomposition, and C and N biomass development in these soils, thus C and N sequestration. To our knowledge, this is the first time that N:P ratios in old growth tropical forests have been examined in relation to differences in above ground vegetation and below ground microbial community composition. Our data show that as the below-ground N:P increases (P is more limited) the below ground biomass increases, and also that this also occurs with as the C:N ratio increases. Moreover, this is occurring in conjunction with decreased amount and diversity of N-fixing vegetation in the PM-L stands, further from the regions dominated by *P. macroloba.* If confirmed, this model has the potential to account for changes in the below ground microbial-related biomass development due to differences in environmental or management conditions that affect the soil N and P, with potential links to C and N sequestration predictions.

5. Conclusion

In these older primary tropical forests, the PM-D stands have a greater amount of and diversity of N-fixing vegetation than the PM-L stands, appearing similar to a forest in a younger stage of succession. Associated with this is a more bacterial-dominant soil biotic community than the PM-L soils, with greater amounts and rates of production of inorganic nutrients, but slower rates of development of below ground C and N biomass, resulting in less DOC and total N than the soils from the PM-L regions. The PM-L soils have far less N-fixing vegetation, yet a strong and healthy vegetation community with similar amounts of coverage and biomass as the PM-D forests, but with different non-N-fixing species involved. Associated with this were less inorganic N, greater amounts of DOC and total N, and greater amounts and rates of production of C and N biomass. This appears to be due to the greater abundance of fungi, linked to their capcity for more complex decomposition, indicated by the greater amounts of laccase activity in the PM-L soils. Thus, the PM-L stand soils are more fungal-dominant, likely resulting in a more complex soil organic C composition, and a more complex microbial community that is making more efficient use of the nutrients provided by the nearby PM-D regions. This probably results in development of more C and N biomass in the PM-L soils, and potentially a greater amount of C and N sequestration in these soils.

The conclusions posed from this work are enticing; however, they also illustrate the tremendous need for much more work to clarify many questions concerning the effect of *P. macroloba* on the soil community and nutrient dynamics and the vegetation community. Perhaps the most striking results to us were how distinct the two microbial and vegetation communities were in the PM-D and PM-L stands. The observed relationships between the C:N and N:P ratios, the evidence for enhanced fungal dominance and soil complexity, and the changes in the vegetation communities warrant more work in the future, and represent major areas of study that should be targeted in order to better understand how this N-fixing tree, that is so critical to these tropical forests, functions to stimulate soil and vegetation ecosystem development and maintenance of a stable and healthy ecosystem.

There are many other questions that remain unanswered concerning the role of this important member of the Fabaceae in these forest ecosystems, which should also serve as targets for future work. How does PM affect the succession of soils in secondary forest regeneration? At what point, over what time frame, and under what conditions does PM enter into a damaged/cleared forest as an early successional tree to recuperate the N in the soil system, re-establish the vegetation, and stimulate development of a "normal" level of both soil and above ground vegetation population dynamics, including C and N sequestration? What conditions are needed for the trees to begin to stimulate the production of inorganic N that stimulates the initial pulse of bacterial activity, which is then needed to stimulate the development of the more simple, then followed by the more complex fungal populations and their associated activities. At what point does the fungal population begin to dominate and how far from the individual trees does this happen—thus, how far from the trees is the stimulating effect of the increased N on the fungal populations, which is balanced out by the inhibitory effect of the inorganic N on fungal populations and laccase activity? It appears as if the PM is important in driving the bacterial-associated N-cycle activity and the provision of inorganic N in these soils, and that this is being used in the immediate area by the bacterial community. However, at some distance away from the PM-D stands (How far?), the fungal inhibition by inorganic N is reduced, and the fungal community is being stimulated and is making use of this inorganic N to build up and maintain their populations and to generate and distribute organic forms of both N and C into the system. Also, if there is an increase in complexity of organic C decomposition occurring at some distance to the trees, is there production of more stable forms of organic C being produced away from the PM-D regions, and if so, at what distance from the PM-D region does this occur, and at what density of PM trees does the this occur?

These are important questions for land managers in the region as there are a variety of strategies of forest regeneration occurring in the region, and it is critical for them to know what the end result forest target should be, so that they know when a secondary forest is re-established as a healthy stable ecosystem, and when might it be least damaging to the ecosystem to selectively re-harvest certain species of trees in that secondary forest. The answers to these questions are also important for developing a more complete understanding of the biotic and abiotic factors that drive the C and N biogeochemical processes in tropical forest ecosystems. More studies should be conducted which examine the impact that critical ecosystem species (keystone perhaps?) have on forest ecosystems, such as an early successional N-fixing tree like *P. macroloba,* as they can provide important information to begin to address these and many more questions concerning the C and N cycle dynamic processes in tropical forest ecosystems, and how they might be affected by anthropogenic changes in the environment.

6. Acknowlegements

This work was supported by funds from NSF DBI grants 0452328, **0453504, 0851792, and** 1034896, as well as the Kean University Presidential Scholars Challenge Grant Program. Also, we wish to thank the Costa Rican Government agency MINAET (Ministerio Del Ambiente, Energia Y Telecomuinicaciones) for permits to work in the region. The authors would like to thank Kurt Schmack and the staff members at the Laguna del Lagarto Lodge in Boca Tapada, Alajuela, Costa Rica for their on-going support of our work, and Dr. Olivier Chassot, Centro Cientifico Tropical, San José, Costa Rica whose efforts have made all our Costa Rica work possible.

7. References

Adair, K.L. & Schwartz, E.(2008). Evidence that ammonia-oxidizing archaea are more abundant than ammonia-oxidizing bacteria in semiarid soils of Northern Arizona, USA. *Microbial Ecology.* 56, 420–426.

Alef, K., & Nannapieri, P. (1995). Enzyme Activities. Methods in Applied Soil Microbiology and Biochemistry. 311–373.

Allison, V.J., Condron, L.M., Peltzer, D.A., Richardson, S.J., & Turner, B.L. (2007). Changes in enzyme activities and soil microbial community composition along carbon and nutrient gradients at the Franz Josef chronosequence, New Zealand. *Soil Biology and Biochemistry.* 39:1770–1781.

Almeida, J.P.F., Hartwig, U.A., Frehner, M, Nösberger, J. & Lüscher, A. (2000). Evidence that P deficiency induces N feedback regulation of symbiotic N_2 fixation in white dover (*Trifolium repens* L.). *Journal of Experimental Botany.* 51:1289-1297.

Anderson, T.H., & Domsch, K.H. (1989). Ratios of microbial biomass carbon to total organic carbon in arable soils. Soil Biology and Biochemistry 21:471-479.

Anderson, J. M. & Ingram J.S.I. (1993). Tropical Soil Biology and Fertility. A handbook of mehods. Second Edition. CAB International, UK,.p65-66.

Anderson, T.H. (2003). Microbial eco-physiological indicators to assess soil quality. *Agriculture, Ecosystems and Environment* 98: 285-293.

Andrews, J. H. & Harris, R.F. (1986). r- and K-selection and microbial ecology. *Advances in microbial ecology.* Plenum Press, New York, 9:pp 99–147.

Aumen, A., Speake, C. & M. Lidstrom. (2001). NifH sequences and nitrogen fixation in type I and type II methanotrophs. *Applied and Environmental Microbiology* 67: 4009-4016.

Bailey, V.L., Smith, J.L. & Bolton, Jr., H. (2002). Fungal-to-bacterial ratios in soils investigated for enhanced C sequestration. Soil Biology and Biochemistry. 34: 997-1007.

Bardgett RD, & McAlister E. (1999). The measurement of soil fungal : bacterial biomass ratios as an indicator of ecosystem self-regulation in temperate meadow grasslands. Biology and Fertility of Soils. 29:282-290.

Bastviken, D., Ejertsson, J., Sundh, I. & Tranvik, L. (2003). Methane as a source of carbon and energy for lake pelagic food webs. *Ecology* 84: 969-981.

Beare, M.H., Hu, S., Coleman, D.C., & Hendrix, P.F. (1997). Influences of mycelial fungi on soil aggregation and organic matter storage in conventional and no-tiillage soils. Applied soil Ecology. 5: 211-219.

Bills, G. F., &. Polishook, J.D. (1994). Abundance and diversity of microfungi in leaf litter of a lowland rain forest in Costa Rica. *Mycologia*, 86:187-198.

Bittman, S. Forge, T.A., & Kowalenko, C.G. (2005). Responses of the bacterial and fungal biomass in a grasslandsoil to multi-year applications of dairy manure slurry and fertilizer. *Soil Biology and Biochemistry*.37:613-623.

Bradford, M.A., Fierer, N., & Reynolds, J.F. (2008). Soil carbon stocks in experimental mesocosms are dependent on the rate of labile carbon, nitrogen, and phosphorus inputs to soil. *Functional Ecology*. 22:964-974.

Brookes, P.C. (1995).The use of microbial parameters in monitoring soil pollution by heavy metals. Biology and Fertility of Soils.19:269-279.

Buckley, D. H. & Schmidt, T. M. (2003). Diversity and dynamics of microbial communities in soils from agro-ecosystems. *Environmental Microbiology* 6.5: 441-452.

Bull, I.D., Parekh, N.R., Hall, G.H., Ineson, P. & Evershed, R.P. (2000). Detection and classification of atmospheric methane oxidizing bacteria in soil. *Nature* 405:175-178.

Cambardella, C.A. a& Elliot, E.T. (1994). Carbon and nitrogen dynamics of soil organic matter fractions from cultivated grassland soils. Soil Science Society of America Journal. 58:123-130.

Campo, J., Maass, M., Jaramillo, V.J., Martinez-Yrizar, A., & Sarukhan, J. (2001) Phosphorus cycling in a Mexican tropical dry forest ecosystem. *Biogeochemistry* 53:161–179.

Carney, K.M., Matson, P.A., & Bohannan, B.J.M. (2004). Diversity and composition of tropical soil nitrifiers across a plant diversity gradient and among land-use types. *Ecology Letters* 7: 684–694.

Carney, K.M., & Matson, P.A. (2005). Plant communities, soil microorganisms, and soil carbon cycling: does altering the world belowground matter to ecosystem functioning? *Ecosystems* 8: 928–940.

Carreiro, M. M.-, Sinsabaugh, R. L., Repert, D. A., & Parkhurst, D.F. (2000). Microbial enzyme shifts explain litter decomposition responses to simulated nitrogen deposition. Ecology 81:2359–2365.

Chaban, B., Ng,S.Y.M. & Jarrell, K.F.(2006). Archaeal habitats—from the extreme to the ordinary. *Canadian Journal of Microbiology* 52:73-116.

Chassot O. & G. Monge (eds.). 2006. Plan de manejo del Refugio Nacional de Vida Silvestre Mixto Maquenque, 2006-2010. Ciudad Quesada, Alajuela, Costa Rica, Área de Conservación Arenal Huetar Norte (ACAHN), Sistema Nacional de Áreas de Conservación (SINAC), Ministerio del Ambiente y Energía (MINAE), Centro Científico Tropical (CCT), 244 p.

Chelius, M.K. & Triplett, E.W. (2001). The diversity of archaea and bacteria in association with the roots of *Zea mays*. *Microbial Ecology* 41:252-263.

Chen, X., Zhu, Y., Xia, Y., Shen, J. & He, J. (2008). Ammonia-oxidizing archaea: important players in paddy rhizosphere soil? Environ. Microbiol. 10, 1978–1987.

Chen, Y., Dumont, M. Cébron, A. & Murrell, J. (2007). Identification of active methanotrophs in a landfill cover soil through detection of expression of 16S rRNA and functional genes. *Environmental Microbiology* 9: 2855-2869.

Chu, K. & Alvarez-Cohen, L. (1999). Evaluation of toxic effects of aeration and trichloroethylene oxidation on methanotrophic bacteria grown with different nitrogen sources. *Applied and Environmental Microbiology* 65: 766-772.

Clark, D.A. (2007). Detecting the responses of tropical forests to global climatic and atmospheric change: current challenges and a way forward. *Biotropica* 39:4-19.

Cleveland, C. C., Townsend, A.R., Schmidt, S.K. & Constance, B.C. (2003). Soil Microbial Dynamics and Biogeochemistry in Tropical Forests and Pastures, Southwestern Costa Rica. Ecological Applications 13:314-326.

Cleveland, C. C., Townsend, A. R., Schimel, D. S., Fisher, H., Howarth, R. W., Hedin, L. O., Perakis, S. S., Latty, E. F., Von Fisher, J. C., Elseroad, A., & Wassan, M. F. (1999). Global patterns of terrestrial biological nitrogen (N_2) fixation in natural ecosystems. *Global Biogeochemical Cycles* 13:623–645.

Cleveland, C.C., Townsend, A.R., Constance, B.C., Ley, R.E.& Schmidt, S.K. (2004). Soil Microbial Dynamics in Costa Rica: Seasonal and Biogeochemical Constraints. *Biotropica* 36: No. 2: pp 184–195.

Cleveland, C.C., & Townsend, A.R. (2006). Nutrient additions to a tropical rain forest drive substantial carbon dioxide losses to the atmosphere. *Proceedings of the National Academy of Sciences of the United States of America.* 103:10316-10321.

Cochrane M.A. & Laurance W.F., (2002). Fire as a large-scale edge effect in Amazonian forests, Journal Of Tropical Ecology, 18:311-325.

Cruz, A.F., Hamel, C., Hanson, K., Selles, F., & Zentner, R.P. (2008). Thirty-seven years of soil nitrogen and phosphorus fertility management shapes the structure an function of he soil microbial community in a Brown Chernozem. *Plant and Soil.* Online August 2008.

De Boer, W., Folman, L.B., Summerbell, R.C., & Boddy, L. (2005). Living in a fungal world: impact of fungi on soil bacterial niche development. FEMS Microbiology Reviews. 29:795-811.

Decaëns, T, Jiménez, J.J., Gioia, C., Measey, G.J. & P. Lavelle. (2006). The values of soil animals for conservation biology. European Journal of Soil Biology 42: S23-S38.

De Vries, F.T., Bloem, J., Van Eekeren, N., Brusaard, L. & Hoffland, E. (2007). Fungal biomass in pastures increases with age and reduced N input. *Soil biology and Biochemistry*. 39:1620-1630.

Dijkstra, F.A, Hobbie, S. E., Knops, J. M. H. & Reich, P. B. (2004). Nitrogen deposition and plant species interact to influence soil carbon stabilization. *Ecology Letters* 7:1192–1198.

Di Stefano, J., Fidler, F., & Cumming, G. (2005). Effect size estimates and confidence intervals: An alternative focus for the presentation and interpretation of ecological data. Pp. 71-102 in A. R. Burk, ed., New trends in ecology research. New York: Nova Science Publishers.

Eaton, W.D. (2001). Microbial and nutrient activity in soils from three different subtropical forest habitats in Belize, Central America during the transition from dry to wet season. *Applied Soil Ecology* 16:219-227.

Eaton, W.D., MacDonald, S., Roed, M., Vandecar, K.L., Hauge, J.B. & Barry, D. (2011). Seasonal and Habitat-Based Variations In The Microbial Community Structure Within Two Soil Types From Old Growth Forests In Costa Rica. In Press *Tropical Ecology* 52: 35-48.

Ehrlich, P.R. & Wilson, E.O. (1991). Biodiversity studies: science and policy. Science 253:758-762.

Eisele, K.A., Schimel, D.S., Kapustka, L.A., & Parton, W.J. (1989). Effects of available P and N-to- P ratios on non-symbiotic dinitrogen fixation in tallgrass prairie soils, *Oecologia*. 79:471–474

Ewing, S.A., Michalski, G., Thiemens, M., Quinn, R.C., Macalady, J.L., Kohl, S., Wankel, S.D., Kendall, C., Mckay, C.P., & Amundson, R. 2007. Rainfall limit of the N cycle on Earth, *Global Biogeochemical Cycles, 21*, GB3009.

Feldpausch, T.R., Rondon, M.A., Fernandes, E.C.M., Riha, S.J., & Wandelli, E. (2004). Carbon and nutrient accumulation in secondary forests regenerating on pastures in central Amazonia. Ecological Applications. 14:S164-S176.

Fierer, N., Bradford, M.A., & Jackson, R.B. (2007). Toward an ecological classification of soil bacteria. *Ecology*, 88: 1354–1364.

Fierer, N, Nemergut, D., Knight, R. & Craine, J.M. (2010). Changes through time: integrating microorganisms into the study of succession. Research in Microbiology, 161:635-642

Galloway, J. N. , Dentener, F. J., Capone, D. G., Boyer, E. W., Howarth, R. W., Seitzinger, S. P., Asner, G. P., Cleveland, C. C., Green, P. A., Holland, E. A., Karl, D. M., Michaels, A. F., Porter, J. H. Townsend, A. R., & Vorosmarty, C. J. (2004). Nitrogen cycles: Past, present, and future. *Biogeochemistry* 70: 153–226.

Gardes, M. & Bruns, T.D.(1993). ITS primers with enhanced specificity for basidiomycetes - application to the identification of mycorrhizae and rusts. *Molecular Ecology* 2: 113-118.

Griffith, G.S., & Bardgett, R.D. (2000). Influence of resource unit distribution and quality on the activity of soil fungi in a particulate medium. *New Phytologist*. 148:143-151.

Guggenberger, G., Zech, W., & Thomas, R.J. (1995). Lignin and carbohydrate alteration in particle-size separates of an Oxisol under tropical pastures following native savanna. Soil Bilogy and Biochemistry. 27:1629-1638.

Guggenberger, G., Christensen, B.T., & Zech, W. (1994). Land-use effects on the composition of organic matter in particle-size separates of soil:I. Lignin and carbohydrate signature. European Journal of Soil Science. 45:449-458.

Guggenberger, G. & Zech, W. (1999). Soil organic matter composition under primary forest, pasture, and secondary forest succession, Regiõn Huetar Norte, Costa Rica. Forest Ecology and Management. 124:93-104.

Hallam, S.J., Mincer, T.J., Schleper, C., Preston, C. M., Roberts, K., Richardson, P. M. & DeLong, E.F. (2006). Pathways pf carbon assimilation and ammonia oxidation suggested by environmental genomic analyses of marine Crenarchaeota. *PLoS Biology*, 4, 520–536.

Hanson, R. & Hanson, T. (1996). Methanotrophic Bacteria. *Microbiological Reviews* 60: 439-471.

Hartshorn, G.S.& Hammel, B.E. (1994). Vegetation types and floristic patterns. In: *La Selva:Ecology and Natural History of a Neotropical Rain Forest*. University of Chicago Press, pp 73-89.

He, Z.L, Yang, X.E., Baligar, V.C., & Calvert, D.V. (2003). Microbiological and biochemical indexing systems for assessing quality of acid soils. *Advances in Agronomy*.78:89-138.

Hedin, L. O., Vitousek, P. M. , & Matson, P. A. (2003). Nutrient losses over four million years of tropical forest development. *Ecology* .84:2231–2255.

Hobbie, S.E. (2008). Nitrogen effects on decomposition: A five-year experiment in eight temperate sites. Ecology. 89: 2633-2644.

Holland, E.A., & Coleman, D.C. (1987). Litter placement effects on microbial and organic matter dynamics in an agroecosystem. *Ecology*.68:425-433.

Ibekwe, A.M., Kennedy, A.C., Halvorson, J.J., & Yang, C-H. (2007). Characterization of developing microbial communities in Mount St. Helens pyroclastic substrate. *Soil Biology and Biochemistry*. 39: 2496–2507

Israel, D.W. (1993). Symbiotic dinitrogen fixation and host-plant growth during development of and recovery from phosphorus deficiency. *Physiologia Plantarum*. 88: 294–300

Jimenez, J.J., Lal, R., Leblanc, H.A., Russo, R.O., & Raut, Y. (2008). The soil C pool in different agroecosystems derived from the dry tropical forest of Guanacaste, Costa Rica. Ecological Engineering. 34, 289-299.

Jobbágy, E.G., Jackson, R.B.(2000). The vertical distribution of soil organic carbon and its relation to climate and vegetation. Ecological Applications. 10:423-436.

Kardol, P. & Wardle, D.A. (2010). How understanding aboveground–belowground linkages can assist restoration ecology. Trends in Ecology and Evolution, 25:670-679

Keller, M., Veldkamp, E., Weitz, A.M. & Reiners, W.A. (1993). Effect of pasture age on soil trace-gas emissions from a deforested area of Costa Rica. Nature 365:244-246.

Kemnitz, D., Kolb, S. & Conrad, R. (2005). Conrad. Phenotypic characterization of Rice Cluster III archaea without prior isolation by applying quantitative polymerase chain reaction to an enrichment culture. *Environmental Microbiology* 7: 553-565.

Kent, A. D., & Triplett, E.W.. (2002). Microbial communities and their interactions in soil and rhizosphere ecosystems. *Annual Review in Microbiology*. 56: 211 236.

Knorr, M., Frey, S.D., & Curtis, P.S. (2005). Nitrogen additions and litter decomposition: a meta-analysis. Ecology 86:3252–3257.

Laurance, W. F., Peres, C.A., Eds. (2006). Emerging threats to tropical forests. University of Chicago Press, Chicago.

Laurance, W.F., Laurance, S.G., Ferreira, L.V., Rankin-De Merona,J.M., Gascon, C. & Lovejoy, T.E. (1997). Biomass collapse in Amazonian forest fragments. Science 278: 1117–1118.

Leininger, S., Urich, T., Schleter, M., Schwark, L., Qi, J., Nicol, G., Prosser, J., Schuster, S., & Schleper, C. (2006). Archaea predominate among ammonia-oxidizing prokaryotes in soils. *Nature*. 442: 806-809.

Litton, C., & Giardina, C.P. (2008). Belowground carbon flux and portioning: global patterns and response to temperature. *Functional Ecology*. 22:941-954.

Lovell, R.D., Jarvis, S.C., & Bardgett, R.D. (1995). Soil microbial biomass and activity in long-term grassland: Effects of management changes. *Soil Biology and Biochemistry*, 27:969-975.

Luis, P., G. Walther, H. Kellner, F. Martin, & Buscot, F. (2004). Diversity of laccase genes from basidiomycetes in a forest soil. *Soil Biology and Biochemistry*, 36:1025-1036.

Malik, K. A.,& Haider, K. (1982). Decomposition of 14C labeled melanoid fungal residues in a marginally sodic soil. Soil Biol. Biochem. 14:457-460.

Malloch, D. W., K. A. Pirozynski, & Raven, P.H. (1980). Ecological and evolutionary significance of mycorrhizal symbioses in vascular plants (a review). *Proceedings of the National Academy of Sciences, USA*, 77:2113-2118.

Mancinelli, R.L. (1995). The regulation of methane oxidation in soil. *Annual Reviews in Microbiology* 49: 581-605.

Martin-Laurent, F., Philippot, L., Hallet, S., Chaussod, R., Germon, J.C., Soulas, G. & Catroux, G. (2001). DNA extraction from soils: old bias for new microbial diversity analysis methods. *Applied and Environmental Microbiology* 67: 2354-2359.

McGroddy, M.E., Daufresne, T., & Hedin, L.O. (2004). Scaling of C:N:P stoichiometry in forests worldwide: implications of terrestrial redfield-type ratios. *Ecology*: Vol. 85, No. 9, pp. 2390-2401.

Monge, G. Chassot, O., Lopez, R. & Chaves, H. (2002). Justificación biológica para el establecimiento del propuesto Parque Nacional Maquenque. San José, Costa Rica: Corredor Biológico San Juan-La Selva / Centro Científico Tropical, 50 p.

Moscatelli, M.C., Lagomarsino, A., Marinari, S., De Angelis, P. & Grego, S. (2005). Soil microbial indices as bioindicators of environmental changes in a poplar plantation. *Ecological Indicators*. 5:171-179.

Murase, J. & Frenzel, P. (2007). A methane-driven microbial food web in a wetland rice soil. *Environmental Microbiology* 9: 3025-3034.

Nelson, D.W. & Sommers, L.E. (1996). Total carbon, organic carbon, and organic matter. In: Methods of Soil Analysis, Part 2, 2nd ed., A.L. Page et al., Ed. Agronomy. 9:961-1010. Am. Soc. of Agron., Inc. Madison, WI.

Nüsslein, K. & Tiedje, J.M. (1999). Soil Bacterial Community Shift Correlated with Change from Forest to Pasture Vegetation in a Tropical Soil. Applied and Environmental Microbiology 65:3622-3626.

O'Brien, H. E., J. L. Parrent, J. A. Jackson, J. M. Moncalvo, & Vilgalys, R. (2005). Fungal community analysis by large-scale sequencing of environmental samples. *Applied and Environmental Microbiology*, 71:5544-5550.

Ochsenreiter, T., Selezi, D., Quaiser, A., Bonch-Osmolovskaya, L. & Schleper, C. (2003). Diversity and abundance of Crenarchaeota in terrestrial habitats studied by 16S RNA surveys and real time PCR. *Environmental Microbiology* 5: 787-797.

Ohtonen, R., Fritze, H., Pennanen, T., Temminghoff, E., & Van Der Lee, J.J. (1999). Ecosystem properties and microbial community structure in primary succession on a glacier forefront. *Oecologia*.119:239-246.

Padmanabhan, P., Padmanabhan, S., Derito, C., Gray, A., Gannon, D., Snape, J.R., Tsai, C.S., Park, W., Jeon, C., & Madsen, E.L. (2003). Respiration of 13C-labeled substrates added to soil in the field and subsequent 16S rRNA gene analysis of 13C-labeled soil DNA. *Applied Environmental Microbiology*. 69:1614–1622.

Pendall, E., Rustad, L., & Schimel, J. (2008). Towards a predictive understanding of belowground process responses to climate change: have we moved any closer? *Functional Ecology*. 22:937-940.

Pons, T.L., Perrejin K., Van Kessel, C., & Werger, M.J.A. (2007). Symbiotic nitrogen fixation in a tropical rainforest: ^{15}N natural abundance measurements supported by experimental isotopic enrichment, *New Phytologist*, *173*(1), 154-167

Powlson, D.S., Brookes, P.C., & Christensen, B.T. (1987). Measurement of soil microbial biomass provides an early indication of changes in total soil organic matter due to straw incorporation. *Soil Biology and Biochemistry*. 19:159–164.

Reed, S.C., Seastedt, T.R., Mann, C.M., Suding, K.N., Townsend, A.R. & Cherwin, K.L.(2007). Phosphorus fertilization stimulates nitrogen fixation and increases inorganic nitrogen concentrations in a restored prairie. *Applied Soil Ecology*.36:238-242.

Reiners, W. A., Bouwman, A.F., Parsons, W.F.J. & Keller, M. (1994). Tropical Rain Forest Conversion to Pasture: Changes in Vegetation and Soil Properties. Ecological Applications 4:363-377.

Rutz, B.A. & Kieft, T.L. (2004). Phylogenetic characterization of dwarf archaea and bacteria from a semiarid soil. *Soil Biology and Biochemistry* 36:825-833.

Sahrawat, K. L. (2008). Factors Affecting Nitrification in Soils. Communications in Soil Science and Plant Analysis. 39:9,1436-1446

Saiya-Cork, K.R., Sinsabaugh, R.L., & Zak, D.R. (2002). The effects of long term nitrogen deposition on extracellular enzyme activity in an *Acer saccharum* forest soil. Soil Biol Biochem 34: 1309–1315.

Schleper, C., Jurgens, G. & Jonuscheit, M. (2005). Genomic studies of uncultivated archaea. *Nature Reviews Microbiology* 3: 479-488.

Schulze, J. (2004). How are nitrogen fixation rates regulated in legumes? *Journal of Plant Nutrition and Soil Science*. 167:125-137.

Schwendenmann, L. & Veldkamp, E. (2006). Long-term CO_2 production from deeply weathered soils of a tropical rain forest: evidence for a potential positive feedback to climate warming. *Global Change Biology* 12:1878-1893.

Simon, H.M., Dodsworth, J.A. & Goodman, R.M. (2000). Crenarchaeota colonize terrestrial plant roots. *Environmental Microbiology* 2: 495-505.

Simon, H.M., Jahn, C.E., Bergerud, L.T., Sliwinski, M.K., Weimer, P.J., Willis, D.K. & Goodman, R.M. (2005). Cultivation of mesophilic soil crenarchaeotes in enrichment cultures from plant roots. *Applied and Environmental Microbiology* 71: 4751-4760.

Sinsabaugh, R.L., Zak, D.R., Gallo, M., Lauber, C., & Amonette, R. (2004). Nitrogen deposition and dissolved organic carbon production in northern temperate forests, Soil Biology & Biochemistry 36:1509–1515.

Sinsabaugh, R.L., Carreiro, M.M., & Repert, D.A. (2002). Allocation of extracellular enzymatic activity in relation to litter composition, N deposition, and mass loss. Biogeochemistry 60:1–24.

Sinsabaugh. R.L.. (2010). Phenol oxidase, peroxidase and organic matter dynamics of soil. Soil Biology & Biochemistry 42 391-404.

Sliwinski, M.K. & Goodman, R.M. (2004). Comparison of consortia inhabiting the rhizospere of diverse terrestrial plants with those in bulk soil in native environments. *Applied and Environmental Microbiology* 70: 1821-1826.

Smit, E. Leeflang, P., Glandorf, B., Van elas, J.D., &. Wenars, K. (1999). Analysis of fungal diversity in the wheat rhizosphere by sequencing of cloned PCR-amplified genes encoding 18S rRNA and temperature gradient gel electrophoresis. *Applied and Environmental Microbiology* 65:2614-2621.

Smith, V.H., Graham, D.W., & Cleland, D.D. (1998). Application of Resource-Ratio Theory to Hydrocarbon biodegradation. *Environmental Science and Technology*. 32:3386-3395.

Smith, V.H. (1992). Effects of nitrogen-to-phosphorus supply ratios on nitrogen fixation in agricultural and pastoral ecosystems, *Biogeochemistry*. 18:19–35.

Sollins, P. (1998). Factors influencing species composition in tropical lowland rain forest: does soil matter? *Ecology* 79:23–30.

Stahl, P.D., Parkin, T.B., & Christensen, M. (1999). Fungal presence in paired cultivated and uncultivated soils in central Iowa, USA. Biology and Fertility of Soils. 29:92-97.

Talbot, J.M., Allison, S.D. &Treseder, K.K. (2008). Decomposers in disguise: mycorrhizal fungi as regulators of soil C dynamics in ecosystems under global change. Functional Ecology 22, 955-963.

Treusch, A.H., Leininger, S., Kletzin, A., Schuster, S.C., Klenk, H.-P. & Schleper, C.(2005). Novel genes for nitrite reductase and Amo-related proteins indicate a role of uncultivated mesophilic crenarchaeota in nitrogen cycling. Environ Microbiol 7: 1985–1995.

Van Der Wal, A., Van Veen, J.A., Smant, W., Boschker, H.T.S., Bloem, J., Kardol, P., Van Der Putten, W.H., & De Boer, W. (2006). Fungal biomass development in a chronosequence of land abandonment. *Soil Biology and Biochemistry*. 38:51-60.

Vitousek, P. M., & Howarth, R. W. (1991). Nitrogen limitation on land and in the sea: how can it occur? *Biogeochemistry* 13:87–115.

Vitousek, P. M., & Farrington, H. (1997). Nutrient limitation and soil development: experimental test of a biogeochemical theory. *Biogeochemistry* 37:63–75.

Waldrop, M.P. & Firestone, M.K. (2006). Response of microbial community composition and function to soil climate change. *Microbial Ecology* 52: 716–724.

Wang, B. & Qiu, Y.L. (2006). Phylogenetic distribution and evolution of mycorrhizas in land plants. Mycorrhiza 16: 299-363.

Wardle, D.A. (2006). The influence of biotic interactions on soil biodiversity. Ecology Letters 9:870-886.

Wardle, D. A., R. D. Bardgett, J. N. Klironomos, H. Setala, W. H. van der Putten, & Wall, D.H. (2004). Ecological linkages between aboveground and belowground biota. Science 304: 1629-1633.

Webster, G., Embley, T.M. & Prosser, J.I. (2002). Grassland management regimens reduce small-scale heterogeneity and species diversity of β-proteobacterial ammonia oxidizer populations. *Applied and Environmental Microbiology* 68: 20-30.

Wessén,E , Nyberg, K., Jansson, J.K., & Hallin. S. (2010). Responses of bacterial and archaeal ammonia oxidizers to soil organic andfertilizer amendments under long-term management. Applied Soil Ecology 45 (2010) 193–200.

Whalen, S.C., Reeburgh, W.S. & Sandbecks, K.A. (1990). Rapid methane oxidation in a landfill cover soil. *Applied and Environmental Microbiology* 56:3405-3411.

Whitfield S.M., Bell K.E., Philippi T., Sasa M., Bolaños F., Chaves G., Savage J.M., II, & Donnelly M.A. (2007) Amphibian and reptile declines over 35 years at La Selva, Costa Rica. Proc. Natl Acad. Sci. USA. 104, 8352–8356.

Wolters V, Silver WL, Bignell DE, Coleman DC, Lavelle P, Van der Putten WH, De Ruiter P, Rusek J, Wall DH, Wardle DA, Brussaard L, Dangerfield JM, Brown VK, Giller KE, Hooper DU, Sala O, Tiedje J, & Van Veen, J.A. (2000). Effects of global changes on above- and belowground biodiversity in terrestrial ecosystems: implications for ecosystem functioning. Bioscience 50:1089–1098.

Worrall, J.J., Anagnost, S.E., & Zabel, R.D. (1997). Comparison of wood decay among diverse lignicolous fungi. Mycologia. 89:199-219.

Zech, W. & Kögel-Knabner, I. (1994). Patterns and regulation of organic matter transformation in soils: litter decomposition and humification. In: Schulze, E.D. (Ed.), Flux Control in Biological Systems. Academic Press, New York, NY, pp. 303-334.

Zeller, V., Bardgett, R.D. & Tappeiner, U. (2001). Site and management effects on soil microbial properties of sub-alpine meadows: a study of land abandonment along a north-south gradient in the European Alps. Soil Biology and Biochemistry. 33:639-649.

Zhang, Q. & Zak, J.C. (1995). Effects of gap size on litter decomposition and microbial activity in a subtropical forest. *Ecology* 76: pp. 2196–2204.

5

Sustainable Solutions for an Environmentally and Socially Just Society*

Sukhmander Singh
Nicholson Family Professor, Civil Engineering, Santa Clara University, Santa Clara, CA USA

1. Introduction

Increasing per capita resource consumption in developed nations on one hand and global poverty, hunger, and poor sanitation on the other hand require efforts directed at both environmental and social sustainability. A model is presented for addressing sustainability in society. In this model, society, environment, and technology are interlocking parameters that dictate the nature and types of efforts for sustainability. Society provides for all its citizens the very basic necessities: potable water, food, education, housing and sanitation. Environment (or ecosystem) includes the natural resources to be exploited economically in service to society, and determines the value of the resource. These factors must be considered simultaneously in developing appropriate and sustainable solutions for an environmentally and socially just society. The model is applicable for any country that seeks sustainability and justice for its citizens. Distinctions between technology and energy-intensive production in developed countries and labour-intensive production in developing countries are made, considering the United States, India, Ghana, and Nicaragua as examples.

It is obvious that the state of a society's economic system influences the production systems that are appropriate to that society, and the production systems and standard of living further impact the ecosystem. The ecosystem contains the resources that support our standard of living. While we depend on it, the ecosystem is autonomous and will continue even as it changes under the pressure of human activity. The ecosystem has been viewed as a commons that can be tapped for production; production often is motivated by the profits to be had, with degradation of the commons ignored to the degree possible.. Increasing per capita consumption of resources, and the resulting pollution, waste and global warming have led to the widespread recognition that we must not deplete and/or pollute the ecosystem to the extent that the ability of future generations to meet their needs is compromised. Such a view accepts human-induced change to the ecosystem, but recognizes that environmental problems are now of such magnitude that they require concerted global efforts. This concern has been expressed and articulated in many ways and through many international conferences (World Council on Economic Development, 1987, Peter et al 1997, Earth Summit, Rio, 1992). Terms such as "sustainability," "renewability," and "sustainable development" are attracting worldwide attention. The key thrust which

---

*This chapter is an extended exposition with new material, of paper published in the Proceedings of Geo Congress-2008 of the American Society of Civil Engineers.

emerges from all these deliberations is the notion of "sustainable development," which is closely tied to "economic development" and thus does not reflect a shift away from the present view that the commons is available to support further development. It is not clear that the standard of living that we have become accustomed to can be supported sustainably.

The calculus of sustainability has not truly entered into decision-making about production and engineering. Therefore, there is a need to change the curricula to explicitly teach engineers to consider sustainability in their design decision-making. Sustainability is a cross-cutting theme that should be present throughout an engineer's education rather than being contained within a single dedicated course. Thus, it is useful to have examples that can carry the essential themes and background to support the emergence of a calculus of sustainability in engineering design. At San Clara University, sustainability concepts are taught using some of the examples that follow in some of the undergraduate courses: Engineering (for juniors), Civil Engineering Materials (for juniors), Green Construction Design (for seniors), and Sustainable Water Resources (a new elective course).

2. Sustainable development

The term "sustainable development" is being interpreted differently by developed nations on one hand and by developing countries on the other. For example, a measure against pollution in a developed country may make sense, but will be a luxury for a developing country. Developing nations may insist on more attention to economic growth than to environmental problems. In developed countries, a check on economic growth to protect ecosystem is often considered a check on freedom and free enterprise. How then to resolve the conflict between a desire to develop, and the need to maintain the integrity of ecosystem?

Figure 1 presents a simple schematic model for a country, which can help guide sustainability decision making. The big box represents a country. As the growth accelerates/cranks faster to further the development of a country, (i) the balance between ecosystem and economic prosperity becomes more delicate for developed nations; and (ii) the balance between economic prosperity and social development becomes more delicate for developing countries. For example if development is linked only to gross national product (i.e. GNP should rise every year), the society may be headed to the depletion of its ecological base, and therefore society may be becoming poorer (Warner, 2006). This means that developed countries must recognize limits to their growth and should look for alternate but sustainable resources or alternate ways of obtaining pollution-free energy.

Mario Belotti, an internationally known professor of economics at Santa Clara University, argues that "resource is a function of technology" (Belotti, 2006). Accordingly, developed nations should focus on advancing technology to find alternate ways to meet the requirements of society while preserving the balance of ecosystem. For example, resources in developed countries should be used to develop bio-based and fusion energy sources, while also improving the efficiency of current renewable energy sources such as wind, hydro, geothermal, and solar technologies. Improved energy technologies would benefit both developing and developed countries.

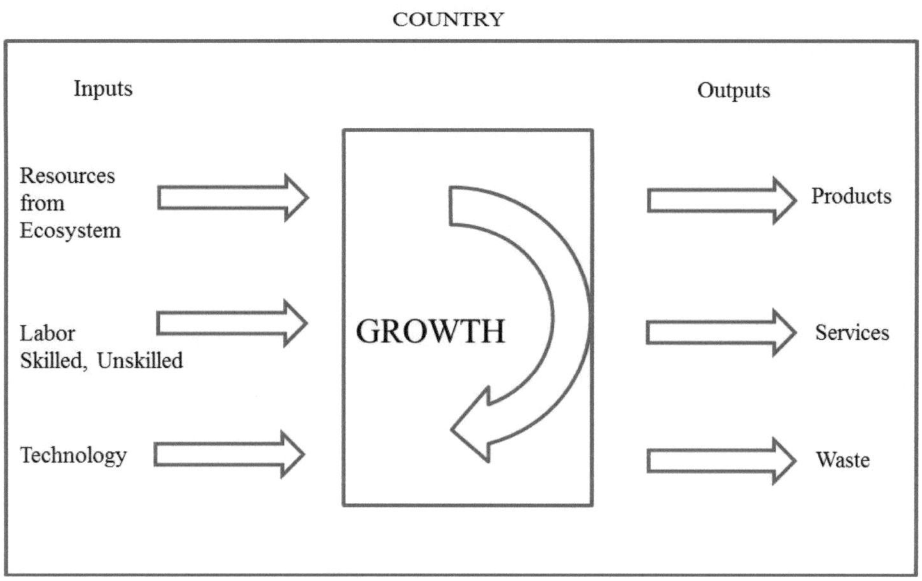

Fig. 1. Schematic model for sustainability decision making.

In developing countries where production is not highly advanced, indiscriminate expansion of production systems may not only lead to an undue burden on the ecosystem, but also to the concentration of wealth in relatively few hands. Sustainable development, however, must consider three dimensions: 1) protection of eco-system, 2) social development and 3) economic prosperity (Earth Summit, 1992). In some poor countries where poverty, hunger and poor sanitation exist, both environmentally acceptable, economically accessible, and socially sustainable solutions must be found to bring about a developed and socially just society. These solutions ought to be simple, inexpensive and environmentally safe. Examples of such solutions are presented below.

3. Mass production versus production by masses

In developing countries, where rice production is the main crop, for example in India and China, large scale burning of rice straw (harvest waste) often creates widespread emissions of $C0_2$ and other pollutants all over the countryside. This straw, however, when shredded and mixed with clay, can be molded into bricks. Thus a mixture of the shredded straw and clay can be used to produce both lightweight and insulating building materials. Shredding, mixing and molding can be achieved using cheap manual labor; with small scale-kilns or ovens employed to bake these bricks. This will provide employment to a large number of persons as well as minimize pollution caused by burning rice straw. Such a labor-intensive process can be termed as "production by masses" in contrast to "mass production". Labor-intensive processes also may be more sustainable; in this case, waste rice straw available at the harvest time is harnessed each year.

Another example is from the upper east region of the country of Ghana. Though Ghana is one of the most stable countries in Sub Sahara Africa, the rural parts of this country are still struggling to push past the poverty line. Many of those struggling people are spending more of their time, energy and money on housing than necessary.

Their construction methods include two types: mud huts, and under block structures with corrugated tin roofing. The most impoverished of the community use the local clay soil for walls and millet straw roofing to build their huts/homes. These huts do not survive multiple seasons of rain fall and must be re-constructed every few years. This is due to the make-up of local soils which are laterites. Laterites are residual soils and are notoriously known for their loss of strength upon reworking and cycles of wetting and drying.

Because of the general poverty, people cannot afford a traditional house made of well baked bricks of good imported soils or of reinforced concrete. Setting up of manufacturing plants for mass production of high quality bricks of local material is an option which, because of extensive availability of local soils, is a sustainable solution. But this may not be an affordable as well as a just solution for the poor people. The concept of production by masses using inexpensive way to produce bricks by local people can not only provide an employment base for the people but it will also be an environmentally and socially just solution.

Over the past two years, students of Santa Clara University undertook senior design projects to produce designs and build houses made up of specially produced bricks of local lateritic soils. The aim was to develop inexpensive procedures for making bricks that could be easily learnt/adapted by local people. Similarly, the design and the construction of the houses were to be simple for easy adaptation of the methodology by local people. The brick-making equipment uses a simple lever system to apply pressure to produce well-compacted bricks. These bricks were sun baked. The design and construction used an arch shape (see Figure 2) to avoid the use of expensive materials such as cement and reinforcing bars not available in that area. In Figure 3, local people can be seen taking part in the construction of the houses. Further details can be found in the report entitled "Sustainable House in Ghana," submitted by Daniel Lawrie and William Sommer, Dept of Civil Engineering, Santa Clara University (May 2011).

There are several NGO's looking for appropriate technology applications to help people in marginal or developing regions of the world. And efforts are often directed towards solutions that are relatively inexpensive and are sustainable. However, it is important that an employment base should also be provided through mass employment of local people who can be trained through a skill-based education in the production of the appropriate technology solution. For example in Nicaragua, in the city of Managua, several teens, even younger children, work in the city dump under extreme un-hygienic and polluted environment for collecting plastic cans, etc. to sell and make a living (Fig 4). NICAHOPE, an NGO has set up a program next to the dump. This program motivates children away from the dump to receive skill-based education to make jewelry (Fig 5), selling of which they make more money in a shorter time than what they make in the dump; whereby they can remain in school also run by the program. This skill-based education leading to production of jewelry by impoverished children is sustained through donations. Such solutions which are both just and sustainable should be encouraged.

Fig. 2. Arch-Shaped Brick House.

An arch shape is used to avoid use of expensive materials such as cement and reinforcing bars not available in the area.

Fig. 3. Community Building Efforts.

Local people of Ghana are seen taking part in the construction of the house.

Fig. 4. City Dump Collecting. Three children collecting bottles and cans at the city dump.

Fig. 5. Activities for Local Children. NICAHOPE helpers training the local children in jewelry-making.

4. Sustainable building materials

Construction materials represent a large percentage of the raw materials used in developed countries. Figure 6 illustrates relative amounts of materials used in the United States, with construction materials by far being the largest of the categories listed. The environmental consequences are significant. For example, in both the United States and globally, the manufacture of Portland Cement for use in concrete accounts for approximately 8% of anthropogenic emissions of CO_2 (Wagner, 2002). Thus, alternative, more sustainable materials must be found.

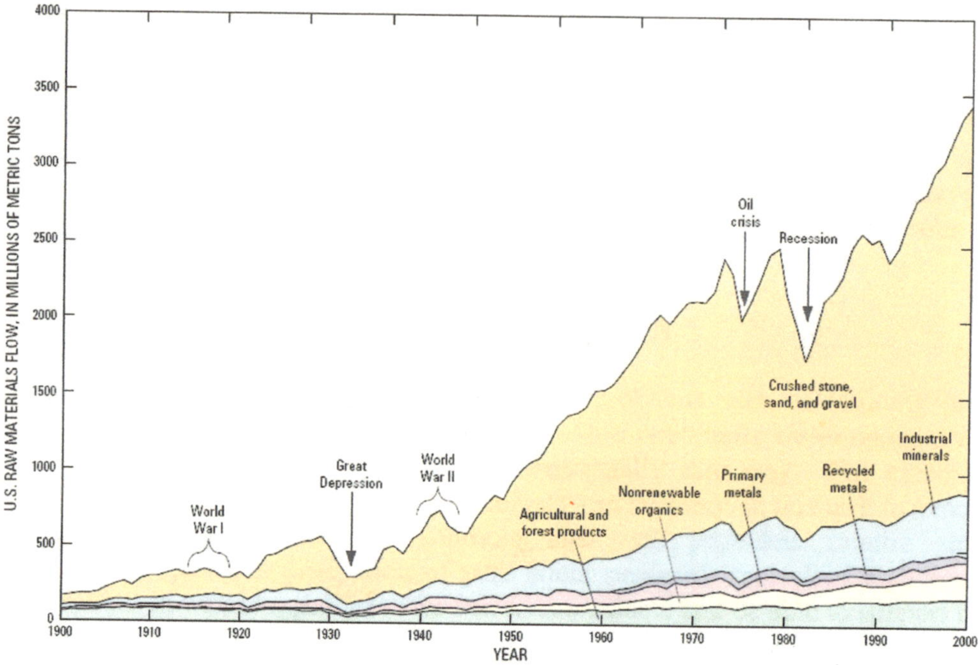

Fig. 6. Raw materials consumed in the United States: 1900-2000. (Wagner, 2002).

In developed countries, high labor costs relative to materials costs tends to result in high performance products that embody significant amounts of energy and technology. Examples include very strong chemical adhesives used for making engineered lumber and plywood. Sustainable alternatives may make use of sophisticated technologies to allow a savings in energy but possibly with some reduction in performance. Thus, it may be feasible to use manufactured biocomposites in place of steel or concrete. An example is the use of engineered lumber or a new alternative in which beams are made from harvested bamboo. The bamboo is attractive because it is stronger than wood and rapidly renewable, since the stalks can be harvested every 3 to 5 years. The bamboo beams may be formed from strips of bamboo assembled together into solid section I-beams or by extruding a beam made from bamboo chips and a bonding agent; the latter being especially appropriate in developed countries where labor is costly, while the former being more applicable in developing countries. Technologies to allow more sustainable adhesives to be used would be especially valuable; lignin resulting from anaerobic digestion is one binder that is currently being explored.

In developing countries, low labor costs and high needs may dictate solutions in which even basic technologies are employed to make production processes more efficient. For example, the firing of clay bricks in developing countries often relies upon relatively simple technologies implemented widely by small businesses. In some cases, the kilns are very crude and inefficient, leading to excessive consumption of wood and emissions of CO_2 and other pollutants. At the same time, there may be little or no quality control and the strengths of the bricks may be far in excess of the strengths required for dwelling construction. Businesses operating at this scale are often unaware of better technologies already in existence which can be employed for production by masses as suggested earlier to reduce pollution as well as to provide employment to local people.

Consequently, the application of relatively simple technologies could dramatically reduce wood consumption, CO_2 emissions, and the cost of masonry construction. Lower costs would allow this form of construction to be more widely used, displacing cheaper alternatives that have a worse history of performance in natural disasters such as earthquakes (Aschheim et al, 2007).

Another renewal and sustainable building material is strawbale. Construction with straw bale is now well developed. The bales of straw are stacked like bricks to form walls, which are then covered with plaster or gunite. Reinforcing mesh used within the plaster confers strength and ductility, while the bales themselves provide ample thermal and sound insulation. The use of good details can provide durability of a hundred years or more, as evidenced by buildings built in Nebraska beginning in the late 1800s. Bales composed of rice or wheat straw are used most often. The straw is a waste product resulting from the production of grain. Since burning in the fields has been prohibited, the bales have to be disposed of. The use of a waste product from food production as a material of construction reduces the need to dedicate land to forestry, and the use of this particular material provides significant savings of energy required to maintain comfortable indoor temperatures. Most strawbale buildings (see figure 7 for example) are approved under the alternative means and methods provisions of the building code, but efforts are underway to have provisions for strawbale construction adopted into mainstream building codes (Aschheim, 2011).

5. Social concerns in geotechnical design

Application of solutions to certain geotechnical problems such as building high earthen enhancements to raise highways that pass thru potentially waterlogged areas can be handled in a both socially and environmentally safe way. For example, although high-technology earth-moving machinery can accomplish this task in a more time efficient way, geotechnical solutions for such problems require slow construction to allow the dissipation of pore water pressures typical of high water table situations in water logged areas for soil to consolidate and be able to support the load of the highway embankment and the traffic. Use of manual labor (usually a large number) for hauling and placing fill on the embankment is a very common practice in India and in many developing countries. This can, not only, be economical on fuel consumption thereby minimizing pollution, but can also be a socially and geotechnically acceptable solution.

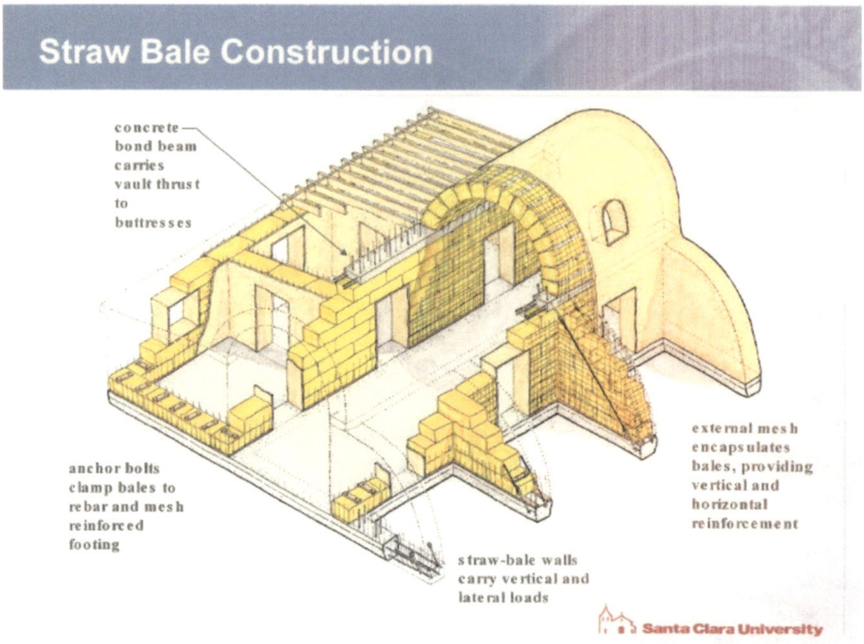

Fig. 7. Strawbale Construction.

Geotechnical engineering often require clearing, excavating, or moving large amounts of earth for the construction of highways, dams, tunnels, and housing construction. Whereas large highly advanced machinery is to accomplish these tasks, it may be possible that the geotechnical construction can be achieved in a more environmentally friendly way. Instead of one large dam two small dams requiring lesser environmental damage may be considered. Use of lightweight construction material for housing construction may help in reducing deep side hill cuts often made to build large view homes. Use of bamboo for housing has already been discussed. Since building in, on, and with earth is what geotechnical engineers do, the aforementioned considerations can help preserve the health of the earth (NRC 2006).

6. Entrepreneurship and social contract

In developing sustainable solutions for developing countries, engineers can play a leading role. But they have a special responsibility in making sure that their innovation or the technology that they have developed should not be exploited by rich /eager/self-made entrepreneurs for money/profit. There is the concept of social contract. According to which, since all entrepreneurships rely on existing infrastructures and existing services in order to function and succeed, it is therefore ethical to expect them to contribute to the society not in a philanthropic way, but in a structured way and that is what is called social contract. Elizabeth Warren of the New York Times has rightfully stated, "There is nobody in this country who got rich on his own. Nobody." She declared, pointing out that the rich can only get rich thanks to the "social contract" that provides a decent functioning society in which they can function (New York Times Sept 23, 2011). This social contract should be so structured as to make it mandatory for the entrepreneurs to train and engage the local community. In some cases it may not be easy and could be time consuming. But it is worth it for the confidence and employment base it will create in the community/people.

Several NGO's over the years have come up with appropriate technology solutions for clean water. Application of these solutions is though for people of developing countries, but they should be guarded against exploitation by self-made entrepreneurs for profits by the above noted concept of "social contracts." Framing of 'social contracts" can vary from country to country to accommodate the social and culture climate of the country, so that the appropriate technology solutions remain sustainable.

At the time of writing this article (2011), no appropriate technology has emerged to clean water of metal impurities of lead and mercury from fertilizers and from industrial wastes dumped in steams.

Several social entrepreneurs are developing affordable technologies to help marginal communities all around the world. These technologies are mostly based on the concept of sustainability. For example in the area of affordable renewable energy to the underserved customers, they are taping sustainable power from the sun and renewable power from biomass to provide cleaner cooking and off-grid lighting. These are the most renewable power sources found everywhere in the world (Center for Science Technology and Society, CSTS, Santa Clara University, SCU 2011). Solar power is well known but power/energy from biomass is relatively less understood. Biomass can be turned into energy in the form of heat by processing it into cooking fuel, or charring it into charcoal or a cleaner cooking ,method can be adopted, for example in India (Fig 8).

Fig. 8a. Old Fashioned Cooking Producing Pollution.

Fig. 8b. Cleaner Cooking Method.

There are several outfits across the world that develop biomass power. There are many more working on solar power to provide off-grid lighting and electricity, cleaner cooking, bio powered engines and generators, portable lanterns, efficient stoves, etc. There are about 40 social entrepreneurs/enterprises which are trying to deliver energy solutions to the underserved populations of the world (Center for Science Technology and Society, CSTS, Santa Clara University, SCU 2011). Some of these are for profit, some for nonprofit and still others are hybrid. These are all highly commendable. But whereas local assets in the form of renewable materials like plants or biomass waste (rice husks, manure, and animal waste) have been utilized, assets of untapped/untrained local human skills have been ignored. Employment can be provided by the entrepreneurs, by training and employing local people to build/make these products. This will further local people's progress towards prosperity or getting out of the poverty line.

7. Overall sustainability in developed nations

With reference to the schematic model for a country (Fig 1), and to repeat that as the growth accelerates in a developed country, to further the development and to raise the standards of living, two possible outcomes can be expected.

First, since the development should not be pushed at the cost of depletion of nation's ecological base, the developed countries must recognize limits to their growth and should look for alternate but sustainable resources of energy. As pointed out earlier, the developed nations should focus on advancing technologies to find alternate ways to meet the requirements of the society while preserving the balance of the ecosystem. Such technologies are to focus on renewable energy resources like wind, hydro, geothermal, and the sun. A lot of progress has been made on these fronts, and efforts are to be directed to improve the efficiency of the technology in using these resources. Bio-based and fusion energy sources can also be developed.

The second outcome is related to the impact of the technology based growth/progress upon the societal values which may not be sustainable due to the onslaught of the technology. Developed nations' ever increasing hunger for energy through technological advances for a more fulfilling life may be a *fallacy*:

A Fallacy:

Back in the early seventies, steel plants in Pittsburg were introducing automation and mechanization and consequently laying off employees who were going on strikes for losing years and years of family tradition. Newspapers, however across the nation carried prominently news headlines that "Three-Day Weekend is coming because we will be producing so much more due to the automation and mechanization, and we will be so much richer that we will not have to work for 5 days." Have we? These new innovations and gadgetry are supposed to relieve us of the drudgery of doing by ourselves so that we will have more free time. Have we? No. Clearly, there is a fundamental fallacy in this line of thought. An average engineer in Silicon Valley is working 60 to 70 hours a week.

In fact the concept of weekend seems to be disappearing. To be successful here as an engineer, one has to optimize not only one's intellectual asset, but asset of family time with one's family, asset of social time with friends and so on. In fact in other walks of life, too, modern society seems to drive us to exhaustion, with continuous frantic activity and busy behavior, leaving hardly time for reflection. Is this then a "moral" problem and not an "energy" problem which ties into reflectiveness? In the reflective vacuum there is an opportunity for fascination with and immersion into all kinds of technology at the expense of human relationships and general well-being. Negative influences seem to have at their root a money-making enterprise. No one makes money on societal values, moral behavior, or ethics, and so the tools of influence (advertising, role models) are not engaged in promoting these perspectives—they are silenced merely by the lack of an individual's profit mongering to promote them. In this milieu when the negative influences out compete the positive, can we sustain societal values. If depletion of the ecosystem requires sustainability efforts, erosion of societal values also require sustainability efforts to guard against the loss of moral values. It is not fair or just to drive a society into a highly technological set up at the cost of its moral values.

Increase competitiveness is creating ruthless climate such that we no longer have colleagues or class fellows, but rivals in all walks of life. And in this ruthless climate, we seem to be more willing to compromise our principles. How, then, can we resolve this fallacy brought about by science and technology? We need to bring reform in our science education. Education in science and technology should include its interface with ethical, moral, social and economic aspects of the society. We must teach the relevance of advances or innovations in science and technology to socio-political as well as cultural aspects of our lives. Science and technology has stridden way ahead and our ability to use these advances wisely has lagged behind. By developing skills and understanding of the interface between science, technology and society, we can minimize negative impacts of science and technology and can sustain societal values.

Since we are never subjected to a feeling of contentment in a climate where catchy commercials are continuously showered on us, inflaming greed or a feeling of left-out if we cannot buy and use the latest, a certain kind of unhappiness creeps in us. Some bad elements in the society may be driven to getting these by unfair means (robbery, stealing, etc). We need to teach a balance between business education and societal values. How to interface these so that each can sustain and go hand in hand to benefit society.

8. Conclusion

- Sustainable solutions to protect ecosystems can vary from country to country and are influenced by economies and social constitution of the country.
- Sustainable development, which includes both environmentally and socially sustainable solutions, for example, can be achieved by incorporating available material, and local labor.
- Solutions in developed countries often involve high technology approaches at times at the cost of erosion in societal values. An understanding of the interface between technology and society can help sustain societal values while developing technology based sustainable solutions.

9. Acknowledgments

Professor Keith Warner of Santa Clara University is acknowledged for his insightful discussions on the subject.

A special thanks to Reggie Antonio for his assistance in helping put this paper together.

10. References

Aschheim, M., Flanagan, S., Harlander, J., Pitt, C., Alfaro, A., Rivas, C., Rodriguez, M. E., (2007) "Enhancing The Sustainability and Earthquake Resistance of Confined Masonry Dwellings in El Salvador," *Journal of Earthquake Engineering.*

Aschheim (2011), "Personal Communication" Santa Clara University

Beloti, Marie, (2006) "Personal Communication" Santa Clara University.

Center for Science, Technology and Society (CSTS), (2011) Santa Clara University, Santa Clara, California.

Daly, H. and Cobb, J. (1989) For the Common Good: Redirecting the Economy toward Community, the Environment and a Sustainable Future, Boston: Beacon Press, 1989.

National Research Council (2006). *Geological and Geotechnical Engineering in the New Millennium-Opportunities for Research And Technological Innovation,* Committee on Geological and Geotechnical Engineering, National Academics Press. Also available from http://www.nap.edu/catalog.php?record_id=11558#toc

Peter M. Vitousek, Harold A. Mooney, Jane Lubchenco, and Jerryv M. Melillo, (1977). "Human Domination of Earth's Ecosystems". Science 277(1997);499.

Wagner, L.A. (2002). *Materials in the Economy-Material Flows,Scarcity, and the Environment,* Circular 1221, United States Geological Survey, 34 pp. Also available from http://pubs.usgs.gov\circ\2002\c1221\.

Warner, Keith Douglas, (2006) "Personal Communication" Santa Clara University.

Warren, Elizabeth (2011), "Social Contract", New York Times, September 23, 2011

World Council on Economic Development (1987). *Our Common Future,* Oxford University Press.

6

Phosphate Solubilization and Mobilization in Soil Through Microorganisms Under Arid Ecosystems

B. K. Yadav* and Arvind Verma
Department of Soil Science, Rajasthan College of Agriculture,
Maharana Pratap University of Agriculture and Technology, Rajasthan
India

1. Introduction

Phosphorous is going to be plant nutrient that will limit the agricultural production in the next millennium. It is a major growth-limiting nutrient, and unlike the case of nitrogen, there is no large atmospheric source that can be made biologically available (Ezawa et al., 2002). As regards the role of P, it stimulate root development and growth, gives plant rapid and vigorous start leading to better tillering, essential for many metabolic processes in plant life and for seed formation and organization of cells, encourages earlier maturity, In most soils, its content is about 0.05% of which only 0.1% is plant available (Achal et al., 2007). The total phosphorous content in arid soils in India ranges from 560 to 900 kg ha^{-1}, the available phosphorous is quite low i.e. 15-25 kg ha^{-1} (Dhir, 1977). The soils being poor in organic matter (1.5- 4.2 mg kg^{-1}), most of these phosphorous is present in inorganic form as calcium and aluminium phosphatase. About 20-25% of total phosphorous in arid soils of India is organic in nature and 68% organic phosphorous in the soil is present as phytin (Yadav & Tarafdar, 2007), which are not directly available to plants. Therefore application of phosphatic fertilizers to the soil is essential to maintain adequate amount of soluble P in the soil solution for optimum plant growth as well as to maintain soils sustainability. Efficiency of P fertilizer throughout the world is around 10-25% (Lindsay, 1979), and concentration of bioavailable P in the soil is very low reaching the level of 1.0 mg kg^{-1} soil (Goldstein, 1994). Phosphorous is taken up from the soil in the form of soluble orthophosphate ions; $H_2PO_4^{-1}$, HPO_4^{-2} and PO_4^{-3} and generally the availability of these ions to the plants is in the order of $H_2PO_4^{-1}$> HPO_4^{-3}>PO_4^{-3}. The type of the orthophosphate ion present in the soil is depending on soil reaction. At the relatively low pH of 4 to 5, orthophosphate usually exist as $H_2PO_4^{-1}$ ions. On increasing pH, first HPO_4^{-2} ions are formed which convert to PO_4^{-3} as the soil reaction become alkaline. Large amount of P applied as fertilizer enters in to the immobile pools through precipitation reaction with highly reactive Al^{3+} and Fe^{3+} in acidic soil and Ca^{2+} in calcareous or normal soils (Gyaneshwar et al., 2002; Hao et al., 2002). Although total P pool is high, only a part is available to plants. So, the release and mobilization of insoluble and fixed forms of P is an important aspect of increasing soil P availability. Soil

---

*Correseponding Author

microorganisms play an important role in mobilizing P mainly by bringing about pH changes in soil microenvironment and producing chelating substances.

2. Forms of phosphorous in soil

Mainly two forms of phosphorous namely organic and inorganic forms occur in soils and both are important to plants as source of phosphorous. The relative amounts of phosphorous in organic and inorganic forms vary gently from soil to soil.

2.1 Organic phosphorous

Organic phosphorous: Generally organic phosphorous represents about 50% of the total P in soils (varies between 4 and 90% in most soils). Most of the organic P compounds are esters of orthophosphoric acid and have been identified primarily as (1) Inositol phosphates, (2) phospholipids and (3) nuclic acids.

1. Inositol phosphate: It represents a series of phosphate esters ranging from monophosphates up to hexaphosphates. Phytic acid, which has an empirical formula $(CH)_6$ $(H_2PO_4)_6$, had six orthophosphate (H_2P) groups attached to each carbon atom in the benzene ring. Phytin (a Ca-Mg salt of phytic acid), is the most abundant of the known organophosphorous compounds in soils. The total proportion of inositol phosphates in soil is 10-50 per cent.
2. Phospholipids: Phospholipids, phosphorous containing fatty compounds, are insoluble in water but are readily utilized and synthesized by soil microorganisms. Some of the most common phospholipids are derivatives of glycerol. The rate of release of phospholipids from organic sources in soil is rapid. Phospholipids constitute 1-5% of total organic P in soils.
3. Nucleic acids: Nucleic acids occurs in all living cells and are produced during the decomposition of residues by soil microorganisms. Two distinct forms of nucleic acids, ribonucleic acid (RNA) and deoxyribonuclic acid (DNA) are release into the soil in greater quantities than inositol phosphates, and they are broken down more quickly. Nucleic acid constitutes 0.2 to 2.5% of total organic P in soil.

2.2 Inorganic phosphorous

Most inorganic phosphorous compounds in soil fall into one of the two groups: (1) those in which calcium is the most dominant controlling cation (calcium phosphate) and (2) those in which iron and aluminium are the controlling cations (iron and aluminium phosphate).

1. Calcium phosphates: The original natural source of phosphorous is the mineral apatite, a calcium phosphate that is nearly insoluble. Apatite minerals may be found in even the more weathered soils, especially in their lower horizons. This fact is an indication of the extreme insolubility and consequent unavailability of the phosphorous contained therein. The simpler compounds of calcium such as mono and dicalcium phosphates are readily available for plant growth. These compounds are present in extremely small quantities only because they easily revert to the more insoluble forms.
2. Iron and aluminum phosphate: In this group the compounds involved are probably hydroxy phosphates such as (a) strengite- iron phosphate and (b) variscite-aluminum

phosphate. Strengite and variscite are too insoluble to contribute much to plant nutrition. The most common P minerals found in soils are presented in Table 1 in order of decreasing solubility.

Minerals	Chemical formula
Acid soils	
Strengite	$FePO_4.2H_2O$
Variscite	$AlPO_4.2H_2O$
Neutral and calcareous soils	
B-tricalcium phosphate	$Ca_3(PO_4)_2$
Dicalcium phosphate	$CaHPO_4$
Dicalcium phosphate dihydrate	$CaHPO_4. 2H_2O$
Fluorapatite	$Ca_5(PO_4)_3 F$
Hydroxyapatite	$Ca_5(PO_4)_3 OH$
Octacalcium phosphate	$Ca_4H(PO_4)_3.2\text{-}5 H_2O$

Table 1. Common phosphorous (P) minerals found in acid, neutral and calcareous soils.

3. Phosphate solubilizing microorganisms

Evidence of the involvement of microorganisms in solubilization of inorganic phosphates was reported as early as 1903 (Kucey et al., 1989; Khan et al., 2007). Since then, extensive studies on the solubilization of mineral phosphates by microorganisms have been reiewed (Goldstein, 1986; Kucey et al., 1989; Tarafdar et al., 2003; Achal et al., 2007; Aseri et al., 2009). Phosphate solubilizing microorganisms (PSMs) are ubiquitous, and their numbers vary from soil to soil. Population of PSMs and organic matter content of some selected arid soils of Rajasthan, India was reported by (Venkateswarul et al., 1984) and presented in Table 2.

S.No.	Site	Soil classification	Organic matter (%)	P-solubilizing microorganisms ($\times 10^3$ g^{-1} soil)
1.	Barmer	Torripsamments	0.16	42.0
2.	Chandan	Calciorthids	0.11	16.0
3.	Churu*	Torripsamments	0.05	5.0
4.	Churu**	Torripsamments	0.10	0.37
5.	Degai	Calci-camborthids	0.74	1300
6.	Gowrisar	Torripsamments	0.11	123.0
7.	Jetpura	Calci-camborthids	0.63	220.0
8.	Jodhpur	Camborthids	0.31	240.0
9.	Masitwali	Torrifluvents	0.25	330.0
10.	Moulasar	Torripsamments	0.33	120.0
11.	Ramgarh*	Torripsamments	0.12	7.0
12.	Ramgarh**	Torripsamments	0.02	0.20

*Stabilized dunes; ** Unstabilized dunes

Table 2. Population of phosphorous solubilizing microorganisms and organic matter content of some selected arid soils of Rajasthan, India (Venkateswarul et al., 1984).

In general among the whole microbial population in soil P solubilizing bacteria constitute 1-50% and P solubilizing fungi 0.1 to 0.5% of the total respective population (Chen et al., 2006). Phosphate solubilizing bacteria generally out-number P solubilizing fungi by 2-150 folds (Banik & Dey, 1982; Kucey, 1983; Kucey et al., 1989; Alam et al., 2002). Most of the PSMs solubilize Ca-P complexes and only a few can able to solubilize Fe-P and Al-P (Kucey et al., 1989).Hence, these PSMs could be effective in calcareous soils in which Ca-P complexes are present, but not in other soils such as Alfisols in which phosphates are complexed with Fe and Al ions. Most P-solubilizing bacteria (Baya et al., 1981; Venkateswarul et al., 1984) and fungi (Venkateswarul et al., 1984; Tarafdar et al., 2003; Tarafdar & Gharu 2005; Achal et al., 2007; Yadav & Tarafdar 2007; Aseri et al., 2009; Yadav & Tarafdar 2010; Yadav & Tarafdar 2011) were isolated from the rhizosphere of various plant and are known to be metabolically more active than those isolated from sources other than rhizosphere. The phosphate solubilizing ability in bacteria was lost upon repeated sub culturing but no such loss has been observed in the case of phosphate solubilizing fungi. (Sperber, 1958; Kucey, 1983). In general, fungal isolates exhibit greater P- solubilizing ability than bacteria in both liquid and solid culture (Banik & Dey 1982; Venkateswarlu et al., 1984; Kapoor et al., 1989).

4. Mechanisms of phosphate solubilization

The potential mechanism for phosphate solubilization might be acidification either by protone extrusion associated with ammonium assimilation (De Freitas et al., 1997; Reyes et al., 1999) or by organic acids production and proton extrusion (Cunningham & Kuiack, 1992; Dutton & Evans, 1996; Nahas, 1996; Jones, 1998). Acid phosphatases and phytases secreated by these microorganisms also have an important role in phosphate solubilization (Richardson et al., 2000; Tarafdar et al., 2003; Aseri et al., 2009). Some phosphatase and phytase producing fungi from arid soils of Rajasthan, India were isolated by Tarafdar et al. (2003) and Aseri et al. (2009). Their phosphatase (acid & alkaline) and phytase secreting efficiency at different time intervals were presented in Table 3& 4 and Table 5.

S.No.	Fungal Species	Acid phosphatases activity (EU ×10^6) g^{-1} dry fungal mat							
		Days after inoculation							
		7		14		21		28	
		I	E	I	E	I	E	I	E
1.	*Aspergillus candidus*	11.7	7.9	32.3	12.4	68.7	28.3	29.8	8.7
2.	*Aspergillus fumigatus**	26.3	10.4	52.7	58.8	69.2	24.9	21.6	10.4
3.	*Aspergillus niger*	13.6	8.7	39.8	29.3	65.6	19.3	14.7	12.3
4.	*Aspergillus parasiticus*	19.3	7.9	56.3	43.8	79.6	25.7	50.3	24.3
5.	*Aspergillus rugulosus*	14.9	11.4	16.3	13.5	27.9	11.1	14.3	7.9
6.	*Aspergillus terreus*	34.6	20.7	57.3	37.2	72.3	25.7	26.9	13.6
7.	*Penicillium rubrum*	10.4	4.3	21.3	12.9	43.6	18.9	13.4	6.9
8.	*Penicillium simplicissimum*	20.4	12.3	39.6	18.7	59.8	21.3	21.3	10.5
9.	*Pseudeurotium zonatum*	16.3	8.9	31.6	17.4	53.8	24.3	22.4	6.7
10.	*Trichoderma harzianum*	22.9	16.8	39.6	21.7	97.3	42.7	26.3	17.4
11.	*Trichoderma viride*	21.7	17.3	38.3	20.6	89.3	40.6	21.7	16.3
	LSD($p<0.05$)	1.2	0.6	2.1	1.0	3.9	2.1	0.8	0.4

*Adopted from Tarafdar et al. (2003), I=Intracellular, E= Extracellular

Table 3. Secretion of acid phosphatase by fungi at different time intervals (Aseri et al., 2009).

S.No.	Fungal Species	Alkaline phosphatases activity (EU × 10^6) g^{-1} dry fungal mat							
		Days after inoculation							
		7		14		21		28	
		I	E	I	E	I	E	I	E
1.	*Aspergillus candidus*	17.3	10.6	32.6	12.7	13.7	5.8	4.8	3.1
2.	*Aspergillus fumigatus**	14.3	5.2	15.5	9.7	9.7	5.7	4.2	2.4
3.	*Aspergillus niger*	5.4	3.2	8.9	5.4	6.6	3.4	4.9	2.6
4.	*Aspergillus parasiticus*	21.6	12.9	38.2	14.3	12.3	6.2	6.9	2.8
5.	*Aspergillus rugulosus*	5.6	3.7	9.1	5.6	6.9	3.2	4.6	2.9
6.	*Aspergillus terreus*	20.6	11.9	30.6	14.6	10.0	6.1	6.8	3.2
7.	*Penicillium rubrum*	4.6	2.8	7.2	4.9	6.1	3.4	3.9	1.2
8.	*Penicillium simplicissimum*	17.9	12.3	29.7	13.9	18.6	8.2	7.2	4.6
9.	*Pseudeurotium zonatum*	18.9	12.8	36.7	9.8	11.3	4.9	6.9	2.1
10.	*Trichoderma harzianum*	28.6	18.3	45.7	18.4	16.2	7.3	8.9	3.6
11.	*Trichoderma viride*	21.9	16.3	42.8	12.8	14.9	7.8	7.6	4.9
	LSD($p<0.05$)	0.5	0.8	1.1	1.3	1.0	0.2	0.3	0.2

*Adopted from Tarafdar et al. (2003), I=Intracellular, E= Extracellular

Table 4. Secretion of alkaline phosphatase by fungi at different time intervals (Aseri et al., 2009).

S.No.	Fungal Species	Phytase activity (EU ×10^6) g^{-1} dry fungal mat							
		Days after inoculation							
		7		14		21		28	
		I	E	I	E	I	E	I	E
1.	*Aspergillus candidus*	0.16	10.1	0.11	6.2	0.04	3.6	0.02	0.9
2.	*Aspergillus fumigatus**	NE	NE	NE	NE	NE	NE	NE	NE
3.	*Aspergillus niger*	0.16	6.2	0.12	5.2	0.08	4.1	0.04	1.2
4.	*Aspergillus parasiticus*	0.20	10.4	0.12	7.1	0.08	4.2	0.03	1.0
5.	*Aspergillus rugulosus*	0.11	6.1	0.09	3.2	0.02	2.1	0.01	0.3
6.	*Aspergillus terreus*	0.19	10.2	0.10	6.9	0.06	4.1	0.01	0.9
7.	*Penicillium rubrum*	0.12	7.2	0.09	4.1	0.06	3.2	0.02	0.8
8.	*Penicillium simplicissimum*	0.14	6.9	0.09	4.0	0.04	2.1	0.02	0.3
9.	*Pseudeurotium zonatum*	0.19	7.0	0.06	3.9	0.02	1.9	0.01	0.4
10.	*Trichoderma harzianum*	0.16	12.8	0.12	8.9	0.08	4.3	0.06	1.2
11.	*Trichoderma viride*	0.15	11.9	0.12	7.6	0.06	3.9	0.01	1.4
	LSD($p<0.05$)	0.02	0.4	0.01	0.3	0.01	0.1	0.001	0.1

*Adopted from Tarafdar et al. (2003); NE-Not estimated, I=Intracellular, E= Extracellular

Table 5. Secretion of phytase by fungi at different time intervals (Aseri et al., 2009).

Phosphorous solubilization is carried out by a large number of saprophytic bacteria and fungi acting on sparingly soluble soil phosphates, mainly by chelating-mediated mechanism (Bajpai & Sundara Rao, 1971; Moghimi et al., 1978; Whitelaw 2000). Inorganic P is solubilized by the action of organic and inorganic acids secreted by PSMs in which hydroxyl and carboxyl groups of acids chelate cations (Ca, Al and Fe) and decrease the pH in basic soils (Kpomblekou & Tabatabai, 1994). The PSMs also dissolved the P through production acids such as (acetate, lactate, oxalate,tartatate,succinate, citrate, gluconate, ketogluconate and glycolate (Banik &Dey, 1982; Goldstein,1986; Cunningham & Kuiack, 1992; Goldstein, 1995; Gyaneshwar et al., 1998; Kim et al., 1997,1998; Deubel et al., 2000) and lowering the pH of the rhiosphere. The pH of the rhizosphere is lowered through biotical production of proton/ bicarbonate release (anion/ cation balance) and gaseous (O_2/CO_2) exchanges. Release of root exudates such as organic ligands can also alter the concentration of P in the soil solution (Hinsinger, 2001).The organic acids produed by PSMs solubilize insoluble phosphates by lowering the pH, chelation of cations and competing with phosphate for adsorption sites in the soil (Nahas, 1996). Inorganic acids such as hydrocholoric acid can also solubilize phosphate but they are less effective compared to organic acids at the same pH (Kim et al., 1997). In certain cases phosphate solubilization is induced by phosphate starvation (Gyaneshwar et al., 1999). Organic phosphates is catalysed through hydrolysis of C-O-P ester bond by phosphatase and phytase released by PSMs (Tarafdar & Rao, 1996; Tarafdar et al., 2001; Yadav & Tarafdar, 2007; Yadav & Tarafdar, 2010; Yadav & Tarafdar, 2011). The efficiency of different phosphatase and phytase secreating fungi isolated from arid soils of Rajasthan, India to hydrolyse different organic P compounds were reported (Tarafdar et al., 2003; Aseri et al., 2009) and presented in Table 6.

		Efficiency µg P release * min^{-1}	
S.No.	Fungal species	Phytin	Glycerophosphate
1.	*Aspergillus candidus*	2.72	4.72
2.	*Aspergillus niger*	1.72	2.46
3.	*Aspergillus parasiticus*	3.21	5.12
4.	*Aspergillus rugulosus*	0.98	1.82
5.	*Aspergillus terreus*	3.09	4.98
6.	*Penicillium rubrum*	2.16	2.90
7.	*Penicillium simplicissimum*	2.21	2.99
8.	*Pseudeurotium zonatum*	2.59	3.62
9.	*Trichoderma harzianum*	3.54	5.89
10	*Trichoderma viride*	3.32	5.46
	LSD($p<0.05$)	0.64	0.71

*Initial P added 500ppm either as phytic acid or as Na- glycerophosphate

Table 6. Efficiency of fungal mat of different fungi to hydrolyze different organic P compounds (Aseri et al., 2009).

5. Solubilization of calcium phosphate (Ca-P)

Soil phosphates mainly the apatites and metabolites of phosphatic fertilizers are fixed in the form of calcium phosphates in alkaline conditions under arid and semi arid region. Many of the calcium phosphates, including rock phosphate ores (fluoroapatite, francolite), are insoluble in soil with respect to the release of inorganic P (Pi) at rates necessary to support agronomic levels of plant growth (Goldstein, 2000). Phosphate solubilizing microorganisms could increase the P nutrition of plants through increased solubility of Ca-phosphates (Sujatha et al., 2004; Vassilev et al., 2006) and their solubility increases with a decrease of soil pH. Phosphate solubilization is the result of combined effect of pH decrease and organic acids production (Fankem et al., 2006). Microorganisms through secretion of different types of organic acids (Deubel et al.,2000; Deubel & Merbach, 2005) and rhizospheric pH lowering mechanisms (He & Zhu, 1988; Hinsinger,2001) dissociate the bound forms of phosphate like $Ca_3(PO_4)_2$. Nevertheless, buffering capacity of the medium reduce the effectiveness of PSMs in releasing P from tricalcium phosphates (Stephen & Jisha, 2009). Acidification of the microbial cell surroundings releases P from apatite by proton substitution / excretion of H^+ (accompanying greater absorption of cations than anions) or release of Ca_2^+ (Illmer & Schinner 1995; Villegas & Fortin 2002). While, the reverse occurs when uptake of anions exceeds that of cations, with excretion of OH^- / HCO_3^- exceeding that of H^+ (Tang & Rengel, 2003). Carboxylic anions produced by PSMs, have high affinity to calcium, solubilize more phosphorus than acidification alone (Staunton & Leprince, 1996). Complexing of cations is an important mechanism in P solubilization if the organic acid structure favors complexation (Fox et al., 1990). It is controlled by nutritional, physiological and growth conditions of the microbial culture (Reyes et al., 2007), but it is mostly due to the lowering of pH alone by organic acids (Moghimi & Tate, 1978) or production of microbial metabolites (Abd- Alla, 1994). Organic anions and associated protons are effective in solubilizing precipitated forms of soil P, chelating metal ions that may be associated with complexed forms of P or may facilitate the release of adsorbed P through ligand exchange reactions (Jones, 1998). Calcium phosphate (Ca-P) release results from the combined effects of pH decrease and carboxylic acids synthesis, but proton release cannot be the single mechanism (Deubel et al., 2000).

6. Solubilization of Iron phosphate / aluminum phosphate (Fe-P / Al-P)

Solubilization of Fe and Al occurs via proton release by PSMs by decreasing the negative charge of adsorbing surfaces to facilitate the sorption of negatively charged P ions. Proton release can also decrease P sorption upon acidification which increases $H_2PO_4^-$ in relation to $HPO_4^{\ 2-}$ having higher affinity to reactive soil surfaces (Whitelaw, 2000). Carboxylic acids mainly solubilized Al-P and Fe-P (Khan et al., 2007; Henri et al., 2008) through direct dissolution of mineral phosphate as a result of anion exchange of PO_4^{3-} by acid anion, or by chelation of both Fe and Al ions associated with phosphate (Omar, 1998). It is through root colonizing pseudomonads with high-affinity iron uptake system based on the release of Fe^{3+} chelating molecules i.e. siderophores (Altomare et al., 1999). Moreover, carboxylic anions replace phosphate from sorption complexes by ligand exchange (Otani et al., 1996; Whitelaw, 2000) and chelate both Fe and Al ions associated with phosphate, releasing phosphate available for plant uptake after transformation. Ability of organic acids to chelate metal cations is greatly influenced by its molecular structure, particularly by the number of carboxyl and hydroxyl groups. Type and position of the ligand in addition to acid strength determine its effectiveness in the solubilization process (Kpomblekou & Tabatabai, 1994).

7. Mineralization of organic phosphate

Mineralization of soil organic P plays an imperative role in phosphorus cycling of a farming system. Organic P may constitute 4-90 % of the total soil P. Almost half of the microorganisms in soil and plant roots possess P mineralization potential under the action of phosphatases (Tarafdar et al., 1988). Mineralization of organic to inorganic phosphate involves processes catalyzed by phosphatase enzymes, which are specifically involved in this conversion. Acid and alkaline phosphatases use organic phosphate as a substrate to convert it into inorganic form (Beech et al., 2001). Principal mechanism for mineralization of soil organic P is the production of acid phosphatases (Hilda & Fraga, 2000). Release of organic anions, and production of siderophores and acid phosphatase by plant roots / microbes (Yadav & Tarafdar, 2001) or alkaline phosphatase (Tarafdar & Claasen, 1988) enzymes hydrolyze the soil organic P or split P from organic residues. Many PSMs produce these enzymes (Tarafdar et al., 1988; Tarafdar et al., 2003; Aseri et al., 2009). In addition, some fungi produce phytase, an enzyme which releases soluble inorganic (PO_4^{3-}) phosphate from organic P compound (inositol hexaphosphate) (Tarafdar & Gharu, 2005; Yadav & Tarafdar 2007; Yadav & Tarafdar 2011). Some heterotrophic micro-organisms are also capable of solubilizing phosphates combined with calcium or magnesium (Atlas & Bartha, 1998). These soluble forms can now be readily taken up by plants, algae, cyanobacteria, and autotrophic bacteria and assimilated into organic cellular components such as DNA, RNA, and ATP. Phosphatase enzymes are present in all organisms but only bacteria, fungi, and some algae are able to secrete them outside of their cells. As exoenzymes, they participate in the dissolution and mineralization of organic phosphate compounds in the environment (Jones, 2002).

8. Response of phosphate solubilizing microorganisms under arid ecosystems

The role of phosphate solubilizing microorganisms in native P solubilization and increasing crop yield under arid ecosystem of India has been studies (Tarafdar et al., 1992; Tarafdar et al., 1995; Tarafdar & Rao, 1996; Tarafdar & Graru, 2005; Yadav & Tarafdar, 2007; Yadav & Tarafdar, 2010; Yadav & Tarafdar, 2011). The work on P solubilization in soil using plant growth test have been carried out under green house conditions (Tarafdar et al., 1992; Tarafdar et al., 1995; Tarafdar & Rao, 1996; Tarafdar & Graru, 2005; Yadav &Tarafdar, 2010), although the results of field studies are also available (Tarafdar & Graru, 2005; Yadav & Tarafdar, 2007; Yadav & Tarafdar, 2010; Yadav & Tarafdar, 2011). The effectiveness of inoculation with phosphate solubilizing microorganisms has been found to vary with the soil physico-chemical properties and the test crop.

Under green house conditions Tarafdar et al. (1992, 1995) and Tarafdar & Rao (1996) studied the effect of different phosphatase- producing fungi on growth and nutrition of mungbean, clusterbean, wheat and chickpea. They reported a significant increase in phosphatase, nitrogenase and dehydrogenase activity after inoculation with different phosphate solubilizing fungi (Table 7 & 8). Inoculation with phosphatase producing fungi significantly increased dry matter production and grain yield (Table 9).

Treatments	Acid phosphatase†	Alkaline phosphatase†	Dehydrogenase‡	Nitrogenase ʃ
Un-inoculated control	7.5	8.6	2.4	0.4
Aspergillus niger	8.1*	9.0*	3.3**	1.0***
Aspergillus fumigatus	8.0*	9.9***	3.1*	1.3***
Aspergillus rugulosus	4.8**	10.8***	5.3***	1.4***
Aspergillus terreus	8.9***	9.4**	5.9***	0.7*

†=n Kat 100g^{-1}soil;‡ =p Kat g^{-1} soil; ʃ= μ mole C_2H_4 h^{-1} plant^{-1};
*p<5% ; ** p<1%; *** p<0.1%

Table 7. Effect of inoculation with phosphatase-producing fungi on rhizosphere enzyme activities in clusterbean (Tarafdar et al., 1995).

Treatments	Dehydrogenase†			Acid phosphatase‡			Alkaline phosphatase‡		
	NR^W	Chickpea	Wheat	NR	Chickpea	Wheat	NR	Chickpea	Wheat
Control	10.6	11.9	12.4	25.0	66.1	38.1	71.1	72.1	71.1
Aspergillus niger	10.4	16.3^c	15.6^c	25.0	78.6	49.1^c	69.1	75.1^a	85.6^c
Aspergillus fumigatus	10.1	14.5^c	14.3^a	24.5	91.6^c	51.1^c	72.6	72.6	87.6^c
Aspergillus rugulosus	11.7	16.3^c	15.0^b	29.0^a	78.6^b	46.1^b	80.6^a	81.6^c	88.6^c
Aspergillus terreus	11.2	15.8^c	14.0^a	25.5	85.6^c	45.1^a	86.1^c	86.6^c	87.1^c

†= μg TPF kg^{-1} per 24 h;‡= μg pNP g^{-1} h^{-1}; ^{W} = soil without plant (a=p<5%; b= p<1%; c =p<0.1%)

Table 8. Dehydrogenase and phosphatase activities as influenced by inoculum with *Aspergillus* in rhizosphere and non-rhizosphere soils. (Tarafdar and Rao, 1996).

Treatments	Yield (g plant^{-1})					
	Clusterbean		Chickpea		Wheat	
	Dry matter	Grain	Dry matter	Grain	Dry matter	Grain
Control	20.8	5.4	2.1	2.0	4.7	1.6
Aspergillus niger	24.1*	6.3*	2.2	2.9*	8.0***	2.1*
Aspergillus fumigatus	26.5***	6.9**	3.2*	3.6***	8.7***	2.3**
Aspergillus rugulosus	27.2***	7.1***	4.0***	3.7***	7.8***	2.7***
Aspergillus terreus	24.4**	6.3*	3.2*	3.6***	6.7***	2.5***

*p<5% ; ** p<1%; *** p<0.1%

Table 9. Effect of inoculation with phosphatase-producing fungi on dry matter production and grain yield of clusterbean, chickpea and wheat (Tarafdar et al., 1995; Tarafdar and Rao, 1996).

They also observed significant improvement in the uptake of major and micronutrients (Table 10 & 11).

Treatments	N†	P†	K†	Ca‡	Mn‡	Na‡
Uninoculated control	19.0	1.8	3.8	86	216	125
Aspergillus niger	26.1**	2.1*	4.2	111**	244*	129
Aspergillus fumigatus	29.7***	2.2*	4.7*	101*	320***	137
Aspergillus rugulosus	31.5***	2.2*	4.2	108**	288***	106
Aspergillus terreus	24.9*	2.1*	4.2	114***	241*	96*

† = mg g-1;‡=mg kg-1; *p<5% ; ** p<1%; *** p<0.1%

Table 10. Effect of inoculation with phosphatase-producing fungi on mineral uptake by clusterbean (Tarafdar et al., 1995).

Treatments	Concentration (mg kg^{-1})			
	Fe	Mn	Cu	Zn
Uninoculated control	3.4	5.0	8.0	6.2
Aspergillus niger	4.9***	5.0	7.2*	8.1*
Aspergillus fumigatus	5.9***	5.5	7.5	8.2*
Aspergillus rugulosus	4.4***	4.9	9.9*	8.7***
Aspergillus terreus	5.1***	5.2	9.8*	8.0*

*p<5%; ** p<1%; *** p<0.1%

Table 11. Effect of inoculation with phosphatase-producing fungi on uptake of some trace elements by clusterbean (Tarafdar et al.,1995).

Further, Yadav et al. (2009) conducted an experiment with *Chaetomium globosum* and organic matter on P mobilization and yield of clusterbean under arid ecosystem. They observed increase in acid phosphatase (15%), alkaline phosphatase (12%) and phytase by 71% more due to inoculation of *C. globosum* with organic matter compared to application of only organic matter. They also reported the seed yield of clusterbean increased by 7.4% with *C. globosum* in the presence of high level of organic matter. In general 6%, 23% and 10% more N, P and K accumulation in plant was recorded due to inoculation of *C. globosum.*

Under field conditions (Tarafdar & Gharu, 2005; Yadav & Tarafdar, 2007, 2010 & 2011) reported significant increased in yield and P content of pearl millet and clusterbean in arid ecosystem after inoculation with different P solubilizing fungi. In pearl millet an increase in dry matter production by 29-39% and P concentration in shoot by 14-19%, in root by 5-7% and seed by 34 to 35% was reported with irrespective inoculation of seed with *Chaetomium globosum, Emericella rugulosa* and *Penicillium purpurogenum.* The experiment conducted by Yadav and Tarafdar (2010) reported that grain yield of clusterbean was increased by 26% and straw yield by 42%, plant P content by 12% and seed P content by 10% due to inoculation of *Emericella rugulosa.* The comparison of results from different experiments is difficult because of the variation in the rainfall of the cropping season, soil properties and phosphate solubilizing microorganisms. Generally, the positive response of the phosphate solubilizing microorganisms have been observed high organic matter content and low P availability under arid ecosystem of Rajasthan, India.

9. Conclusion

The arid ecosystem extended over the earth from the tropical to the sub-alpine zones and mean sea level to above 3000 m. In Indian arid zone average rainfall varies from as low as 100 to 400 mm with different drought intensities. Annual crop can hardly be cultivated under moderate to severe drought conditions. With increase in both human and population in Indian arid zone, the demand for grain, fodder and fuel wood is increasing. But the agricultural production is low owing to the poor soil fertility and occurrence of frequent droughts. So, the enhancement of soil fertility thus assumes a great significance for sustained agriculture production in drought prone areas. Given the socio-economic conditions of the farmers, the extensive use of chemical fertilizers to augment the crop production is a risky proposition. Besides, use of only chemical fertilizers may result in degradation of soil productivity/health as has happened in other ecosystems. The use of P solubilizing microorganisms as a P biofertilizer under arid ecosystem may improve the soil fertility and increase the crop production to fulfill the requirement. Further, the efficiency of these microorganisms to meet P requirement of crops will depend greatly on their impact under practical farming conditions.

10. References

Abd-Alla, M.H. (1994). Phosphatases and the Utilization of Organic P by *Rhizobium leguminosarum* Biovar Vicear. *Letters in Applied Microbiology*, Vol.18, No.5, (May 1994), pp. 294-296, ISSN 1472-765X

Achal, V.; Savant, V.V. & Sudhakara Reddy, M. (2007). Phosphate Solubilization by Wide Type Strain and UV-induced Mutants of *Aspergillus tubingensis*. *Soil Biology and Biochemistry*, Vol. 39, No.2, (February 2007),pp. 695-699,ISSN 0038-0717

Alam, S.; Khalil, S.; Ayub, N. & Rashid, M. (2002). In *vitro* Solubilization of Inorganic Phosphate by Phosphate Solubilizing Microorganism (PSM) for Maize Rhizosphere. *International Journal of Agriculture and Biology*, Vol. 4, No.4, (December 2002), pp. 454-458, ISSN 1560-8530

Altomare, C.; Noevell, W.A.; Björkman, T. & Harman, G.E. (1999). Solubilization of Phosphates and Micronutrients by the Plant Growth Promoting and Biocontrol Fungus *Trichoderma harzianum* Rifai 1295-22. *Applied and Environmental Microbiology*, Vol.65, No. 7, (July 1999), pp. 2926-2933, ISSN 0099-2240

Aseri, G.K.; Jain, N & Tarafdar, J.C. (2009). Hydrolysis of Organic Phosphate forms by Phosphatases and Phytase Producing Fungi of Arid and Semi-arid Soils of India. *American-Eurasian Journal of Agriculture and Environment Science*, Vol. 5, No.4, (December 2009), pp. 564-570, ISSN 1818-6769

Atlas, R.M. & Bartha, R. (1988). *Microbial Ecology. Fundamentals and Applications*, Benjamin Cummings Publishing Company Inc, ISBN 10:0805306563, USA

Bajpai, P.D. & Sundra Rao, W.V.B. (1971). Phosphate Solubilizing Bacteria. Part 2. Extracellular Production of Organic Acids by Selected Bacteria Solubilizing Insoluble Phosphate. *Soil Science and Plant Nutrition*, Vol. 17, No.2, (April 1971), pp. 44-45, ISSN 1747-0765

Banik, S. & Day, B.K. (1982). Available Phosphate Content of an Alluvial Soil as Influenced by Inoculation of Some Isolated Phosphate Solubilizing Microorganisms. *Plant and Soil*, Vol. 69, No.3, (September 1982), pp. 353-364, ISSN 1573-5036

Baya, A.M.; Robert, S.B. & Ramos C.A. (1981). Vitamin Production in Relation to Phosphate Solubilization by Soil Bacteria. *Soil Biology and Biochemistry*, Vol. 13, No.6, (June 1981), pp. 527-532, ISSN 0038-0717

Beech, I.B.; Paiva, M.; Caus, M. & Coutinho, C. (2001). Enzymatic Activity and within Biofilms of Sulphate-Reducing Bacteria. In: *Biofilm Community Interactions: change or necessity?*, P.G. Gilbert; D. Allison; M. Brading; J. Verran & J. Walker (Eds.), 231-239, Boiline, ISBN 9780521812405 , Cardiff, U.K.

Chen, Y.P.; Rekha, P.D.; Arunshen, A.B.; Lai, W.A. & Young, C.C. (2006). Phosphate Solubilizing Bacteria from Subtropical Soil and their Tricalcium Phosphate Solubilizing Abilities. *Applied Soil ecology*, Vol. 34, No.1, (November 2006), pp. 33-4, ISSN 0929-1393

Cunningham, J.E. & Kuiack, C. (1992). Production of Citric and Oxalic Acids and Solubilization of Calcium Phosphates by *Penecillium Bilaii*. *Applied and Environmental Microbiology*, Vol. 58, No. 5, (May 1992), pp. 1451-1458, ISSN 0099-2240

De Freitas, J.R.; Banerjee, M.R. & Germida, J.J. (1997). Phosphate Solubilizing Rhizobacteria Enhance the Growth and Yield but not Phosphorous Uptake on Canola (Brassica *napus* L.). *Biology and Fertility of Soils*, Vol. 24, No. 4, (May 1997), pp. 358-364, ISSN 1432-0789

Deubel, A.; Gransee, A. & Merbach, W. (2000). Transformation of Organic Rhizodeposits by Rhizoplane Bacteria and its Influence on the Availability of Tertiary Calcium Phospate. *Journal of Plant Nutrition and Soil Science*, Vol. 163, No.4, (August 2000), pp. 387-392, ISSN 1436-8730

Dhir, R.P. (1977). Desertification and Its control, ICAR Monograph, New Delhi, pp.102.

Dutton, V.M. & Evans, C.S. (1996). Oxalate Production by Fungi: It's Role in Pathogenicity and Ecology in the Soil Environment. *Canadian Journal of Microbiology*, Vol.42, No. 9, (September 1996), pp. 881-895, ISSN 0008-4166

Ezawa, T.; Smith, S.E. & Smith, F.A. (2002). P Metabolism and Transport in AM Fungi. *Plant and Soil*, Vol. 244, No. 1-2, (July 2002), pp.221-230, ISSN 1573-5036

Fankem, H.; Nwaga, A.; Dueubel, A.; Dieng, L.; Merbach, W. & Etoa, F.X. (2006). Occurrence and Functioning of Phosphate Solubilizing Microorganisms from Oil Palm Tree (*Elaeis guineensis*) Rhizosphere in Cameroon. *African Journal of Biotechnology*, Vol. 5, No. 24, (December 2006), pp. 2450-2460, ISSN 1684-5315

Fox, R.; Comerford, N.B. & Mcfee, W.W. (1990). Phosphorous and Aluminium Release from a Spodic Horizon Mediated by Organic Acids. *Soil Science Society of America Journal*, Vol. 54, No.6, (December 1990), pp. 1763-1767, ISSN 0361-5995

Golstein, A.H. (1986). Bacterial Phosphate Solubilization: Historical Perespective and Future Prospects. *Amercian Journal of Alternative Agriculture*, Vol. 1, No.2, (June 1986), pp. 57-65, ISSN 0889-1893

Goldstein, A.H. (1994). Involvement of the Quinoprotein Glucose Dehydrogenises in the Solubilization of Exogenous Phosphates by Gram-Negative Bacteria. In: *Phosphate in Microorganisms: Cellular and Molecular Biology*,A. Torriani Gorini; E. Yagil and S. Silver (Eds.), 197-203, Washington, ISBN 1-55581-265-1, D.C.

Goldstein, A.H. (1995). Recent Progress in Understanding the Molecular Genetics and Biochemistry of Calcium Posphate Solubilization by Gram-negative Bacteria.

Biological, Agriculture and Horticulture, Vol. 12, No. 1 (March 1995), pp. 185-193, ISSN 0144-8765

Goldstein, A.H. (2000). Bioprocessing of Rock Phosphate Ore: Essential Technical Considerations for the Development of a Successful Commercial Technology, Proceedings of 4th International Association for Technical Conference, pp. 220-242, ISBN 1-57498-079-3, New Orieans, USA, October 1-4, 2000

Gyaneshwar, P.; Kumar, N.J.; Pareka, L.J. & Podle, P.S. (2002). Role of Soil Microorganisms in Improving P Nutrition of Plants. *Plant and Soil*, Vol. 245, No. 1, (August 2002), pp. 83-93, ISSN 1573-5036

Gyaneshwar, P.; Pareka, L.J.; Archana, G.; Podle, P.S.; Collins,M.D.; Huston, R.A. & Naresh, K.G. (1999). Involvement of a Phosphate Stravation Inducible Glucose Dehydrogenase in Soil Phosphate Solubilization by *Enterobacter asburiae. FEMS Microbiology Letters*, Vol. 171, No. 2, (February 1999), pp. 223-229, ISSN 0378-1097

Gyaneshwar, P.; Naresh Kumar, G. & Parekh, L.J. (1998). Effect of Buffering on the Phosphate Solubilizing Ability of Microorganisms. *World Journal of Microbiology and Biotechnology*, Vol. 14, No. 5, (October 1998), pp. 669-673, ISSN 0959-3993

Hao, X.; Cho, C.M.; Racz, G.J. & Chang, C. (2002). Chemical Retardation of Phosphate Diffusion in an Acid Soil as Affected by Liming. *Nutrient Cycling in Agro ecosystems*, Vol. 64, No. 3, (September 2002), pp. 213-224, ISSN 1385-1314

He, Z.L. & Zhu, J. (1988). Microbial Utilization and Transformation of Phosphate Adsorbed by Variable Charged Minerals. *Soil Biology and Biochemistry*, Vol. 30, No. 7, (July 1998), pp. 917-923, ISSN 0038-0717

Henri, F.; Laurette, N.N.; Annertte, D.; John, Q.; Wolfgang, M.; Francois-Xavier, E. & Dieudonné (2008). Solubilization of Inorganic Phosphates and Plant Growth Promotion by Strains of *Pseudomonas fluorescens* Isolated From Acidic Soils of Cameron. *African Journal of Microbiology Research*, Vol. 2, No.7, (July 2008), pp. 171-178, ISSN 1996-0808

Hilda, R. & Fraga, R. (1999). Phosphate Solubilizing Bacteria and their Role in Plant Growth Promotion. *Biotechnology Advances*, Vol. 17, No. 4-5, (October 1999), pp.319-339, ISSN 0734-9750

Hinsinger, P. (2001). Bioavailability of Soil Inorganic P in the Rhizosphere as Affected by Root-induces Chemical Changes: A Review. *Plant and Soil*, Vol. 237, No.2, (December 2001), pp. 173-195, ISSN 0032-0079X

Illmer, P. & Schinner, F. (1995). Solubilization of Inorganic Calcium Phosphates-Solubilization Mechanisms. *Soil Biology and Biochemistry*, Vol. 27, No. 3, (March 1995), pp. 257-263, ISSN 0038-0717

Jones, D.L. (1998). Organic Acids in the Rhizosphere- A Critical Review. *Plant and Soil*, Vol. 205, No. 1, (August 1998), pp. 25-44, ISSN 0032-0079X

Jones, R.D. (2002). Phosphorus Cycling In: *Manual of environmental microbiology*, C.J Hurst, (Ed.), 450-465, Washington, ISBN 13:978-1-5558-1525-7, DC

Kapoor, K. K.; Mishra, M. M. & Kukreja, K. (1989). Phosphate Solubilization by Soil Microorganisms. *Indian Journal of Microbiology*, Vol. 29, No. 2, (June 1989), pp.119-127, ISSN 0046-8991

Khan, M.S.; Zaidi, A. & Wani, P.A. (2007). Role of Phosphate-Solubilizing Microorganisms in Sustainable Agriculture-A Review. *Agronomy for Sustainable Development*, Vol. 27, No. 1, (March 2007), pp. 29-43, ISSN 1774-0746

Kim, K.V.; Jordan, D. & McDonald, G.A. (1998). Effect of Phosphate Solubilizing Bacteria and Vesicular-Arbuscular Mycorrhizae on Tamoto Growth and Soil Microbial Activity. *Biology and Fertility of Soils*, Vol.26, No. 2, (December 1997), pp. 79-87, ISSN 1432-0789

Kim, K.Y.; Jordan, D. & McDonald, D. (1997) Solubilization of Hydroxyapatite by Enterobacter Agglomerans and Cloned Escherichia in Culture Medium. *Biology and Fertility of Soils*, Vol. 24, No. 4, (May 1997), pp. 347-352, ISSN 1432-0789

Kpomblekou, K. & Tabatabai, M.A. (1994). Effect of Organic Acids on Release of Phosphorous from Phosphate Rocks. *Soil Science*, Vol. 158, No. 6, (December 1994), pp. 442-453, ISSN 0038-075X

Kucey, R.M.N. (1983). Phosphate Solubilizing Bacteria and Fungi in Various Cultivated and Virgin Albreta Soils. *Canadian Journal of Soil Science*, Vol. 63, No. 4, (December 1983), pp. 671-678, ISSN 0008-4271

Kucey, R.M.N.; Janzen, H.H. & Legget, M.E. (1989). Microbial Mediated Increases in Plant Available Phosphorous. *Advances in Agronomy* Vol. 42, pp.199-288, ISSN 0065-2113

Lindsay, W.L. 1979). *Chemical Equilibrium in Soils*, Willy, ISBN 0471027049, New Work

Moghimi, A.; Tate, M. E. & Oades, J.M. (1978). Characterization of Rhizospheric Products Espically 2- Ketogluconic Acid. *Soil Biology and Biochemistry*, Vol. 10, No. 4, (April 1978),pp. 283-287, ISSN 0038-0717

Nahas, E. (1996). Factors Determining Rock Phosphate Solubilization by Micro-organisms. *World Journal of Microbiology and Biotechnology*, Vol. 12, No. 6, (November 1996), pp. 567-572, ISSN 0959-3993

Omar, S.A. (1998). The Role of Rock Phosphate Solubilizing Fungi and Vesicular Arbuscular Mycorrhizae (VAM) in Growth of Wheat Plants Facilitated with Rockphosphate. *World Journal of Microbiology and Biotechnology, Vol.* 14, No. 2, (March 1998), pp. 211-218, ISSN 0959-3993

Otani,T.; Ae, N. & Tanaka, H. (1996). Phosphorous (P) Mechanisms of Crops Grown in Soils with Low P Status.II. Significance of Organic Acids in Root Exudares of Pigeonpea. *Soil Science and Plant Nutrition*, Vol. 42, No.3, (June 1996), pp. 553-560, ISSN 1747-0765

Reyes, I.; Bernier, L.; Simard, R.R. & Anroun, H. (1999). Effect of Nitrogen Source on the Solubilization of Different Inorganic Phosphates by an Isolate of *Penicillium Rugulosum* and Two UV-induced Mutants. *FEMS Microbiology Ecology*, Vol. 28, No. 3, (March 1999), pp. 281-290, ISSN 0168-6496

Reyes, I.; Valery, A. & Valduz, Z. (2007). Phosphate Solubilizing Microorganisms Isolated from Rhizospheric and Bulk Soils of Colonizer Plants at an Abandoned Rock Phosphate Mine. In: *First International Meeting on Microbial Phosphate Solubilization*, E. Valazquez & C. Rodriguez-Barrueco (Eds.), 69-75, Salamnca, ISBN 978-1-4020-4019-1, Spain

Richardson, A. E.; Hadobas, P.A. & Hayes, J.E. (2000). Acid Phosphomonoesterase and Phytase Activities of Wheat (*Triticum aestivum* L.) Roots and Utilization of Organic Phosphorous Substrates by Seedlings Grown in Sterile Culture. *Plant, Cell and Environment*, Vol. 23, No.4, (April 2000), pp. 397-40, ISSN 0140-7791

Sperber, J.I. (1957) Solution of Mineral Phosphates by Soil Bacteria. *Nature (London),* Vol. 180 (November 1957), pp. 994-995, ISSN 0-415-31128-4

Sperber, J.I. (1958). Solution of Apatite by Soil Microorganisms Producing Organic Acids. *Australian Journal of Agricultural Research*, Vol. 9, No. 6, (December 1958), pp. 778-781, ISSN 0009-9409

Staunton, S. & Leprince, F. (1996). Effect of pH and Some Organic Anions on the Solubility of Soil Phosphate: Implications for P Bioavailability. *European Journal of Soil Science*, Vol. 47, No. 2, (June 1996), pp. 231-239, ISSN 1351-0754

Stephen, J. & Jisha, M.S. (2009). Buffering Reduces Phosphate Solubilizing Ability of Selected Strains of Bacteria. *World Journal of Agricultural Sciences*, Vol. 5, No. 1, (January 2009), pp. 135-137, ISSN 1817-3047

Stevenson, F.J. (2005). *Cycles of Soil: Carbon, Nitrogen, Phosphorus, Sulfur, Micronutrients*. John wiley and Sons, ISBN 0-471-32071-4, New York.

Sujatha, S.; Sirisham, S. & Reddy, S.M. (2004). Phosphate Solubilization by Thermophilic Microorganisms. *Indian Journal of Microbiology*, Vol. 44, No. 2, (June 2004), pp. 101-104, ISSN 0046-8991

Tang, C. & Rengel, Z. (2003). Role of Plant Cation/Anion Uptake Ratio in Soil acidification. In: *Hand book of Soil Acidity*, Z. Rengel (Ed), 57-81, Marcel &Marcel, Dekker, ISBN 0824742591, New York

Tarafdar, J.C. & Gharu A. (2005). Mobilization of Organic and Poorly Soluble Phosphate by *Chaetomium globosum*. *Applied Soil Ecology*, Vol. 32, No. 3, (July 2006), pp. 273-283, ISSN 0929-1393

Tarafdar, J.C. & Claassen, N. (1988). Organic Phosphorus Compounds as a Phosphorus Source for Higher Plants through the Activity of Phosphatases Produced by Plant Roots and Microorganisms. *Biology and Fertility of Soils*, Vol. 5, No. 4, (January 1988), pp. 308-312, ISSN 1432-0789

Tarafdar, J.C.; Rao, A.V. & Bala, K. (1988). Production of Phosphatases by Fungi Isolated from Desert Soils. *Folia Microbiology*, Vol. 33, No. 6, (December 1988), pp. 453-457, ISSN 1874-9356

Tarafdar, J.C.; Yadav, R.S. & Meena, S.C. (2001). Comparative Efficiency of Acid Phosphatase Originated from Plant and Fungal Sources. *Journal of Plant Nutrition and Soil Science*, Vol. 164, No. 3, (June 2001), pp. 279-282, ISSN 1522-2464

Tarafdar, J.C.; Bareja, M. & Panwar, J. (2003). Efficiency of Some Phosphatase Producing Soil-Fungi. *Indian Journal of Microbiology*, Vol. 43, No. 1, (March 2003), pp. 27-32, ISSN 0046-8991

Tarafdar, J.C., Rao, A.V. & Kumar, P. (1992). Effect of Different Phosphatase Producing Fungi on Growth and Nutrition of Mung Beans [*Vigna radiata* (L.) Wilczek] in an Arid Soil. *Biology and Fertility of Soils*, Vol.13, No. 1, (March 1992), pp. 35-38, ISSN 1432-0789

Tarafdar,J.C.; Rao, A.V. & Kumar, P.(1995). Role of Phosphate Producing Fungi on the Growth and Nutrition of Clusterbean (*Cyamopsis tetragonoloba* (L.) Taub.). *Journal of Arid Environments*, Vol. 29, No. 3, (March 1995), pp. 331-337, ISSN 0140-1963

Tarafdar. J.C. & Rao, A.V. (1996). Contribution of *Aspergillus* Strains in Acquisition of Phosphorous to Wheat (*Triticum aestivum* L.) and Chick pea (*Cicer arietinum* L.) Grown in a Loamy Sand Soil. *Applied Soil Ecology*, Vol. 3, No. 2, (March 1996), pp. 109-114, ISSN 0929-1393

Vassilev, N.; Vasileva, M.A. & Nikolaeva, L. (2006). Simultaneous P Solubilizing and Biocontrol Activity of Microorganisms: Potentials and Future Trends. *Applied*

Microbiology and Biotechnology, Vol. 71, No. 2, (March 2006), pp. 137-144, ISSN 0175-7598

Venkateswarlu, B.; Rao, A.V. & Raina, P.(1984). Evaluation of Phosphorous Solubilization by Microorganisms Isolated from Arid Soils. *Journal of the Indian Society of Soil Science*, Vol. 32, No. 3, (September 1984), pp. 273-277, ISSN 0019-638X

Villegas, J. & Fortin, J. A. (2002). Phosphorous Solubilization and pH Changes as a Result of the Interactions between Soil Bacteria and Arbuscular Mycorrhizae Fungi on a Medium Containing NO_3 as Nitrogen Source. *Canadian Journal of Botany*, Vol. 80, No. 5, (May 2002), pp. 571-576, ISSN 0008-4026

Whitelaw, M.A. (2000). Growth Promotion of Plants Inoculated with Phosphate Solubilizing Fungi. *Advances in Agronomy*, Vol. 69, (December 1999), pp. 99-151, ISSN 0065-2113

Yadav, R.S. & Tarafdar, J.C. (2001). Influence of Organic and Inorganic Phosphorous Supply on the Maximum Secretion of Acid Phosphatase by Plants. *Biology and Fertility of Soils*, Vol. 34, No. 3, (September 2001), pp.140-143, ISSN 1432-0789

Yadav, B.K. & Tarafdar, J.C. (2007). Ability of *Emericella rugulosa* to Mobilize Unavailable P Compounds during Pearl Millet (*Pennisetum glaucum* L.) Crop under Arid Condition. *Indian Journal of Microbiology*, Vol. 41, No. 1, (March 2007), pp. 57-63, ISSN 0046-8991

Yadav, B.K. & Tarafdar, J.C. (2010). Studies on Phosphatase Activity and Clusterbean Production as Influenced by the P Mobilizing Organism *Emericella Rugulosa*. *Legume Research*, Vol. 33, No. 2, (June 2010), pp.118-220, ISSN 0250-5371

Yadav, B.K. & Tarafdar, J.C. (2011). *Penicillium Purpurogenum*, Unique P Mobilizers in Arid Agro-Ecosystems. *Arid Land Research and Management*, Vol. 25, No. 1, (January 2011), pp. 87-99, ISSN 1532-4982

Yadav, B.K.; Niwas, R.; Yadav, R.S. & Tarafdar, J.C. (2009). Effect of *Chaetomium Globosum* Inoculation and Organic Matter on Phosphorus Mobilization in Soil and Yield of Clusterbean. *Annals of Arid Zone*, Vol. 48, No. 1, (March 2009), pp. 41-44, ISSN 0570-1791

7

Human Impacts on a Small Island Ecosystem: Lessons from the Lucayans of San Salvador, Bahamas for This Island Earth

Jeffrey P. Blick
Georgia College and State University
USA

1. Introduction

The issues of prehistoric impact on the environment, resource overexploitation, and overfishing have emerged as major themes over the last few decades as scholars have come to realize that indigenous peoples, even those living at relatively simple levels of technology, such as land-based hunter-gatherers and "maritime hunter-gatherers" (Yesner, 1980), appear to have had the capability of significantly altering their local and regional environments by reducing biodiversity through customary subsistence practices such as horticulture, hunting, shellfish collecting, and fishing (Keegan, Portell, & Slapcinsky, 2003; Krech, 1999; Newsom & Wing, 2004; Redman, 1999; Wing, 1989, 2001; Wing & Wing 2001). Modern zooarchaeological techniques and accompanying statistical analyses are allowing us to address these important themes by providing quantitative measures of prehistoric human-environmental interaction, but zooarchaeological data must be multivariate (and demonstrate consistent trends), and must be understood within the broader context of human social behavior, modes of prehistoric subsistence, and human adaptation to ancient environments. Multifaceted evidence from numerous zooarchaeological studies across the Greater Caribbean illustrates that the Taíno and related cultures of the Antilles, including the Lucayans of the Bahamas, indeed had significant, wide-reaching, and profound impacts on the terrestrial, intertidal, and marine environments of the Greater and Lesser Antilles and the Bahamian Archipelago (Blick, 2006; Blick & Murphy, 2005; Carlson, 1999; Carlson & Keegan, 2004; Keegan, Portell, & Slapcinsky, 2003; Newsom & Wing, 2004; Steadman & Stokes, 2002; Wing, 2001; Wing, deFrance, & Kozuch, 2002; Wing & Wing, 2001). In fact, prehistoric impact on the environment appears to be one of the main emerging themes of island and coastal archaeology, particularly in the Caribbean, of the last three decades or so (Erlandson & Fitzpatrick, 2006; Newsom & Wing, 2004; Wing, 1989; Wing & Reitz, 1982; Wing & Scudder, 1983).

The issues of marine resource exploitation and fishing pressure, sometimes referred to as overfishing, appear to have reached a critical mass in the scientific literature in recent years (Baum et al., 2003; Coleman et al., 2004; Conover & Munch, 2002; Hawkins & Roberts, 2004; Jackson et al., 2001; Mumby et al., 2006; Pandolfi et al., 2003; Pauly et al., 1998; Sibert et al., 2006; Worm et al., 2006). Recreational, artisanal, and industrial fishing pressures on coastal and marine ecosystems clearly have implications for the conservation, establishment, and management of modern no-take reserves and marine protected areas. Based upon an examination of the scientific literature, characteristics of overfishing include: targeting and depletion of apex predators (Baum et al., 2003; Estes, et al., 2011; Frank, et al., 2005; Hawkins & Roberts, 2004; Jackson et al., 2001) such as sharks, groupers, snappers, jacks, and larger individuals (Conover & Munch, 2002; Hawkins & Roberts, 2004; Pandolfi et al., 2003; Pauly et al., 1998; Sibert et al., 2006); truncated or decreasing size and age distributions of targeted species ("growth overfishing") (Coleman et al., 2004; Conover & Munch, 2002; Pauly et al., 1998); genetic changes in somatic growth rates (Conover & Much, 2002); declining fisheries biomass (Hawkins & Roberts, 2004; Sibert, 2006); altered community composition or functioning (Bjorndal & Jackson, 2003; Coleman et al., 2004; Estes, et al., 2011; Frank, et al., 2005; Hawkins & Roberts, 2004; Jackson et al., 2001; Mumby et al., 2006; Worm et al., 2006); decline in the mean trophic level of fisheries landings (Pauly et al., 1998); reduction of biodiversity (Jackson et al., 2001; Pandolfi et al., 2003; Worm et al., 2006); and regional extirpations or global extinctions (Hawkins & Roberts, 2004; Jackson et al., 2001; Pandolfi et al., 2003). Several of these phenomena are observable in the data presented in this chapter (see also Blick, 2006; 2007; Blick & Kjellmark, 2006).

Regarding human impact on coastal and marine ecosystems, the good news appears to be that current rates of overfishing are reversible, at least on local and regional scales (Myers et al., 1995; Worm et al., 2006), and biologists are now aware of the pitfalls of focusing on only one key species (Baum et al., 2003) or substrate (e.g., sea grasses, reefs) (Mumby et al., 2006) for conservation purposes. The bad news is that "humans have been disturbing marine ecosystems since they first learned how to fish" (Jackson et al., 2001:629) and that "Changes in ecosystem structure and function occurred as early as the late aboriginal [stage]" (Jackson et al., 2001:636) and will likely continue unabated into the near future when global fisheries are projected to collapse by ca. A.D. 2050 (Pauly et al., 1998; Worm et al., 2006). Finally, the question of whether or not the pre-Columbian Lucayans of San Salvador practiced "sustainable" harvesting of coastal and marine organisms will be addressed. The answer to this question has serious implications for the concept of ecological sustainability in conservation circles and has grave implications regarding human population size, resource use, and environmental impacts in the future.

The main purpose of this chapter is to present new data from the archaeological record of the island of San Salvador, Bahamas, on the nature of prehistoric impact on the environment and to examine the nature of pre-Columbian resource overexploitation and overfishing. The findings derive from two roughly contemporaneous archaeological sites, Minnis-Ward (SS-3) and North Storr's Lake (SS-4), located on opposite sides of San Salvador with somewhat different ecological and archaeological characteristics (see Background, below). Remains of key invertebrate and vertebrate fauna are used to illustrate significant trends of decline in abundance and size of prehistoric subsistence items, several of which were highly ranked or preferential food resources of the Lucayan Indians of the Bahamas (see Keegan, 1992:132, Table 6.1). Even though the pre-Columbian population of the island was relatively minor, the evidence presented herein indicates that the Lucayan population of San Salvador had significant environmental impacts on the terrestrial, intertidal, and marine environments of this small island ecosystem.

2. Background

2.1 The study area: San Salvador Island, Bahamas

San Salvador Island is one of some 700 islands in the Commonwealth of the Bahamas located at approximately 24°03'10.44"N, 74°29'26.78"W (GoogleEarth, 2006) (Figure 1). San Salvador is best known as the leading contender for the site of the historic Columbus landfall in October of 1492 (Long Bay Site, SS-9, white triangle in Figure 2; Blick, 2011a; Brill & Hoffman, 1985; Brill et al., 1987; Hoffman, 1987; Keegan, 1992; Obregón, 1987, 1989). Prior to Columbus's arrival that fateful year, San Salvador and the rest of the Bahamian Archipelago had been occupied since about A.D. 600 by a prehistoric Arawakan-speaking people known as the Lucayans, the "island people" of the Bahamas (Albury, 1975; Berman & Gnivecki, 1995; Granberry & Vescelius, 2004; Keegan, 1992).

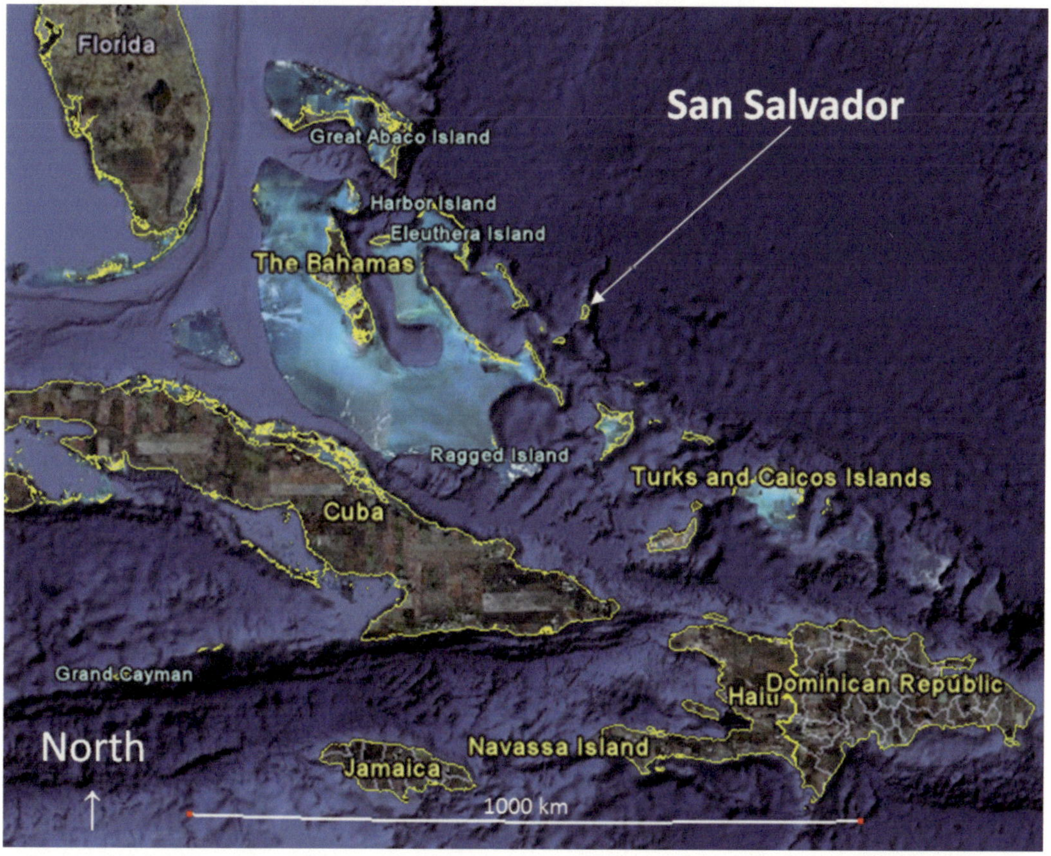

Fig. 1. Map of the Commonwealth of the Bahamas. San Salvador is located at approximately 24°N 74°30'W in the central eastern Bahamas. Scale is 1000 km (622 miles).

The Lucayans were a seafaring, fishing, and horticultural society with origins ultimately on the South American mainland (Rouse, 1992; S. Wilson, 1999). They were master dugout canoe makers and wood carvers (Albury, 1975; Berman, 1999) and they traded with cultures as far away as Central and South America (Johnson, 1980). Lucayan subsistence was based on a mixed economy of fishing in the semi-tropical waters of their islands and on horticulture of such plants as manioc, cocoyam, maize, other starchy vegetables, gourds or squashes, chili peppers, fruits, cotton, tobacco, etc. (Berman & Pearsall, 2000, 2005, 2008; Keegan, 1992; Perry, et al. 2007). There is substantial plant macrofossil, phytolith, and pollen evidence from San Salvador and Andros Island that indigenous horticultural practices altered the terrestrial landscapes of the islands of the Bahamas during the pre-Columbian period (Berman & Pearsall, 2000; Blick & Kjellmark, 2006; Kjellmark, 1996).

Some of the earliest evidence for human occupation in the Bahamas comes from San Salvador (Berman & Gnivecki, 1995). Dated to about A.D. 600, these peoples had ceramic ties with nearby islands of the Greater Antilles such as Cuba and Hispaniola (Berman & Gnivecki, 1995; Granberry & Vescelius, 2004; Hoffman, 1967; Winter & Gilstrap, 1991). Over the ca. 900 years of occupation on the island of San Salvador, the Lucayans founded, and thrived at, some 40 prehistoric sites across the island (Blick, Hopkins, & Oetter, 2011; Hopkins, Oetter, & Blick, 2011), most of which were located on dune ridges near the coastline and the ocean resources they had come to rely upon for their sustenance (Keegan, 1997).

This study is based upon archaeological work performed at two sites on San Salvador, the Minnis-Ward site (SS-3) and the North Storr's Lake site (SS-4) (Figure 2). Generally speaking, these two sites are roughly contemporaneous, located on opposite sides of the island, and have somewhat different ecological and archaeological characteristics. Calibrated radiocarbon dates from Minnis-Ward span the period ca. A.D. 880-1490 (2-sigma) (Blick & Dvoracek 2011; Blick, Creighton, & Murphy 2006; Winter, 1981, 1997) while calibrated pre-Columbian dates from North Storr's Lake range from ca. A.D. 860-1520 (2-sigma) (Blick & Dvoracek, 2011; Blick, Creighton, & Murphy 2006; Delvaux, Fry, & Murphy, 2006; Shaklee, Fry, & Delvaux 2007). The Minnis-Ward site is a linear village on a dune ridge located on the western or leeward side of San Salvador near a patch reef complex and is situated about 200 m from the ocean. North Storr's Lake is also a linear village on a dune ridge and adjacent lakeshore on the eastern or windward side of the island about 180 m from the ocean near a linear fringing reef. Archaeological deposits differ, with Minnis-Ward characterized by a high density midden with abundant Scaridae (parrotfishes), moderate amounts of Cheloniidae (sea turtle), and a few high status artifacts including imported pottery, an imported sandstone tablet, a rock crystal, and a serpentine *cohoba* (*Anadenanthera peregrina*) hallucinogenic snuff grinder (Blick, et al., 2009). North Storr's Lake appears to have a medium-to-high density midden with abundant Serranidae (grouper) and Cheloniidae (Blick & Murphy, 2005; Blick, Creighton, & Murphy, 2006), and several likely high status artifacts including copper fragments, greenstone nodules, a quartz crystal, a petaloid ax (Delvaux, Fry, & Murphy, 2006), and an incised shell inlay fragment for a wooden pre-Columbian idol known as a *zemi* ("spirit figure") (Blick, Creighton, & Murphy, 2006:29, Fig. 23a).

More specifically, the Minnis-Ward site is located on the northwest corner of San Salvador Island, approximately 1 km south of Rocky Point some 200 m east (inland) from the ocean and some 100 m west of Triangle Pond (Figure 2). The site's coordinates are 24°06'2.22"N, 74°31'5.5"W (Winter, 1997). Dimensions of the site are approximately 100 x 100 m, covering an elongated elliptical area of ca. 8400 m^2 (Winter, 1997) along a southwest-northeast trending dune ridge. The Minnis-Ward site (aka Ward site, Ward/Minnis site, etc.) was first reported by Ruth Wolper in the late 1950s and tested by John Goggin in 1960 (Hoffman, 1967; Winter, 1997). Testing and excavations by John Winter (1980, 1981, 1997) have yielded large quantities of vertebrate and invertebrate faunal remains (Winter & Wing, 1995) dominated by Scaridae and Serranidae, including relatively abundant quantities of Cheloniidae, some of which have been identified to the species *Caretta caretta* (Linnaeus, 1758) (loggerhead turtle) (Winter, 1980, 1981). Prehistoric radiocarbon dates reported by Winter (1981, 1997) and Blick, Creighton, & Murphy (2006; see also Blick & Dvoracek, 2011) span the period ca. A.D. 880-1490 (calibrated, 2-sigma). Winter (1980:3) also reports three post molds (impressions of posts) in the "yellow-white coral sand" of the site at about 51-53 cm below the surface, perhaps indicative of a pre-Columbian structure at the site. Blick's (2003) preliminary shovel testing program at the site also revealed evidence for the presence of multiple prehistoric houses at the site based on the spatial distribution of artifacts recovered in shovel tests. Winter (1981:5) classifies the site as an "open village site," an interpretation supported by Blick (2003, 2004). The archaeological materials from Minnis-Ward discussed in this chapter derive from Blick's May 2004 excavation of a 5x5 m unit in a high density area of artifacts near a pre-Columbian structure designated as Household 1 (Blick, 2004). Additional work at Minnis-Ward, now a threatened site, continued in 2009 and 2010.

The North Storr's Lake site (SS-4) is a prehistoric Lucayan archaeological site located on the northeastern side of the island of San Salvador (Figure 2) contained within an area of approximately 300 meters north-south by 150 m east-west between the northeast shoreline of northern Storr's Lake and the east facing foredune paralleling the Atlantic Ocean. Based on current data available from GoogleEarth (2006), the coordinates of the approximate center of the site are 24°05'00.00"N, 74°26' 19.00"W. Like many of the archaeological sites on San Salvador, the North Storr's Lake site was discovered in the early 1980s by John Winter (Winter, 1981, 1982). Winter (1981) reports that he excavated six test pits at the North Storr's Lake site and recovered prehistoric Lucayan Palmetto Ware pottery from two of the six test pits. "The two pits which produced sherds are located in that region of the lake which is covered with swamp vegetation" (Winter, 1981:7). Also according to Winter (1982:8), in January 1982, "five test pits were dug at the Storrs Lake Site ... in order to better evaluate the size and nature of the site." Although a portion of the site appears to have been disturbed at that time by cultivation and bulldozer operations, "cultural material came from a depth of between 29-35 cm" (Winter, 1982:8). At the time of the 1982 investigation, Winter reported the dimensions of the site to be approximately 111 m long by 56 m wide. According to Winter (1982:8), "Since parts of the site seem undisturbed, and since it is larger than many other sites presently known in the Bahamas, excavation in the future may be warranted." As far as the author's knowledge is concerned, the North Storr's Lake site subsequently lay untouched by archaeologists until 1996 when Dr. Gary Fry of Youngstown State University (YSU) began a multi-year project that lasted until 2001. Previous work performed under the auspices of YSU under the supervision of Dr. Gary Fry and Tom Delvaux has been recently described by Delvaux, Fry, & Murphy (2006) and Shaklee, Fry, & Delvaux (2007). Radiocarbon dates reported by Shaklee, Fry, & Delvaux (2007:34, Table 1) for the North Storr's Lake site span the time range ca. A.D. 860-1650 (calibration not reported; the later date is most probably non-Lucayan) with new prehistoric dates reported by Blick, Creighton, & Murphy (2006) and Blick & Dvoracek (2011) on a different portion of the site covering the period ca. A.D. 1300-1520 (calibrated, 2-sigma). The archaeological materials from North Storr's Lake discussed in this chapter derive from Blick's May 2005 2x2 m excavation in a high density household midden area (Blick & Murphy, 2005) associated with abundant fish remains and the latter radiocarbon dates.

Fig. 2. Map of the island of San Salvador, Bahamas showing the locations of the Minnis-Ward site (SS-3) (NW corner of the island) and the North Storr's Lake site (SS-4) (NE corner of the island) (sites indicated by yellow triangles). Scale is 10 km (6.22 miles). The Long Bay site (SS-9) (white triangle) is the probable site of the 1492 Columbus landfall (Blick, 2011a).

2.2 Pre-Columbian resource overexploitation in the Greater Caribbean and Bahamas

Over approximately the last three decades, zooarchaeologists working in the Greater Caribbean and Bahamas have recorded numerous archaeological sites where quantities and sizes of crab, mollusks, and fishes have declined over time from earlier to later deposits. One of the first cases in which such a trend was noticed was the decline in abundance of Gecarcinidae (land crab) remains on Puerto Rico by Rainey (1940) which he interpreted as the transition from a "crab culture" to a "shell culture," suggesting two different cultural groups with varying subsistence strategies. Rainey's model has since been shown to be incorrect since pottery styles (thought to be indicative of different cultures or migrations) did not change during this transition (Goodwin, 1980; Rouse, 1992). Various explanations have been put forth to explain this shift from an abundance of crab in earlier cultural periods to a greater reliance on mollusks in later cultural periods, including depletion of crab by human predators, climate change, and alteration of the terrestrial landscape by habitat-destroying human agricultural practices (Jones, 1985; Carlson & Keegan, 2004). The declines in abundance and size of Gecarcinidae have since been noticed on many Caribbean islands such as Antigua, Nevis, Puerto Rico, Saba, St. John, St. Martin, St. Thomas, etc. (Jones, 1985; Quitmyer, 2003; Wing, 2001; Wing, deFrance, & Kozuch, 2002), and is today recognized as a widespread phenomenon in the archaeological record of the Caribbean (Serrand, 2002). The most common explanation for this widespread pattern of decline in Gecarcinidae remains is overexploitation by human collectors (see also LaPilusa & Heilveil, 2011).

Mollusks, such as *Cittarium pica* (Linnaeus, 1758) (West Indian top shell), *Strombus gigas* (Linnaeus, 1758) (queen conch), etc., are also reported to exhibit changes in frequencies and sizes over time in the archaeological record of the Caribbean on such islands as Jamaica, Nevis, St. John, St. Thomas, and others (Keegan, Portell, & Slapcinsky, 2003; Quitmyer, 2003; Wing, 2001; Wing, deFrance, & Kozuch, 2002). A preliminary field study by Blick (unpublished data) of *S. gigas* indicates total length of the body of the queen conch was significantly longer on specimens from pre-Columbian middens (e.g., Barker's Point Shell Midden, SS-37) and shorter on specimens from modern conch middens (e.g., modern conchs along the western side of Dump Point site, SS-28/35). Some scholars have suggested that as crab becomes scarce, pre-Columbian shellfish gatherers switch to other protein sources to make up for the lack of others (e.g., Wing, 2001). Keegan (1992), on the other hand, lists *S. gigas* and *C. pica* as high ranking food resources which would have been targeted first due to their desirable protein and calorie yields, while low value species such as chitons "should have been among the last items added to the diet" (Keegan,1992:130). Chitons, such as *Chiton tuberculatus* (Linnaeus, 1758) (common West Indian chiton) and *Acanthopleura granulata* (Gmelin, 1791) (fuzzy chiton), appear to be poorly studied and under-reported in the archaeological literature of the Caribbean (see for example Keegan, 1992; Newsom & Wing, 2004), so their inclusion in this chapter may motivate others to look more carefully at chitons as potential indicators of human overexploitation of the environment. Reported changes in frequencies and sizes of mollusks from earlier to later deposits, to local disappearance of the resource, are generally taken to reflect stress and intensive exploitation as a result of human predation leading to overfishing and eventual local extirpations (Keegan, Portell, & Slapcinsky, 2003; Newsom & Wing, 2004; Wing, 2001; Wing, deFrance, & Kozuch, 2002).

Perhaps the most significant finding to have emerged from Caribbean archaeology in recent decades is the recognition of several trends of decreasing abundance and size of various fishes and changing composition of tropical fish communities, especially notable among reef dwellers and coral reef communities. Islands where such studies have been performed include Antigua, Grand Turk, Haiti, Jamaica, Nevis, Puerto Rico, Saba, St. Martin, and St. Thomas, among others (Carlson & Keegan, 2004; Jones, 1985; Wing, 2001; Wing, deFrance, & Kozuch, 2002; Wing & Wing, 2001). These studies show clear and convincing evidence of decreasing abundance of a variety of types of fishes, decreasing size of those fishes as measured by widths of atlases and vertebral centra (which can then be converted to body weights using allometric formulae - see Methods, below), and decline in top predators present in the faunal assemblages with a concomitant increase in herbivores and omnivores (primarily parrotfishes). Only rarely do similar studies show opposing or somewhat different trends (Carder, Reitz, & Crock, 2007; Whyte, Berman, & Gnivecki, 2005). The recognition of similarities in these trends between pre-Columbian fisheries and modern global fisheries has major conservation implications regarding fisheries management, sustainability, and human population impact on the environment.

Many of the same studies cited above also demonstrate convincing patterns of decline in the number of taxa identified archaeologically which is taken to represent decline in taxa richness, or biodiversity. In earlier deposits, there are numerically more (and more diverse) taxa than in later archaeological deposits. At the same time, a number of these studies also reports declines in the average trophic levels over time of the total aquatic fauna in the archaeological deposits of various Caribbean islands from the Greater to the Lesser Antilles (Puerto Rico, St. Thomas, St. Martin, Saba, Nevis, etc.) (Newsom & Wing, 2004; Wing, 2001; Wing, deFrance, & Kozuch, 2002; Wing & Wing, 2001). The phenomena mentioned in the paragraphs above are attributed to human-induced resource overexploitation and overfishing of local environments near prehistoric habitation sites.

3. Methods

All cultural and faunal remains from the Minnis-Ward and North Storr's Lake sites were collected by the author and students in archaeological sifters or screens using window screen with 1.59 mm mesh. The archaeological excavation at the Minnis-Ward site (SS-3) was a 5x5 m excavation located near Household 1 at E10N5 excavated in three levels to a depth of ca. 40 cm (Blick, 2004). The excavation at North Storr's Lake (SS-4) was a smaller 2x2 m excavation unit located in an area of household midden deposit at 81E17N which was dug in five levels to a depth of ca. 70 cm (Blick & Murphy, 2005). All soils were passed through window screen allowing recovery of very small artifacts such as 1-2 mm shell disc beads, fish vertebrae, and other miniscule bones. This method ensured excellent sampling and collection of small faunal materials that might otherwise be lost. Excavations were conducted within natural sedimentary layers, and excavators attempted to stay within 10 cm thick arbitrary levels within natural strata. Due to the size of the 5x5 m excavation at Minnis-Ward, volume of earth excavated was ca. 10.00 m^3 yielding 31,408 artifacts and ecofacts; the smaller 2x2 m excavation at North Storr's Lake revealed 11,927 artifacts and ecofacts in ca. 2.80 m^3 of earth. Details of the excavations are available in Blick (2004) and Blick & Murphy (2005).

Vertebrate and invertebrate faunal materials were identified using the comparative collections at the University of Florida Museum of Natural History (FMNH), Gainesville and the University of Georgia (UGA) Museum of Natural History, Athens and various published guidebooks (see Blick & Murphy, 2005 for details). Taxonomic nomenclature follows Turgeon, et al., (1998) for invertebrates (mollusks) and Nelson, et al., (2004) for vertebrates (fishes). Zooarchaeological methods employed the techniques developed by Reitz & Wing (1999), with additional identifications and assistance by Nanny Carder (then conducting research at UGA), Irvy R. Quitmyer (FMNH), and Elizabeth J. Reitz (UGA). Although additional faunal analyses from both sites continue, all efforts were made to identify specimens discussed in this report to the lowest taxonomic level. Faunal material was counted and weighed and recorded on preprinted data recording forms and entered into computer spreadsheets and/or databases for statistical analyses. Because the purpose of these excavations was to investigate changing subsistence patterns and changing exploitation of subsistence resources over time, all key specimens were analyzed, weighed (in g), quantified, measured (in mm), and tabulated separately according to their respective archaeological excavation levels. Weights were recorded using an Ohaus Scout Pro SP601 digital scale with 0.1 g accuracy. Linear measurements were taken on all complete vertebrate specimens presented in this report using a Helios needle nose dial caliper with an accuracy of 0.05 mm. All statistical analyses were performed using SYSTAT® (SPSS Inc., 1997) Version 7.0 and verified with a university statistician. For the purposes of this chapter, only five faunal categories are reported - three invertebrate and two vertebrate taxa - due to sample size and other limitations. The five taxa, the particular variables investigated for each, and other measurements based upon them, are described briefly, below.

Gecarcinidae (land crab) shell fragments, chelipeds and dactyls (claws), etc. were quantified, weighed and tabulated separately for each excavation level of the two archaeological sites. There are two varieties of Gecarcinidae on San Salvador today, *Cardisoma guanhumi* (Latreille, 1825) (great land crab) and *Gecarcinus ruricola* (Linnaeus, 1758) (black land crab) (Diehl, et al., 1988). Archaeological specimens of Gecarcinidae are identifiable typically only to the family level (Jones, 1985; Newsom & Wing, 2004; Wing, 2001; Wing, deFrance, & Kozuch, 2002) and are notable for their creamy coloration and eggshell-like texture and weight. Modern specimens that retain more life-like colors (gray or maroon, depending on the species) were excluded from analysis. It has been demonstrated throughout archaeological deposits in the West Indies that "land crabs were intensively harvested" (Newsom & Wing, 2004:196; see also Quitmyer, 2003; Serrand, 2002; Wing, 2001), especially during the Saladoid period (ca. 500 B.C.-A.D. 600). Intensive land crab harvesting was also practiced in later prehistory on San Salvador, as shall later be demonstrated. LaPilusa & Heilveil (2011) report similarly heavy exploitation of land crabs on modern Andros Island, Bahamas for feasting purposes.

C. pica shell fragments (very few complete specimens were found) were quantified, weighed and tabulated separately for each excavation level. *C. pica* is readily identifiable by its notable black-and-white coloration and its top-like (or turban-like) shape (Lawson, 1993; Robertson, 2003). Fragmented specimens of *C. pica* are easily identifiable based on color and sometimes morphology; in the absence of coloration, the shiny whitish-yellow nacre inside the shell may sometimes be used as an additional diagnostic indicator (Blick

& Murphy, 2005). *C. pica* has been called, "the third most important marine invertebrate species eaten by man in the West Indies.... [Today] Its stocks are depleted at many places, large animals having been selectively removed by fishermen and others" (Robertson, 2003; see also Wing, 2001). Depletion of this stock will also be demonstrated for pre-Columbian San Salvador.

Dorsal plates and dorsal plate fragments of chitons (*C. tuberculatas* and/or *A. granulata*) were quantified, weighed and tabulated separately for each excavation level. Chitons are readily identifiable by their eight keeled, overlapping, hinged dorsal plates (Diehl, et al., 1988). Unfortunately, this creature has not been well reported archaeologically (see Keegan, 1992; Newsom & Wing, 2004), and there is generally no attempt to distinguish the two similar species in the archaeological record (Jones, 1985; Keegan, Portell, & Slapcinsky, 2003; Newsom & Wing, 2004). It is said that the dorsal plate of the *C. tuberculatus* is "heavily keeled" (Diehl, et al., 1988:52) in comparison to that of *A. granulata*. Chitons are considered intertidal creatures that "inhabit rock faces near the high tide line" (Diehl, et al., 1988:52), so their presence inland (away from the rocky swash zone) on an archaeological site is considered to be indicative of human collection, transport, and eventual consumption (Blick, 2003; Keegan, 1992).

Sparisoma viride (Bonnaterre, 1788) (stop light parrotfish) premaxillae, dentaries, and upper and lower pharyngeal grinding mills are readily identifiable in the archaeological record (Reitz & Wing, 1999:55, Fig. 3.11) and are relatively easy to assign to left or right sides allowing for quick calculation of MNI (minimum number of individuals, the smallest number of individuals necessary to account for all skeletal or other preserved elements of a given species). The lower pharyngeal grinding mill (LPGM) width was taken as a lateral measurement at the greatest medio-lateral breadth (aka greatest width) of the grinding mill from the right posterior edge to the left posterior edge of the grinding plate surface (Blick, 2007; see also Bellwood, 1994:25, Fig. 12; Carder, Reitz, & Crock, 2007:591, Fig. 3). LPGM length was taken as an antero-posterior measurement from the anterior edge to the posterior edge of the grinding plate surface (Blick, 2007). See Bellwood (1994:25, Fig. 12e) and Reitz & Wing (1999:55, Fig. 3.11d) for illustrations of the *S. viride* LPGM (but note that Bellwood and Reitz & Wing label anterior differently in these diagrams). The technique for taking these measurements was modeled after Reitz's lateral and antero-posterior measurements (personal communication, July 2004). Fragmentary LPGMs were not measured unless at least one of the two measurements was possible. Parrotfishes have been reported as a major subsistence item of the Lucayans on San Salvador (Hoffman, 1967; Whyte, Berman, & Gnivecki, 2005; Wing, 1969; Winter & Wing, 1995). The large sample size of *S. viride* LPGMs (MNI = 249) at Minnis-Ward made this bony element particularly amenable to statistical analyses to investigate change over time of one heavily utilized marine (coral reef) resource.

Serranidae (sea bass/grouper) atlases are particularly diagnostic due to the presence of an anterior medial protuberance and dual superior articular facets which articulate with the paired exoccipitals of the basioccipital portion of the neurocranium (Nanny Carder, personal communication, July 2004; Carder, Reitz, & Crock 2007). The serranids most likely present in and around the waters of San Salvador are the groupers, especially *Epinephelus morio* (Valenciennes, 1828) (red grouper) and *E. striatus* (Bloch, 1792) (Nassau grouper) and certain members of the genus *Mycteroperca* such as *M. tigris* (Valenciennes, 1833) (tiger grouper) and

M. venenosa (Linnaeus, 1758) (yellowfin grouper). In fact, most of the grouper remains identified at the Minnis-Ward and North Storr's Lake sites appear to be those of *E. morio* based upon morphology (Florida Museum of Natural History disarticulated comparative specimen). The serranid atlas width was taken as a lateral measurement at the greatest medio-lateral breadth on the anterior portion of the vertebral centrum (Morales & Rosenlund, 1979; Reitz, 1994; see also Reitz & Wing, 1999:176, Fig. 7.3 for an illustration of where the atlas width is taken). The serranid atlas height was taken as the greatest dorso-ventral height of the vertebral centrum (Morales & Rosenlund, 1979; see also Blick, 2006, 2007).

Based upon the serranid atlas width measurement described above, an allometric formula (Reitz & Wing, 1999; Wing, 2001; Wing & Wing, 2001) was used to calculate serranid body weight from the atlas width. An allometric formula is an "equation [that] describes the line that statistically best fits the data and is calculated by the method of least squares" (Reitz & Wing, 1999:70). Using the atlas widths of a particular fish, one can correlate "dimensions of fish skeletal elements and total weight or length" (Reitz & Wing, 1999:70) of that particular class or family of fish (there are separate formulae for different categories of fishes, so one must chose the formula one uses with care). In this case, for the serranid atlases analyzed for the present chapter, the formula utilized was one derived from a sample of teleost fish atlas widths utilized in other similar studies by Wing (2001) and Wing & Wing (2001:3, Table 2):

$$\log Y = \log a + b (\log X) \tag{1}$$

where:

X = the measurement of the skeletal element (atlas width, mm)
Y = the estimated body weight (g)
log a = the Y-intercept (.872)
b = the slope of the line (2.53)

Since sample size is largest at the Minnis-Ward site, the total number of identified vertebrate taxa (a measure of "richness") was calculated for each archaeological level in the 5x5 m excavation unit. This simply amounted to a presence/absence analysis of whether a vertebrate taxon was present or not in a particular archaeological level and does not take into account "abundance" (how many individuals or specimens are present within a particular taxon). For each archaeological level, from earliest (deepest) to most recent, the number of taxa present were summed to yield the total number of identified taxa per archaeological level to assess whether or not there was a generally observable decline in taxa richness as one might expect as resources are continuously exploited over time (Newsom & Wing, 2004; Pauly et al., 1998; Reitz & Wing, 1999; Wing, 2001; Wing & Wing, 2001). Prehistoric vertebrate fauna identified in the Minnis-Ward materials include a number of predominantly tropical reef taxa and other fishes, and generally identified Cheloniidae. Intrusive vertebrate fauna include the post-Columbian European introductions of *Rattus* sp. (Linnaeus, 1758) (rat) and Caprinae (possible Barbadan sheep; Reitz, personal communication, July 2004) which have an associated AMS radiocarbon date of A.D. 1690-1900 (UGAMS-17156, calibrated, 2-sigma). The post-Columbian introductions were excluded from the analysis of prehistoric change in the total number of identified taxa. The complete list of identified vertebrate taxa at Minnis-Ward is presented in Table 1.

Family or Taxon	Genus (& species if possible)	Trophic Level	Level 3	Level 2	Level 1	L3TL	L2TL	L1TL
Belonidae	*Tylosurus* (?)	4.46						
Serranidae			1	1	1			
	Epinephelus cf. *morio*	3.60	1	1	1	3.60	3.60	3.60
	Epinephelus sp.	3.60	1	1	1	3.60	3.60	3.60
Carangidae	*Caranx* cf. *latus*	4.36	1			4.36		
	Caranx sp.	4.36	1			4.36		
Lutjanidae				1				
Haemulidae	*Haemulon* (?)	3.27						
Sparidae	*Calamus* cf. *bajonado*	3.17	1	1	1	3.17	3.17	3.17
Labridae			1	1	1			
	Halichoeres cf. *radiatus*	3.26	1	1	1	3.26	3.26	3.26
	Halichoeres sp.	3.26	1		1	3.26		3.26
Scaridae	*Scarus* cf. *coelestinus*	2.02	1	1	1	2.02	2.02	2.02
	Scarus cf. *vetula*	2.03	1	1	1	2.03	2.03	2.03
	Scarus sp.	2.02	1	1	1	2.02	2.02	2.02
	Sparisoma cf. *rubripinne*	2.00	1			2.00		
	Sparisoma viride	2.00	1	1	1	2.00	2.00	2.00
Acanthuridae	*Acanthurus coeruleus*	2.01	1			2.01		
	Acanthurus sp.	2.01	1			2.01		
Scombridae			1					
Balistidae	*Balistes* cf. *vetula*	3.40	1	1	1	3.40	3.40	3.40
Diodontidae	*Diodon*	3.37		1			3.37	
UID Fish			1	1	1			
Cheloniidae	(*Chelonia mydas* ? - herbivore)	2.00	1	1	1	2.00	2.00	2.00
UID Bird (Aves)			1					
Mammal	*Rattus* (UID Mammal?)*			1				
Mammal	Caprinae (Barbadan sheep?)**				1			
	TOTAL		21	16	15	2.82	2.77	2.76
	*possible intrusive/post 1492	from						
	**possible historic/post 1780	fishbase.org						

Table 1. Complete list of identified vertebrate taxa at Minnis-Ward. Trophic levels for each taxon are provided as is taxon presence/absence information for each of the archaeological Levels 3-1. Total numbers of identified taxa are provided in the Total row, as are the average trophic level figures for each archaeological level. Taxa not positively identified, indicated by (?), and historic/intrusive taxa, have been left out of the pre-Columbian trophic level calculations. Trophic levels are based on Froese & Pauly (2005). Note: L3TL = Level 3 trophic level, etc.

Each of the vertebrate taxa identified at Minnis-Ward was assigned a trophic level which is basically an animal's place in the food chain. These trophic levels are based on Pauly et al. (1998) and Froese & Pauly (2005) who rank primary producers as 1 and apex predators near 5. Trophic level values (and their explanations) for each of the identified taxa are available on the Internet at www.fishbase.org (Froese & Pauly, 2005). The trophic level analysis of Wing (2001; Wing & Wing, 2001) was inspirational and subsequently modified for specific application to the available archaeological dataset from the Minnis-Ward site for the purposes of this chapter. Utilizing the presence/absence analysis of the number of identified taxa (described above), each taxon was assigned a 1 or a 0 if it was present or absent, respectively, in a particular archaeological level. The trophic levels for each individual taxon were then multiplied by a 1 if the taxon was present or a 0 if the taxon was absent, summed within each archaeological level, and an average or mean trophic level was then calculated for each of the archaeological strata or excavated levels (Table 1). This allowed for a general, yet revealing, analysis of prehistoric change in trophic level across time as represented by each of the archaeological strata from earliest (deepest) to most recent.

4. Results

Beginning with our invertebrate fauna, there were significant declines over time in the weight of Gecarcinidae (land crab) remains at both the Minnis-Ward site (SS-3) and North Storr's Lake (SS-4). In the three archaeological levels at Minnis-Ward on the northwestern corner of San Salvador, weight of land crab remains declines linearly from 56.1 g in Level 3 to 19.1 g in Level 1 (Figure 3). The approximate time frame of this decline is ca. A.D. 880-1490 (calibrated, 2 sigma) based on previously reported radiocarbon dates for the Minnis-Ward site (Blick, Creighton, & Murphy, 2006, Appendix 1; Blick & Dvoracek, 2011; Winter, 1981, 1997). The significance of this decline was measured using a likelihood ratio test (Myers et al., 1995) with Poisson distribution yielding $p = 0.00001$. The likelihood ratio test was utilized rather than linear correlation due to the sample size of archaeological levels involved ($n = 3$), although it should be noted that there are hundreds of crab fragments per archaeological level that contributed to the total weight for each level (total of 687 land crab fragments in all three levels combined). This likelihood ratio test assumes that the average weight per fragment per level is equal, an assumption validated through the construction of confidence intervals for the counts and weights of Gecarcinidae fragments for each archaeological level (Jason Stover, personal communication, April 2005). At the North Storr's Lake site on the northeastern side of San Salvador, weight of Gecarcinidae remains declines in near-linear fashion from 50.0 g in Level 5 to 15.1 g in Level 1 (Figure 3). The approximate time frame of this decline is ca. A.D. 1300-1520 (calibrated, 2-sigma) based on recent AMS radiocarbon dates from North Storr's Lake (Blick & Dvoracek, 2011; Blick, Creighton, & Murphy, 2006, Appendix 1). The significance of this decline was measured using a likelihood ratio test with Poisson distribution yielding $p < 0.001$. The likelihood ratio test was utilized rather than linear correlation due to the sample size of archaeological levels involved ($n = 5$), although it should be noted that there are hundreds of crab fragments per archaeological level that contributed to the total weight for each level (total of 1320 land crab fragments in all five levels combined).

Weight of the mollusk *C. pica* remains declines significantly at the Minnis-Ward site from 402.4 g in Level 3 to 97.5 g in Level 1 (Figure 4). The approximate time frame of this decline is ca. A.D. 880-1490 (calibrated, 2-sigma) based on previously reported radiocarbon dates for the Minnis-Ward site (Blick, Creighton, & Murphy, 2006, Appendix 1; Blick & Dvoracek, 2011; Winter, 1981, 1997). The significance of this decline was measured using a likelihood ratio test with Poisson distribution yielding $p = 0.012$. The likelihood ratio test was utilized rather than linear correlation due to the sample size of archaeological levels involved ($n = 3$) although it should be noted that there are tens to hundreds of *C. pica* fragments per archaeological level that contributed to the total weight for each level (total of 277 *C. pica* fragments in all three levels combined). Another mollusk, *S. gigas*, also exhibits a dramatic decline in raw count at the Minnis-Ward site (Blick, 2004:16, Tables 3 and 4), indicative of general decline in mollusks over time.

Similarly, weight of chiton (*C. tuberculatus/A. granulata*) remains declines significantly at the Minnis-Ward site from 131.9 g in Level 3 to 64.3 g in Level 1 (Figure 5). The approximate time frame of this decline is ca. A.D. 880-1490 (calibrated, 2-sigma) based on previously reported radiocarbon dates for the Minnis-Ward site (Blick, Creighton, & Murphy, 2006, Appendix 1; Blick & Dvoracek, 2011; Winter, 1981, 1997). The significance of this decline was measured using a likelihood ratio test with Poisson distribution yielding p = 0.0211. The likelihood ratio test was utilized rather than linear correlation due to the sample size of archaeological levels involved (n = 3) although it should be noted that there are dozens of chiton fragments per archaeological level that contributed to the total weight for each level (total of 143 chiton fragments in all three levels combined).

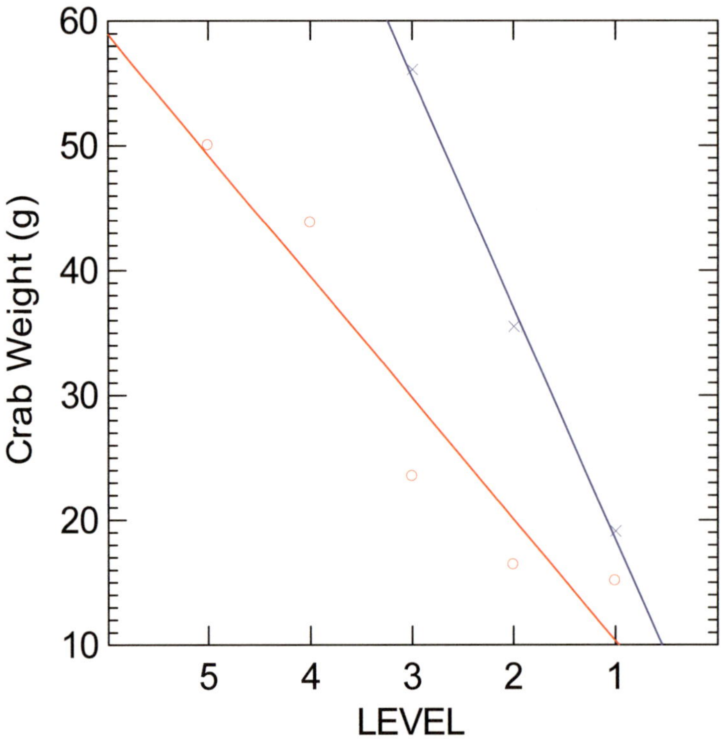

Fig. 3. Decline in weight of Gecarcinidae (land crab) remains at the Minnis-Ward site (x, blue) from Level 3 to Level 1, ca. A.D. 880-1490 (likelihood ratio test with Poisson distribution, n = 687, p = 0.00001, least squares line: y = -18.5 x + 0.60) and decline in weight of land crab remains at the North Storr's Lake site (o, red) from Level 5 to Level 1, ca. A.D. 1300-1520 (likelihood ratio test with Poisson distribution, n = 1320, p < 0.001, least squares line: y = -8.725 x + 6.375). Open circles represent total weights of crab remains per stratigraphic level.

Moving from the invertebrate to the vertebrate fauna, analysis of the width of the LPGM of the parrotfish *S. viride* reveals a statistically significant decline in size over time from an average of 13.102 mm in Level 3 to 12.106 mm in Level 1 (Figure 6). Once again, the approximate time frame of this decline is ca. A.D. 880-1490 (calibrated, 2-sigma) based on previously reported radiocarbon dates for the Minnis-Ward site (Blick, Creighton, &

Murphy, 2006, Appendix 1; Blick & Dvoracek 2011; Winter, 1981, 1997). The significance of this decline was measured using analysis of variance by examining the average width of LPGMs per archaeological level (n = 249, F-ratio = 3.747, p = 0.025). Likewise, analysis of the length of the LPGMs of *S. viride* reveals a moderately significant decline in size over time from an average of 11.210 mm in Level 3 to 10.313 mm in Level 1 over the same time frame mentioned above. The significance of this decline was measured using analysis of variance by examining the average length of LPGMs per archaeological level (n = 192, F-ratio = 2.361, p = 0.097, i.e., there is < 10 percent probability that the observed difference is due to random chance). Therefore, both the width and the length of the *S. viride* LPGM exhibit statistically significant declines in average size over time, reflecting decline in the average size of the entire organism. "Animals under intense predation show a decline in size as a result of overexploitation" (Southerland, 1990 in Reitz & Wing, 1999:67; see also Coleman et al., 2004; Conover & Munch, 2002; Pauly et al., 1998). Similar declines in the width and length of *S. viride* LPGMs have also been observed at the North Storr's Lake site (SS-4) during the period ca. A.D. 1300-1520 (calibrated, 2-sigma), although current sample sizes are not statistically adequate to verify this trend (Blick, 2007).

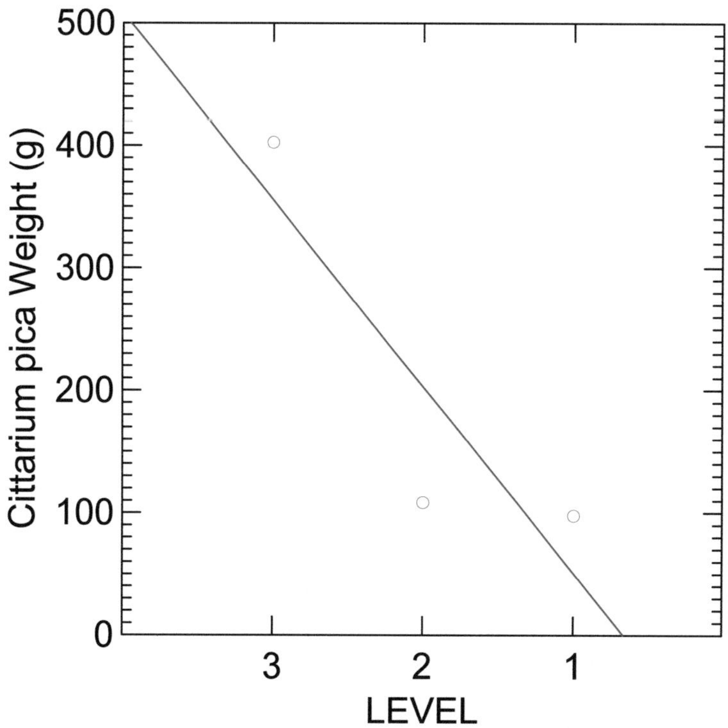

Fig. 4. Decline in weight of *Cittarium pica* (West Indian top shell) remains at the Minnis-Ward site from Level 3 to Level 1, ca. A.D. 880-1490 (likelihood ratio test with Poisson distribution, n = 277, p = 0.012, least squares line: y = -152 x + -104). Open circles represent total weights of *C. pica* remains per stratigraphic level.

Measurements of the widths of Serranidae atlases from the Minnis-Ward site reveal a bimodal distribution likely indicative of the exploitation of different sized serranids and/or the use of different fishing techniques (Figure 7), a finding similar to that of Reitz (1994) for grouper on Montserrat. For example, it is generally the case that nets or traps are used to catch smaller fish and that hook and line are more effective in catching larger fish (Froese & Pauly, 2005; Keegan, 1986; Wing & Reitz, 1982), especially predatorial serranids like *E. morio* and *E. striatus* that are more likely to take bait from a hook (Froese & Pauly, 2005; Ostrander & Brocksmith, 1997). The bimodal distribution of Serranidae atlases in Figure 7 was shown to be statistically significant through the use of a Kolmogorov-Smirnov (K-S) test for normality to determine if two different populations are represented in the sample. The results of the K-S test were based on a sample size of n = 30 (D = 1.000, $p < 0.0001$); a t-test also indicated a significant difference between the two serranid populations (n = 30, t = -11.294, $p < 0.0001$). These results were also confirmed with hierarchical cluster analysis and K-means cluster analysis (Figure 8) (F-ratio = 122.248, $p < 0.0001$) that divided the serranids into two natural clusters with a clear divide in atlas widths at ca. 15 mm.

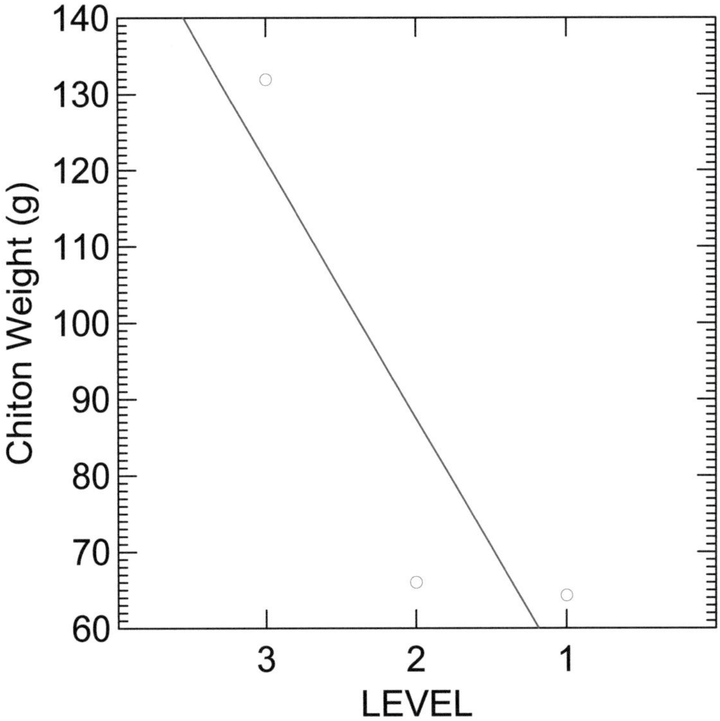

Fig. 5. Decline in weight of chiton remains at the Minnis-Ward site from Level 3 to Level 1, ca. A.D. 880-1490 (likelihood ratio test with Poisson distribution, n = 143, p = 0.0211, least squares line: $y = -34.2x + 18.6$). Open circles represent total weights of chiton remains per stratigraphic level.

Using the allometric formula to calculate fish weight from the atlas width (Wing, 2001; Wing & Wing, 2001), two groups of serranids result: Size Group 1 or the smaller serranids with an average atlas width of 11.219 mm (range 9.20-13.20 mm) estimated to weigh an average of 3470.13 g (7.65 lb) (range 2043.54-5093.92 g), and Size Group 2 or the larger serranids with an average atlas width of 18.047 mm (range 15.75-21.60 mm) with an estimated body weight averaging 11,456.66 g (25.26 lb) (range 7963.79-17,707.94 g) (Figure 9). It is therefore hypothesized that the smaller serranids were likely caught in traps while the larger serranids were more likely taken with hook and line. Additional analyses of the widths and heights of Serranidae atlases at Minnis-Ward reveal a general decline in size over time during the period ca. A.D. 880-1490, although these declines are not statistically significant at the 0.05 level due to small sample size (especially per-level sample size) (Figure 10).

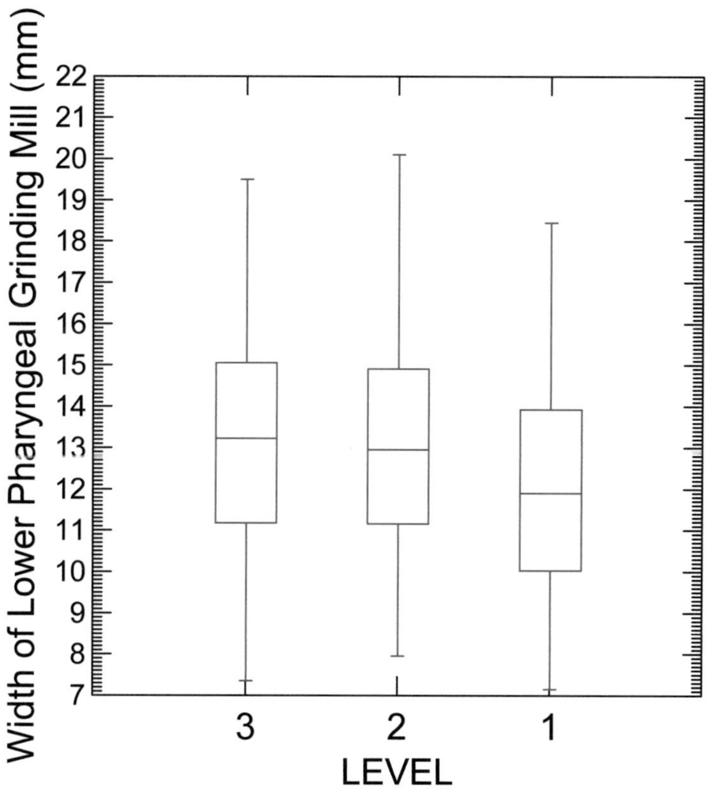

Fig. 6. Decline in the width of the lower pharyngeal grinding mill of *Sparisoma viride* (stoplight parrotfish) at the Minnis-Ward site from Level 3 to Level 1, ca. A.D. 880-1490 (ANOVA, n = 249, F-ratio = 3.747, p = 0.025). Each of the box plots shows median LPGM width (horizontal line within box), interquartile range (box), and range (whiskers) for each stratigraphic level.

The total number of identified taxa shows a notable decline over time from 21 total taxa in Level 3 to 14 taxa in Level 1 (Figure 11). The approximate time frame of this decline is ca. A.D. 880-1490 (calibrated, 2-sigma) based on previously reported radiocarbon dates for the Minnis-Ward site (Blick, Creighton, & Murphy, 2006, Appendix 1; Blick & Dvoracek, 2011; Winter, 1981,1987). This decline is not amenable to statistical analyses of the sort presented previously in this chapter due to the way the data were lumped into three archaeological levels yielding a sample size of n = 3. If one were to attempt linear correlation analysis to describe the strength of the decline in the total number of identified taxa in Level 3 to Level 1, one would find $r = 0.952$ and $r^2 = 0.855$, but linear correlation is not amenable to a sample size of three levels. It is clear from the slope of the line in Figure 11 that the decline in identified taxa is dramatic, especially between Levels 3 and 2. Based on an AMS radiocarbon date of A.D. 1000±40 (UGAMS-17157, calibrated, 2-sigma) for Level 3 (Blick & Dvoracek, 2011; Blick, Creighton, & Murphy, 2006, Appendix I), it is clear that total number of identified taxa (taken as a proxy for taxa richness) is higher during the earlier pre-Columbian period (ca. A.D. 900-1150) than in later pre-Columbian times (ca. A.D. 1300-1490) at Minnis-Ward. Decline in ecosystem richness is generally interpreted as resulting from human overexploitation, in this case fishing pressure, although other ecological factors may also be responsible (e.g., climate change, see Discussion, below).

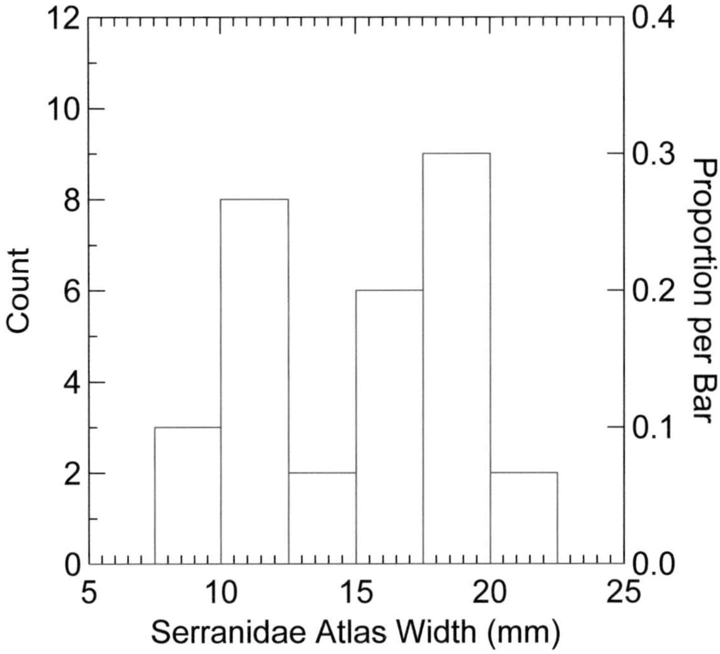

Fig. 7. The bimodal distribution of Serranidae (sea bass/grouper) atlases at the Minnis-Ward site is statistically significant based on the results of a Kolmogorov-Smirnov test ($n = 30$, $D = 1.000$, $p < 0.0001$) and a t-test ($n = 30$, $t = -11.294$, $p < 0.0001$). This morphological dichotomy suggests different sized serranids were caught using different fishing techniques.

Likewise, the average trophic levels for archaeological Levels 3 through 1 at Minnis-Ward decline over time from a mean of 2.819 to 2.760 (on a scale of 1 to 5) (Figure 12). This decline in trophic level is similar in scale to that reported by Wing & Wing (2001, Fig. 6) for Nevis, Puerto Rico, Saba, St. Martin, and St. Thomas during the period ca. A.D. 100-1390. The approximate time frame of this decline at Minnis-Ward is ca. A.D. 880-1490 (calibrated, 2-sigma). The decline in trophic levels is not amenable to statistical analyses of the sort presented previously in this report due to the way the data were lumped into three archaeological levels yielding a sample size of $n = 3$. If one were to attempt linear correlation analysis to describe the strength of the decline in the average trophic levels from Levels 3 through Level 1, one would find $r = 0.933$ and $r^2 = 0.871$, but linear correlation is not amenable to a sample size of three. Based upon the same taxa dataset, yet calculated in a different manner (Table 1), the decline in average trophic level is notable, especially between

Levels 3 and 2. It is clear, therefore, that the average trophic level is higher in earlier pre-Columbian times (ca. A.D. 900-1150) and had declined substantially by late prehistoric times (ca. A.D. 1300-1490). Such a decline in trophic level is generally interpreted as a result of overfishing (Pauly et al., 1998; Wing & Wing, 2001) and is symptomatic of the phenomenon known as "fishing down the marine food web" (Pauly et al., 1998). Not only does the average trophic level decline over time from Level 3 to Level 1 at the Minnis-Ward site, but a change in community composition was observed over time. For example, in Level 3 dated to A.D. 1000±40 (calibrated, 2-sigma), the average trophic level is 2.819 and there are both more carnivores and more herbivores among the identified taxa (Figure 13). The higher level carnivores consist of positively identified *Caranx latus* (Agassiz in Spix & Agassiz, 1831) (horse-eye jack) and a more generally identified member of the genus *Caranx* sp. (jack). Other fishes in the middle of the trophic scale in archaeological Level 3 include *E. morio* and a more generally identified member of the genus *Epinephelus*, *Balistes vetula* (Linnaeus, 1758) (queen triggerfish), *Halichoeres radiatus* (Linnaeus, 1758) (puddingwife) and *Halichoeres* sp., and positively identified *Calamus bajonado* (Bloch & Schneider, 1801) (jolthead porgy).

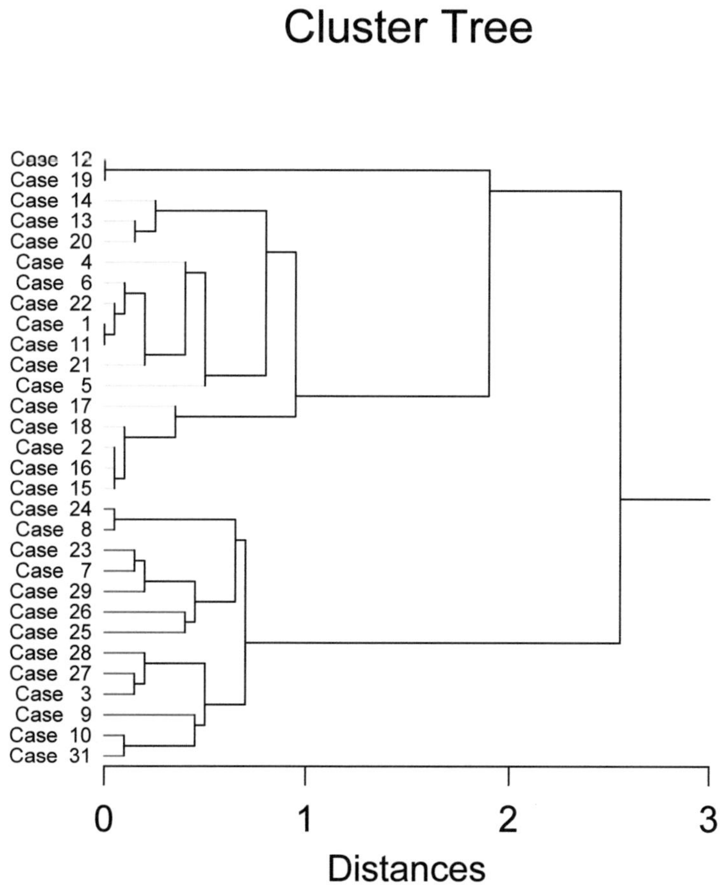

Fig. 8. K-means cluster analysis (n = 30, F-ratio = 122.248, $p < 0.0001$) divides the Serranidae at the Minnis-Ward site into two natural clusters with a clear divide in atlas widths at ca. 15 mm (compare to Figures 7 and 9).

Fishes and other organisms ranked as herbivores in Level 3 include a variety of Scaridae, Acanthuridae, and even Cheloniidae, probably *Caretta caretta* (Linnaeus, 1758) (loggerhead turtle) (Table 1). In Level 2, the carnivorous *Caranx* spp. are no longer present, the mid-level *Halichoeres* sp. disappears, and the herbivorous *Sparisoma rubripinne* (Valenciennes, 1840) (yellowtail parrotfish) and *Acanthurus* spp. (tangs) are missing. Level 2 gains a mid-level feeder with the presence of *Diodon*, probably *D. hystrix* (Linnaeus, 1758) (porcupinefish), while Level 1 has the same taxa as Level 2 with the exception of *Diodon* and the reappearance of *Halichoeres* sp. Clearly, the decline in average trophic level and the change in community composition are more pronounced from Level 3 to Level 2 and continue to a lesser degree in Level 1. The significance of this alteration in community composition as a response to human exploitation will be addressed in more detail below.

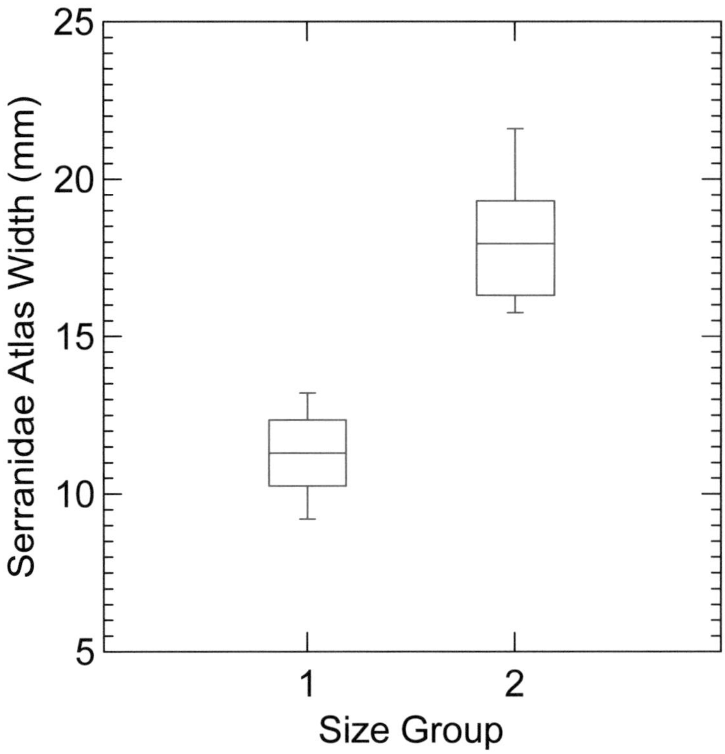

Fig. 9. Two size groups of Serranidae at the Minnis-Ward site ($n = 30$, $t = -11.294$, $p < 0.000001$). An allometric formula to calculate fish body weight from atlas width (see text) indicates that the smaller serranids, with an average atlas width of 11.387 mm, have an estimated mean body weight of 3470.13 g (7.65 lb); the larger serranids, with an average atlas width of 18.047 mm, have an estimated mean body weight of 11,456.66 g (25.26 lb). Each of the box plots show median atlas width (horizontal line within box), interquartile range (box), and range (whiskers) for each stratigraphic level.

5. Discussion

The time period of the declines of the weights and sizes of the various organisms discussed in this chapter basically covers the majority of the pre-Columbian occupation on the island of San Salvador. Prehistoric deposits at the Minnis-Ward site (SS-3) date to ca. A.D. 880-1490 (calibrated, 2-sigma) with Level 3 dating to ca. A.D. 900-1150, Level 2 corresponding to ca. A.D. 1150-1300, and Level 1 representing the latest pre-Columbian period, ca. A.D. 1300-1490. Historic era radiocarbon dates from Minnis-Ward dating to the English colonization and Loyalist periods, ca. A.D. 1670-1900 and later, are associated with historically introduced material culture remains not considered in this chapter. Thus, the changes observed in the invertebrate and vertebrate fauna at Minnis-Ward described above occurred during the ca. 600-year period between ca. A.D. 880 to just prior to the arrival of Columbus in the Bahamas. Deposits from Blick's 2005 excavation at North Storr's Lake (SS-4) appear to date fairly late in the prehistoric sequence and cover a range from ca. A.D. 1300-1520 (Blick & Dvoracek 2011; Blick & Murphy, 2005; Blick, Creighton, & Murphy, 2006, Appendix I). Level 3 in the middle of the stratigraphic column at North Storr's Lake dates clearly to A.D. 1300-1370 (UGAMS-17152, calibrated, 2-sigma) while the more superficial Level 2 dates to A.D. 1390-1490 (UGAMS-17151, calibrated, 2-sigma). Therefore it is apparent that the deposit excavated by Blick at North Storr's Lake represents the last ca. 200 years before the arrival of Columbus. Yet at both Minnis-Ward and North Storr's Lake, despite their varying durations of occupation, it is possible to detect prehistoric human impact on the ancient environment through the analysis of changing weights, quantities, and sizes of biological organisms utilized by the Lucayans as food resources.

The statistically significant diminution in the abundance of Gecarcinidae at both Minnis-Ward (SS-3) and North Storr's Lake (SS-4) is reminiscent of the decline in land crab abundance first noted by Rainey (1940) on pre-Columbian sites in Puerto Rico. Likewise, this decline in land crab remains has been noted on many Caribbean islands such as Antigua (Jones, 1985), Nevis, Puerto Rico, Saba, St. Martin, St. Thomas (Wing, 2001; Wing, deFrance, & Kozuch, 2002), St. John's (Quitmyer, 2003), St. Kitts (Goodwin, 1980), and several islands of the French Lesser Antilles (Serrand, 2002). Today, this trend is recognized as a widespread phenomenon in the archaeological record of the Greater Caribbean (Serrand, 2002). Factors such as climate change, habitat destruction, and human overexploitation have all been proposed as mechanisms for this pan-Caribbean decline in Gecarcinidae remains. Carlson & Keegan (2004:88), for example, have suggested that the decline in land crab remains (and subsequent replacement by mollusks in the diet) might have been due to "increased aridity at the end of the Saladoid period" (ca. A.D. 500). The climate change explanation for the decline in land crab abundance fails simply due to the fact that the decline in crab is observed at different time periods from relatively early in the Saladoid sequence (ca. A.D. 100-400) into later prehistory (ca. A.D. 1000-1500) (e.g., Newsom & Wing, 2004). It is possible that habitat destruction due to agricultural practices (Jones, 1985) may have played a role in the decline of Gecarcinidae, but the most widely accepted explanation for this decline now appears to be human overexploitation of land crabs (Blick, 2006, 2007; Newsom & Wing, 2004; Serrand, 2002; Wing, 2001; Wing, DeFrance, and Kozuch, 2002). Hill (2001:6) gives us an idea of the intensity of this crab exploitation (in modern times): "Throughout the Bahamas and Caribbean, *Cardisoma guanhumi* is intensively exploited as a food resource. Harvesters of wild populations ... have reported that as many as 400 crabs per harvester per night can be collected even during the months of lowest catch" (see also LaPilusa & Heilveil, 2011 for modern festival-related crab harvesting on Andros Island).

The statistically significant decline of *C. pica* at the Minnis-Ward site resembles declines in abundance reported for other islands in the Caribbean such as Jamaica (Keegan, Portell, &

Slapcinsky, 2003), Saba, St. John, St. Martin, St. Thomas, and Nevis (Quitmyer, 2003; Wing, 2001; Wing, deFrance, & Kozuch, 2002). At rare sites, there has been confusion regarding natural vs. human deposits of *C. pica* (Scudder & Quitmyer, 1998), but the majority of archaeological sites seems to provide evidence of *C. pica* as a substantial contributor to the pre-Columbian Caribbean diet (Jones, 1985; Keegan, 1992; Keegan, Portell, & Slapcinsky, 2003; Newsom & Wing, 2004; Wing, 2001; Wing, deFrance, & Kozuch, 2002). Rainey (1940) indicated that with the decline of the Gecarcinidae there was a concomitant rise in the abundance of mollusks, especially *C. pica*. If measured by weight, however, there appears to be a synchronous decline in land crab as well as *C. pica* (see for example Wing, 2001:115, Table 2). *C. pica* is also recorded to decline in size at many Caribbean sites where aperture measurements have been taken (Quitmyer, 2003; Wing, deFrance, & Kozuch, 2002). Therefore, *C. pica*, throughout the Greater Caribbean, appears to have been subject to the same intensive harvesting pressures as the terrestrial Gecarcinidae, despite *C. pica*'s rocky intertidal habitat, to the point of local extirpation in some cases (Keegan, Portell, & Slapcinsky, 2003). Decline in the quantity of *S. gigas* at Minnis-Ward mirrors the decline in *C. pica* due to high levels of exploitation throughout the Greater Caribbean (Keegan, 1992; Keegan, Portell, & Slapcinsky, 2003; Robertson, 2003). Patterns of decline in abundance in both of these organisms (*C. pica* and *S. gigas*) reflect a major reliance upon mollusks as preferred high-ranking protein sources (Keegan, 1992:132, Table 6.1).

The statistically significant decline of *Chiton tuberculatus/Acanthopleura granulata* at Minnis-Ward echoes the trends of decline in abundance of Gecarcinidae and other edible mollusks (*C. pica, S. gigas*) previously reported. Although apparently neither well studied nor well reported, the decline in chiton is best explained by human overexploitation. The steady decline of chiton quantity and weight from Level 3 to Level 1 suggests depletion of the resource or its gradual abandonment as a dietary item. Chiton's presence in an archaeological context is definitely a sign of human transport from its rocky intertidal habitat, and these organisms were likely consumed in prehistoric times as they are today by some of the inhabitants of San Salvador (Blick, 2003). Although similar in kilocalories per kg and grams of protein per kg to the more preferred *S. gigas* and *C. pica* (Keegan, 1992:132, Table 6.1), Keegan (1992:130) considers chiton to be a "low value" food "exploited during periods of food shortage" and "among the last items added to the diet." It is interesting to note that the decline in chiton is synchronous with the declines in abundance and size of many other organisms reported herein. However, rather than a non-preferred starvation food as implied by Keegan, it is likely that chiton was being exploited like any other food resource. The generally low numbers of chiton throughout the deposits at Minnis-Ward (SS-3) and North Storr's Lake (SS-4) suggest that chiton occurred as a normal part of the Lucayan diet, although in very small quantities (Blick & Murphy, 2005).

The statistically significant declines in both the width and length of *S. viride* LPGMs from Level 3-1 at Minnis-Ward provide some of the most striking evidence in this chapter of the effects of pre-Columbian resource overexploitation. Similar declines are also noted at North Storr's Lake, although the sample sizes are too small to yield valid statistics (Blick, 2007). Declining sizes of parrotfishes have been reported for other islands in the Caribbean such as Nevis, Puerto Rico, Saba, St. Martin, and St. Thomas (Newsom & Wing, 2004; Wing, 2001; Wing, deFrance, & Kozuch, 2002, Wing & Wing, 2001). Declining size is one of the chief patterns that indicate that *S. viride* (and other reef fishes) was heavily exploited in prehistoric times (Wing & Wing, 2001). The statistically significant decline in the size of the LPGM is taken as representative of decline in size of the overall organism. In the words of Newsom & Wing (2004:54), "Overexploitation can manifest itself in many ways and cause population changes. Chief among these in terms of faunal resources is size change. ... Overexploitation by artisanal or subsistence fishermen and gatherers can result in a decline in the body size of territorial organisms," such as coral reef dwelling parrotfishes (Wing & Wing, 2001). This size (and age structure) reduction is a result of what some scholars have termed growth overfishing - essentially catching fish before they have time to grow to maturity (Newsom & Wing, 2004; Russ, 1991). More on growth overfishing will be discussed in the next section on Serranidae atlas measurements.

The two clusters of Serranidae atlases demonstrate two significantly different size groups of serranids (sea basses/groupers) at the Minnis-Ward site likely caught using two different capture techniques: nets or traps for the smaller (ca. 3470.13 g, 7.65 lb) serranids, and hook and line for the larger (ca. 11,456.66 g, 25.26 lb) serranids (Reitz, 1994; Wing 2001). For the purposes of the present discussion concerning overfishing, the most important result is that both serranid atlas width (and height) show declines in size over time from Level 3 to Level 1 at Minnis-Ward (Figure 10), although sample sizes are relatively small (n = 30), especially per-level sample size. These declines in serranid average atlas widths and heights mirror the declines reported for Serranidae on St. Thomas and Nevis (Wing & Wing, 2001). Serranidae atlases also become less common in the deposits at Minnis-Ward, a trend noted for St. Thomas and Saba (Wing & Wing, 2001:5): "Serranids that dominated the early samples [periods] were less dominant or absent in the later samples." The trend of declining atlas size for the serranids suggests three possibilities: 1) there was a gradual decline in serranid size over time; 2) serranids targeted in the earlier time period (Level 3) were slightly larger than targeted serranids in the later time periods (Levels 2 and 1); and/or, 3) the different size groups were caught with different harvesting technologies during the different time periods (sample size is currently too small to rigorously test these three hypotheses). Whichever the case, the decline in size of the serranid atlases over time suggests, once again, the phenomenon of growth overfishing (especially notable from Levels 3 to 2), with fishing selection pressure possibly resulting in some somatic size recovery in the latest pre-Columbian period, Level 1 (Conover & Munch, 2002). The decline in Serranid abundance (as reflected by the number of atlases) in later levels at Minnis-Ward indicates earlier targeting of higher level predatorial feeders and eventual altered composition of the coral reef community due to the removal of higher level predators by selective fishing practices (Wing, 2001; Wing & Wing, 2001) (more on the alteration of community composition due to overfishing will be discussed below). The decline of the predatorial serranids and the persistence of the herbivorous parrotfishes is likely due to the fact that, "Large predators such as groupers (Serranidae) ... are more vulnerable to overfishing than reef herbivores" (Wing, 2001, citing Russ, 1991).

It should be noted at this point that the evidence from both the *S. viride* LPGMs and the Serranidae atlases is strongly suggestive of growth overfishing (Russ, 1991; Wing, 2001; Wing, deFrance, & Kozuch, 2002; Wing & Wing, 2001). To find similar evidence in two such disparate groups of fishes is certainly striking. While both reef dwelling fishes, clearly the herbivorous *S. viride* and the carnivorous Serranidae both exhibit smaller average body sizes (as indicated by their particular skeletal elements) in the later deposits than in the earlier ones. Even though reef fishes, such as parrotfishes, are usually caught in traps and more aggressive predators are typically caught with hook and line (Wing, deFrance, & Kozuch, 2002), both the *S. viride* and the Serranidae were being subjected to significant harvesting pressures at the same time. Growth overfishing is typically indicated by decline in body size of the targeted taxon as larger, older fish are removed from the population while smaller, younger fish escape harvest. Eventually, larger older fish become rare, and smaller younger fish become more common, resulting in declining average body size of the targeted taxon over time. While some might choose to debate whether the size changes described above are due to growth overfishing or some other factor, zooarchaeologists point to such changes as clear evidence of stress in the exploited species (Wing, 2001; Wing, deFrance, & Kozuch, 2002; Wing & Wing, 2001).

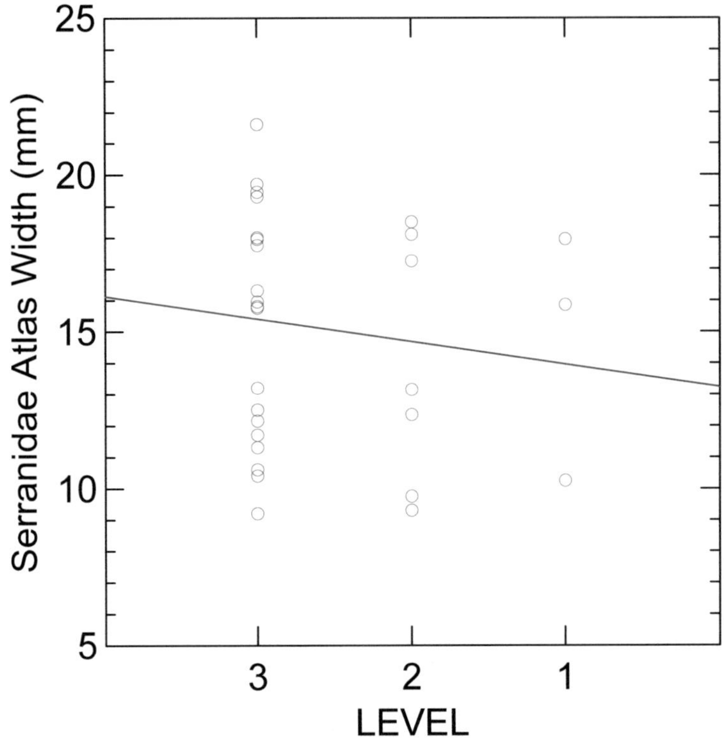

Fig. 10. Decline in Serranidae atlas widths at the Minnis-Ward site from Level 3 to Level 1, ca. A.D. 880-1490, suggestive of reduction in fish size over time (n = 30, p = 0.097, least squares line: y = -0.725 + 16.1). Open circles represent individual atlas widths.

The decline in total number of identified taxa at the Minnis-Ward site is a rough measure of decline in taxa richness, or biodiversity. This decline is indeed dramatic and exhibits a drop of 33 percent (from 21 to 14) in identified taxa from the earliest pre-Columbian level to the latest. It is a known fact that island ecosystems are especially vulnerable to human-induced environmental change, including reduction in the number of endemic species (Bahn & Flenley, 1992; Burney, 1997; Newsom & Wing, 2004; Steadman, 1995). It is interesting to note that a rare bone of Aves (bird) appears in the vertebrate fauna of the earlier Level 3 but not in later pre-Columbian levels - island birds are known to be highly susceptible to human predation (Steadman, 1995). The pattern observed in the Minnis-Ward marine vertebrate assemblage is not simply due to growth overfishing. This trend represents a transition from a fairly abundant pre-Columbian vertebrate fauna (mostly fishes and sea turtle) to a significantly reduced fauna most likely as the result of overexploitation and possibly even local extirpation of certain taxa. For example, Scombridae (tuna family), *Caranx* sp. and *C. latus*, all typically ranked as high-level predators, are present in the earlier level but absent in both later levels. This finding is in line with Wing & Wing's (2001) observation of decline in abundance of predatory fishes on several Caribbean island sites since predators are more susceptible to overfishing compared to reef herbivores (Wing, 2001). Also present in the earlier Level 3 but missing from the later two levels are the herbivorous reef fishes *S. rubripinne*, *Acanthurus* sp., and *A. coeruleus*, most likely captured with traps. Typically, in a situation in which the predatory taxa are removed from the ecosystem, herbivores and omnivores flourish and increase in number (Wing, deFrance, & Kozuch, 2002). The demonstrated decline in size for *S. viride*, another obligate reef dweller, indicates that growth overfishing, or at a minimum fishing pressure, was being exerted on Scaridae and likely other reef dwelling taxa. This type of stress might explain the absence of *S. rubripinne* and the Acanthuridae from the later deposits at the Minnis-Ward site. Overall, *Caranx* sp., *C. latus*, *S. rubripinne*, *Acanthurus* sp., *A. coeruleus*, Scombridae, and Aves characterize the earlier pre-Columbian fauna at Minnis-Ward while absent in the later pre-Columbian levels. While some of this variation could be due to sampling error from level to level, the general pattern of declining biodiversity over time is striking indeed.

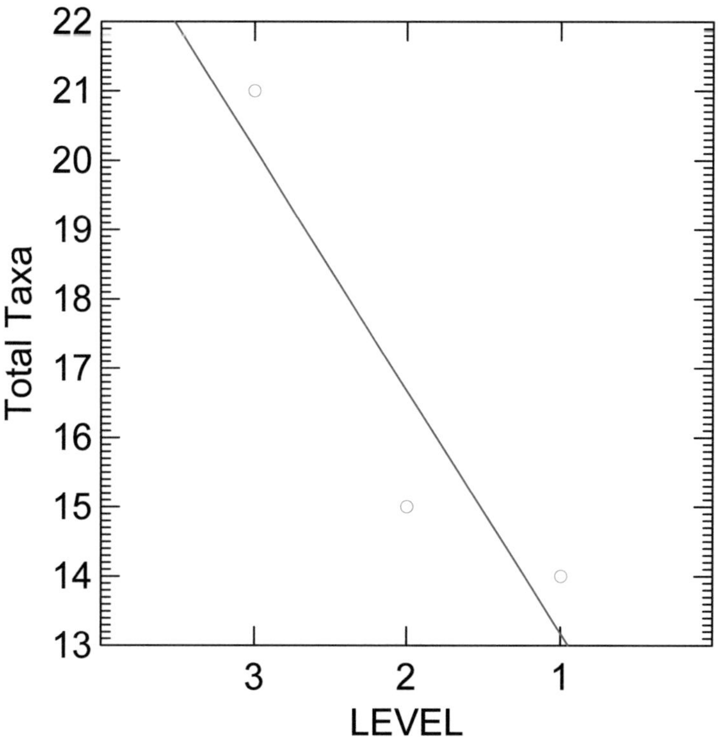

Fig. 11. Total number of identified vertebrate taxa at the Minnis-Ward site declines dramatically from 21 total taxa in Level 3 to 14 taxa in Level 1, indicative of a decline in biodiversity over time, ca. A.D. 880-1490 ($n = 50$, $p < 0.001$, least squares line: $y = -3.50\,x + 9.67$). Open circles represent individual identified taxa.

The observable decline in average trophic level from Level 3 to Level 1 at Minnis-Ward mirrors that found by Wing (2001) and Wing & Wing (2001) for five Caribbean islands including Nevis, Puerto Rico, Saba, St. Martin, and St. Thomas during the period ca. A.D. 100-1390. Although the method utilized for the purposes of this chapter is not the same as that used by Wing (albeit modeled upon her work), the fact that similar results were obtained is encouraging and significant. The general hypothesis explaining declining trophic levels in fish catches over time is that high trophic level taxa - the predatorial, carnivorous fishes (in this case, represented by *Caranx* sp. and *C. latus*) are targeted and eventually extirpated first, followed by a shift to, or greater emphasis upon, lower trophic level taxa (in this case, represented primarily by the herbivorous and varied Scaridae) (see Pauly et al., 1998). According to Wing (2001:123), "Analysis of mean trophic levels is a way of examining these and other changes in marine organisms in the sites;" she goes on to note that in her study of five island archaeological sites, "Without exception the mean trophic levels of reef fishes declines [sic] in the later deposits." Explaining these changes for the Tutu site on St. Thomas, Wing, deFrance, & Kozuch (2002:165) describe the mechanism driving declining trophic levels: "Overexploitation at the top of the food chain reduces this segment of the fauna…. This allows the populations of species at lower trophic levels to … become more available for further human exploitation." Although referring to contemporary fisheries landings, Pauly et al. (1998:860) note that, "Fishing down food webs (that is, at lower trophic levels) leads at first to increasing catches, then to a phase transition associated with stagnating or declining catches. These results indicate that [these] exploitation patterns are unsustainable." The concept of sustainability will be revisited in the Conclusions.

Related to declining biodiversity and declining average trophic level is another observable trend in the faunal materials from the Minnis-Ward site: a change in the coral reef community composition over time. The primary notable changes are the absences of the Carangidae, *S. rubripinne*, and the Acanthuridae in the two later levels. Change in community composition is generally taken to represent the result of human exploitation that typically targets apex predators first (Baum et al., 2003; Caddy et al., 1998; Conover & Munch, 2002; Hawkins & Roberts, 2004; Jackson et al., 2001; Wing, 2001). Depletion of predatory species (such as groupers, jacks, etc.) then causes a cascade effect, altering community composition as a result of harvesting practices that eventually switch to lower level feeders in the absence of carnivores (Coleman et al., 2004; Hawkins & Roberts, 2004; Jackson et al., 2001; Mumby et al., 2006; Pandolfi et al., 2003; Pauly, Froese, & Christensen, 1998; Pauly et al., 1998; Wing, 2001; Wing, deFrance, & Kozuch, 2002); this phenomenon is generally referred to as "fishing down the marine food web" (Pauly et al., 1998). Coleman et al. (2004:1959) report that, "Commercial and recreational fishing have similar demographic and ecological effects on fished populations. They truncate size and age structures, reduce biomass, and alter community composition. … All these fishery removals can cause cascading trophic effects that alter the structure, function, and productivity of marine ecosystems." Hawkins & Roberts (2004:15), in their study of fishing pressure on six Caribbean islands, have shown that, "intensive artisanal fishing has transformed Caribbean reefs" (Hawkins & Roberts, 2004:215) by targeting larger-bodied individuals, reducing fishery biomass, and causing cascading effects and eventual depletion and reduction in size of predatory groupers, snappers, and reef herbivores - especially parrotfishes. This appears to be what happened in the case of the Minnis-Ward vertebrate fauna (composed mostly of fishes). The findings of Hawkins & Roberts (2004:225) indicate that, "[modern] artisanal fisheries have transformed coral reefs in ways that seriously compromise their ecological and economic value." It would appear, based upon the evidence presented above, that the same was true for the prehistoric subsistence fishery of the Lucayans on San Salvador.

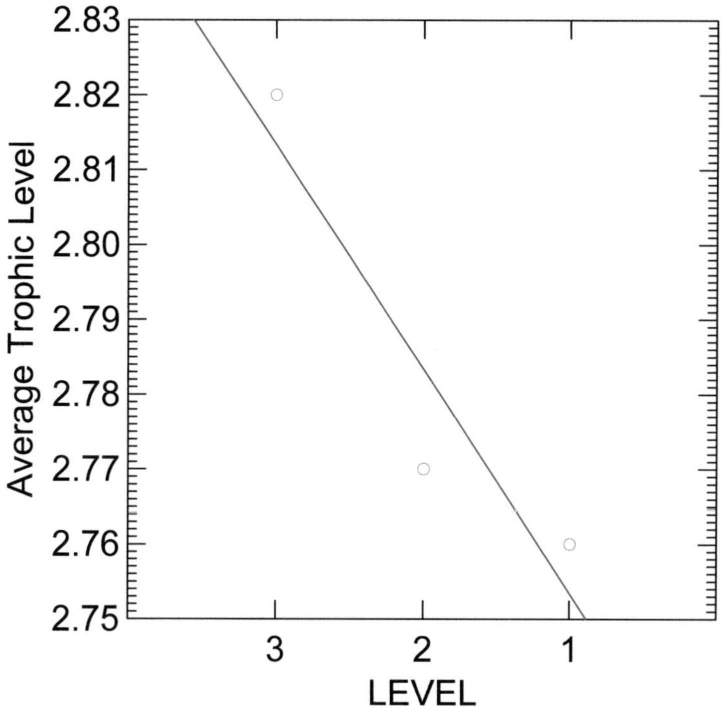

Fig. 12. Average trophic level at the Minnis-Ward site declines notably from a mean of 2.819 in Level 3 to 2.760 in Level 1, indicative of fishing down the marine food web over time, ca. A.D. 880-1490 ($n = 50$, $p < 0.001$, least squares line: $y = -2.95 x + 2.72$). Open circles represent individual trophic levels calculated for each stratigraphic level.

Finally, some last words should be said about the possible effects of climate change on the trends observed in the invertebrate and vertebrate fauna at both the Minnis-Ward and North Storr's Lake sites. It should be noted that the time period represented by the various archaeological levels reported herein, ca. A.D. 900-1500, marks a time of major transition between two widely recognized global climate phenomena: the Medieval Warm Period (MWP), ca. A.D. 800-1250, and the Little Ice Age (LIA), ca. A.D. 1300-1850 (Archer, 2007; Broecker, 2001; Cook et al., 2004; Cronin, 1999; deMenocal et al., 2000; Fagan, 2000, 2004; Goudie, 1992; Hardy, 2003; Keigwin, 1996; Osborn & Briffa, 2006; Sridhar et al., 2006; R. Wilson, Drury, & Chapman, 2000). At the time of the MWP, sea surface temperatures (SSTs) are recorded as being ca. 1°C warmer than today, while during the LIA SSTs were ca. 1°C cooler than today (Keigwin, 1996). In fact, based on a study of archaeological *Codakia orbicularis* (Linnaeus, 1758) (tiger lucine) dated to A.D. 1440-1530 (calibrated) from the Pigeon Creek site on San Salvador, nearby Pigeon Creek winter surface temperatures are estimated to have been 1.65-2.33°C cooler than at present (Cerajewski, 2002). Clearly, cooler SSTs in an otherwise semi-tropical region would likely have a negative effect on fisheries richness and abundance. For example, the Atlantic cod fishery declined notably during the late LIA due to colder temperatures and resulting lower productivity (Rose, 2004). Whether a shift to a slightly cooler climate can explain the decline in the terrestrial Gecarcinidae as well as the decline in average trophic level (which would suggest that higher level fishes were somehow affected disproportionately compared to lower level feeders) seems doubtful. It is asserted that the declines in weights, sizes, number of identified taxa, and trophic levels that have been described in this research are most parsimoniously attributed to human predation and overexploitation of the various resources under consideration (Carlson, 1999; Carlson & Keegan, 2004; Keegan, Portell, & Slapcinsky, 2003; Newsom & Wing, 2004; Quitmyer, 2003; Wing, 2001; Wing, deFrance, & Kozuch, 2002; Wing & Wing, 2001).

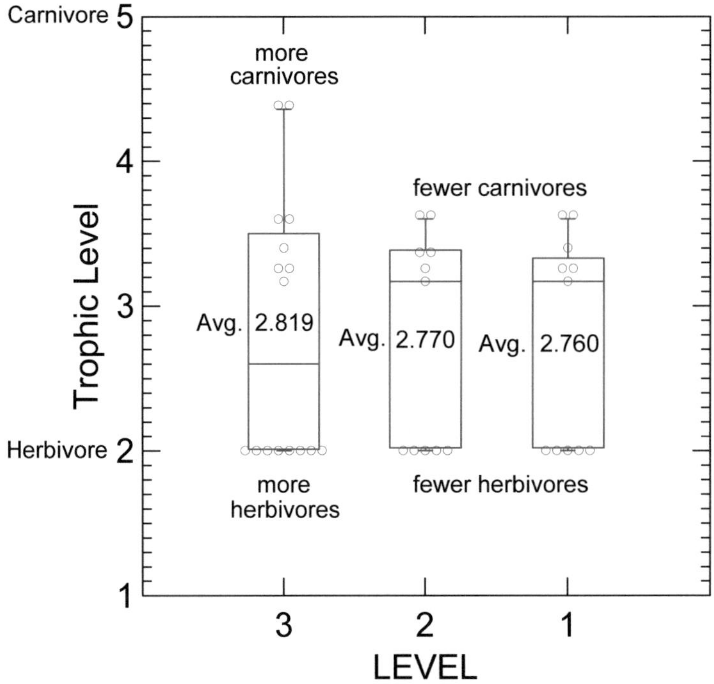

Fig. 13. Decline in average trophic level and alteration of community composition of marine vertebrate fauna at the Minnis-Ward site (SS-3), A.D. 880-1490. Note that there are both more carnivores and more herbivores (trophic levels 5 and 2, respectively) in the earlier Level 3 compared to the later levels. Average trophic level is indicated numerically in the center of each box plot. Each of the box plots show average and median trophic level (horizontal line within box), interquartile range (box), and range (whiskers) for each stratigraphic level. Open circles represent individual identified marine taxa.

6. Conclusions

Confidence in the results presented above is bolstered by the fact that similar trends are apparent in multivariate datasets involving terrestrial (Gecarcinidae), intertidal (*C. pica*,

Chiton tuberculatus/*Acanthopleura granulata*), and marine (*S. viride* and Serranidae) fauna from two different archaeological sites and slightly different time periods: the Minnis-Ward site (SS-3) dating to ca. A.D. 880-1490, and the North Storr's Lake site (SS-4) with deposits dating to ca. A.D. 1300-1520. Trends of decline in abundance and size of exploited resources appear to have occurred during both earlier (ca. A.D. 900-1150) and later (ca. A.D. 1300-1490) pre-Columbian periods on San Salvador, regardless of the length of duration of the deposits (ca. 600 years for Minnis-Ward and ca. 200 years for North Storr's Lake). These trends involve declines in abundance (as measured by weight in g), declines in sizes (as measured in mm on key skeletal elements), and declines in number of identified taxa (reflecting declining biodiversity) and trophic levels (reflecting the elimination of higher level feeders and alteration of community composition) across time as recorded in the various archaeological levels of the two sites. The reliability of the present research is verified by the fact that other scholars have recorded analogous findings using similar techniques on multiple Caribbean islands on deposits ranging in age from ca. 500 B.C.-A.D. 1500. In the words of Worm et al. (2006:787), "when the available experimental data are combined, they reveal a strikingly general picture" - and that is a picture of prehistoric overexploitation of terrestrial and intertidal resources, such as land crabs and various mollusks, and overfishing of marine coral reef dwellers such as Scaridae and Serranidae.

This brings us to the issue of sustainability, particularly regarding the sustainability and maximum sustainable yield (MSY) of coastal and maritime organisms. Sustainability refers to the idea that an "activity can be maintained without exhaustion or collapse" (Harper, 2004:276). Another widely accepted definition of sustainability is that an activity "meets the needs of the present without compromising the ability of future generations to meet their own needs" (World Commission on Environment and Development, 1987). MSY is often defined as "maximum use that a renewable resource can sustain without impairing its renewability through natural growth or replenishment" (European Environment Agency, 2006). Equipped with the above definitions, and based on the evidence at hand, it is now possible to answer the question, "Did the Lucayans of San Salvador practice sustainable harvesting/fishing practices?" The answer would appear to be "No." Lucayan harvesting of land crabs, intertidal mollusks, and fishes in the nearby coral reefs appears to have resulted minimally in the exhaustion of the resources under consideration. By over-harvesting and fishing down the local marine food web, Lucayans were engaging in opportunistic subsistence practices that, in our opinion, would have eventually compromised the ability of future generations to meet their needs (this lifestyle was put to a premature end with the arrival of Columbus and subsequent Spanish slave raiders who exterminated the Lucayans between 1492 and 1513; see Sauer, 1966). It is also clear that the Lucayans did not practice MSY since the evidence suggests that they were engaged in growth overfishing of certain marine organisms including *S. viride* and certain Serranidae. Continuation of these harvesting and fishing practices certainly would have impaired the renewability of terrestrial, intertidal, and local marine (coral reef) ecosystems due to the interruption of natural growth and replenishment processes by Lucayan subsistence practices.

It should be noted that the prehistoric Lucayan population of San Salvador is estimated to have been relatively low - in the range of 500-1000 people (Blick, 2011b; Kelley, 1992), although some scholars (based upon erroneous extrapolations of the data, in our opinion) argue for a much larger total population of 2250 with a range of 1000-4400 people (Pickering, 1997). Based upon the number of archaeological sites on San Salvador and throughout the Bahamas per square kilometer of land area (Keegan, 1992, 1997), the total population of the Bahamas is estimated to have been ca. 46,800-87,750 (Blick, 2011b), a range that corresponds well with Spanish conquest documents (Craton & Saunders, 1992; Pickering, 1997; Sauer, 1966). Based on our calculations, the total population of San Salvador (land area ca. 101 km^2, total area 163 km^2) would have been ca. 535-1008 people (Blick, 2011b) with a population density of ca. 5.30-9.98 per km^2. This estimated prehistoric population range is near the modern fluctuating population of San Salvador, 465-970 people (White, 1985; see also Government of the Bahamas, 2005), suggesting some sort of upper limit for the carrying capacity of the island. The prehistoric Minnis-Ward site probably contained some 15-18 houses (Blick, 2003, 2004) with an estimated population of ca. 120-144 people (at 8 persons per ordinary house; see Ramcharan 2004:83-84, Table 5.3). According to Columbus' *Diario*, villages were small in the Bahamian Archipelago with some 120-225 people living in approximately 12-15 houses per village (Dunn & Kelley, 1989; Rouse, 1992). The late Lucayan village at North Storr's Lake was probably similar in size. All of these figures suggest that the environmental impact recorded in the faunal remains of pre-Columbian San Salvador was likely caused by a relatively small human population with a low, hunter-gatherer/horticulturalist type population density.

The evidence presented in this chapter indicates that relatively minor, low density human populations can have significant environmental impacts, especially on small, vulnerable island ecosystems. In the words of Quitmyer (2003:131-132), "Island ecosystems are particularly fragile and susceptible to human disturbances because their biological reservoirs are small and not easily replenished. … human contact, even at very low levels of cultural complexity, can degrade the environment … The most dramatic evidence appears in the faunal records of island midden deposits" (the subjects of the excavations at the Minnis-Ward and North Storr's Lake archaeological sites). Although once considered "Ecological Indians" (Krech, 1999), it has become clear that even small indigenous populations have had serious impacts on their local environments via horticultural and animal exploitation practices (Redman, 1999). The findings presented in this chapter have major implications regarding the sustainability of fisheries, the effect of human population size and resource pressure, and human-induced environmental impacts in the past, present, and future.

7. Acknowledgments

The generosity of the Dr. Donald T. and Kathy D. Gerace Foundation is recognized for its support for students and assistants and for general support for work at the Minnis-Ward and North Storr's Lake sites from 2003-2010 (Figure 14). The Bahamian Antiquities, Monuments and Museums Corporation is thanked for its permission to analyze vertebrate faunal remains and radiocarbon samples from Minnis-Ward and North Storr's Lake. This research was made possible by the Georgia College & State University International Studies office and 2003-2010 Faculty Development Funds through the Department of Government &

Sociology at Georgia College & State University. Supplies and equipment used in this research were provided by funds through a 2003 Faculty Research Grant from Georgia College & State University, a 2004 Dean's Award from the Dean of the School of Liberal Arts & Sciences at GC&SU, and by a 2004 Council on Undergraduate Research grant to Jeffrey P. Blick and David Brinson. Dr. Doug Dvoracek of University of Georgia's Center for Applied Isotope Studies is much appreciated for facilitating the processing of AMS radiocarbon dates. Funding for radiocarbon dating was provided by a 2006 Faculty Research Grant to Jeffrey P. Blick from Georgia College & State University. An abbreviated version of a similar paper appeared in the *Journal for Nature Conservation* (Blick, 2007).

Fig. 14. Archaeological fieldwork at the Minnis-Ward site (SS-3), June 2010, San Salvador, Bahamas. Dr. Blick instructs students in the drawing of a soil profile.

8. References

Albury, P. (1975). *The Story of the Bahamas*, Macmillan Caribbean, ISBN 978-0333171325, London

Archer, D. (2007). *Global Warming: Understanding the Forecast*, Blackwell, ISBN 978-1405140393, Malden

Bahn, P., & Flenley, J. (1992). *Easter Island, Earth Island*, Thames & Hudson, ISBN 978-0500050651, New York

Baum, J.K., Myers, R.A., Kehler, D.G., Worm, B., Harley, S.J., & Doherty, P.A. (2003). Collapse and Conservation of Shark Populations in the Northwest Atlantic, *Science*, 299, pp. 389-392, ISSN 0036-8075

Bellwood, D.R. (1994). A Phylogenetic Study of the Parrotfishes, Family Scaridae (Pisces: Labroidei), with a Revision of Genera. *Records of the Australian Museum, Suppl. 20*, pp. 1-86, Australian Museum, ISBN 0731036638, Sydney

Berman, M.J. (1999). Wooden Artifacts from the Deadman's Reef Site (GB4), Grand Bahama, Commonwealth of the Bahamas, Report of Investigations to the National Geographic Society, Scientific Research Grant 6374-98, Permission to Cite Granted by the Author

Berman, M.J., & Gnivecki, P.L. (1995). Colonization of the Bahama Archipelago: A Reappraisal, *World Archaeology*, 26, 3, pp. 421-441, ISSN 0043-8243

Berman, M.J., & Pearsall, D. (2000). Plants, People, and Culture in the Prehistoric Central Bahamas: A View from the Three Dog Site, an Early Lucayan Settlement on San Salvador Island, Bahamas, *Latin American Antiquity*, 11, 3, pp. 219-239, ISSN 1045-6635

Berman, M.J., & Pearsall, D.M. (2005). At the Crossroads: The Role of Starch Grain Analysis in Lucayan Prehistory, Paper Presented at the 11th Symposium on the Natural History of the Bahamas, Gerace Research Centre, San Salvador, June 2005

Berman, M.J., & Pearsall, D.M. (2008). At the Crossroads: Starch Grain and Phytolith Analyses in Lucayan Prehistory, *Latin American Antiquity*, 19, 2, pp. 181-203, ISSN 1045-6635

Bjorndal, K.A., & Jackson, J.B.C. (2003). Roles of Sea Turtles in Marine Ecosystems: Reconstructing the Past, In: *The Biology of Sea Turtles, Volume II*, P.L. Lutz, J.A. Musick, & J. Wyneken, (Eds.), pp. 259-273, CRC Press, ISBN 978-0849311239, Boca Raton

Blick, J.P. (2003). *Systematic Shovel Testing at the Minnis-Ward site (SS-3), San Salvador, Bahamas: Archaeological Evidence for Pre-Columbian Households and Subsistence Patterns*, Preliminary Report Presented to the Gerace Research Centre, San Salvador, Bahamas

Blick, J.P. (2004). *Report on the 2004 Archaeological Investigations at Barker's Point (SS-15, SS-37) and Minnis-Ward (SS-3), San Salvador, Bahamas, with Commentary on the Nature of Fire-Cracked Rock*, Preliminary Report Presented to the Gerace Research Centre, San Salvador, Bahamas

Blick, J.P. (2006). Recent Archaeological and Paleoecological Investigations on San Salvador: Evidence for Pre-Columbian Impact on Terrestrial and Marine Resources in the Bahamas, Paper Presented at the International Council for Archaeozoology (ICAZ), Mexico City, Mexico, August 2006

Blick, J.P. (2007). Pre-Columbian Impact on Terrestrial, Intertidal, and Marine Resources, San Salvador, Bahamas (A.D. 950-1500), *Journal for Nature Conservation*, 15, 3, pp. 174-183, ISSN 1617-1381

Blick, J.P. (2011a). The Case for San Salvador as the Site of the 1492 Columbus Landfall: Principles of Historical Archaeology Applied to the Current Evidence, Paper Presented at the 14th Symposium on the Natural History of the Bahamas, Gerace Research Centre, San Salvador, Bahamas, June 2011

Blick, Jeffrey P. (2011b). Paleodemographic Reconstruction of the Pre-Columbian Population of Guanahaní (San Salvador, Bahamas). In: *Proceedings of the Thirteenth Symposium on the Natural History of the Bahamas,* J. Baxter and E. Cole, (Eds.), pp. 200-212, Gerace Research Centre, San Salvador, Bahamas

Blick, J.P., & Dvoracek, D. (2011). Forty New Radiocarbon Dates from Recent Archaeological Research on San Salvador, Bahamas: Implications for Island Chronology and Settlement, Paper Presented at the 14th Symposium on the Natural History of the Bahamas, Gerace Research Centre, San Salvador, Bahamas, June 2011

Blick, J.P., & Kjellmark, E. (2006). Recent Archaeological and Paleoecological Investigations in the Bahamas: Evidence for Pre-Columbian Impact on the Environment, Paper Presented at the 83rd Annual Meeting of the Georgia Academy of Science, Georgia Perimeter College, Atlanta, March 2006

Blick, J.P., & Murphy, B. (2005). *Report on the 2005 Archaeological Investigations at the North Storr's Lake Site (SS-4), San Salvador,Bahamas (Preliminary Report, 2nd edition),* Preliminary Report Presented to the Gerace Research Centre, San Salvador

Blick J.P., Creighton, A., & Murphy, B. (2006). *Report on the 2006 Archaeological Investigations at the North Storr's Lake Site (SS-4), San Salvador, Bahamas: Stratigraphic Excavations and the Role of the Sea Turtle in Lucayan Subsistence (Preliminary Report, Revised),* Preliminary Report Presented to the Gerace Research Centre, San Salvador, Bahamas

Blick, J.P., Hopkins, J.H., & Oetter, D. (2011). The Prehistoric Settlement Pattern of San Salvador, Bahamas, *Journal of Island and Coastal Archaeology* 6, 3, pp. 421-441, ISSN 1556-4894

Blick, J.P., Jackson, C.C., Thacker, F.O, & Pittman, J.M. 2009. *Archaeological Excavations at the Minnis-Ward Site (SS-3), and the Discovery of the Mary Ann Blick Site (SS-41), May-June 2009,* Research Report Presented to the Gerace Research Centre, San Salvador, Bahamas, Georgia College & State University, Milledgeville

Brill, R.H., & Hoffman, C.A. (1985). Some Glass Beads Excavated on San Salvador Island in the Bahamas, *Annales du 10e Congrès de l'Association Internationale pour l'Histoire du Verre,* Madrid-Segovia, September 1985, pp. 373-400

Brill, R.H., Barnes, I.L., Tong, S.S.C., Joel, E.C., & Murtaugh, M.J. (1987). Laboratory Studies of Some European Artifacts Excavated on San Salvador Island, In: *Columbus and His World,* D.T. Gerace, (Ed.), pp. 247-292, Bahamian Field Station, ISBN 978-0935909234, San Salvador

Broecker, W.S. (2001). Paleoclimate: Was the Medieval Warm Period Global?, *Science,* 291, pp. 1497-1499, ISSN 0036-8075

Burney, D.A. (1997). Tropical Islands as Paleoecological Laboratories: Gauging the Consequences of Human Arrival, *Human Ecology,* 25, 3, pp. 437-457, ISSN 1402-6902

Caddy, J.F., Csirke, J., Garcia, S.M., & Grainger, R.J.R. (1998). How Pervasive is "Fishing Down Marine Food Webs"?, Technical Comments, *Science*, 282, p. 1383a, ISSN 0036-8075

Carder, N., Reitz, E.J, & Crock, J. (2007). Fish Communities and Populations During the Post-Saladoid Period (AD 600/800-1500), Anguilla, Lesser Antilles, *Journal of Archaeological Science*, 34, pp. 588-599, ISSN: 0305-4403

Carlson, L.A. (1999). *Aftermath of a Feast: Human Colonization of the Southern Bahamian Archipelago and its Effect on the Indigenous Fauna*, Ph.D. thesis, University of Florida

Carlson, L.A. & Keegan, W.F. (2004). Resource Depletion in the Prehistoric Northern West Indies, In: *Voyages of Discovery: The Archaeology of Islands*, S.M. Fitzpatrick, (Ed.), pp. 85-107, Praeger, ISBN 978-0275979478, Westport

Cerajewski, R.J. (2002). *Paleoclimate Reconstruction Using Isotopic Analysis of Tropical Bivalves from the Pigeon Creek Archaeological Site, San Salvador, Bahamas*, Master's thesis, University of Georgia

Coleman, F.C., Figueira, W.F., Ueland, J.S., & Crowder, L.B. (2004). The Impact of Recreational Fisheries on Marine Fish Populations, *Science*, 305, pp. 1958-1960, ISSN 0036-8075

Conover, D.O., & Munch, S.B. (2002). Sustaining Fisheries Yields over Evolutionary Time Scales, *Science*, 297, pp. 94-96, ISSN 0036-8075

Cook, E.R., Woodhouse, C.A., Eakin, C.M., Meko, D.M., & Stahle, D.W. (2004). Long-term Aridity Changes in the Western United States, *Science*, 306, pp. 1015-1018, ISSN 0036-8075

Craton, M., & Saunders, G. (1992). *Islanders in the Stream: A History of the Bahamian People, Volume I: From Aboriginal Times to the End of Slavery*, University of Georgia Press, ISBN 978-0820321226, Athens

Cronin, T.M. (1999). *Principles of Paleoclimatology*, Columbia University Press, ISBN 978-0231109550, New York

deMenocal, P., Ortiz, J., Guilderson, T., & Sarnthein, M. (2000). Coherent High- and Low-Latitude Climate Variability During the Holocene Warm Period, *Science*, 288, pp. 2198-2202, ISSN 0036-8075

Delvaux, T.R., Fry, G.F., & Murphy, B. (2006). A Review of the Storr's Lake Site (SS4), San Salvador Island, Bahamas, 1996-2005, Paper Presented at the 71st Annual Meeting of the Society for American Archaeology, San Juan, Puerto Rico, April 2006

Diehl, F.A., Mellon, D., Garrett, R.H., & Elliott, N. (1988). *Field Guide to the Invertebrates of San Salvador Island, Bahamas*, Bahamian Field Station, San Salvador

Dunn, O., & Kelley, J.E. (1989). *The Diario of Christopher Columbus's First Voyage to America, 1492-1493*, University of Oklahoma Press, ISBN 978-0806123844, Norman

Estes, J.A., Terborgh, J., Brashares, J.S., Power, M.E., Berger, J., Bond, W.J., Carpenter, S.R., Essington, T.E., Holt, R.D., Jackson, J.B.C., Marquis, R.J., Oksanen, L., Oksanen, T., Paine, R.T., Pikitch, E.K., Ripple, W.J., Sandin, S.A., Scheffer, M., Schoener, T.W., Shurin, J.B., Sinclair, A.R.E., Soulé, M.E., Virtanen, R., & Wardle, D.A. (2011). Trophic Downgrading of Planet Earth, *Science*, 333, 6040, pp. 301-306, ISSN 0036-8075

Erlandson, J.M., & Fitzpatrick, S.M. (2006). Oceans, Islands, and Coasts: Current Perspectives on the Role of the Sea in Human Prehistory, *Journal of Island and Coastal Archaeology*, 1, pp. 5-32, ISSN: 1556-4894

European Environment Agency. (2006). *EEA Glossary*, Date of Access 29 July 2011, Available from: http://glossary.eea.europa.eu/terminology/concept_html?term=maximum%20sustainable%20yield

Fagan, B.M. (2000). *The Little Ice Age: How Climate Made History (1300-1850)*, Basic Books, ISBN 978-0465022724, New York

Fagan, B.M. (2004). *The Long Summer: How Climate Changed Civilization*, Basic Books, ISBN 978-0465022823, New York

Frank, K.T., Petrie, B., Choi, J.S., & Leggett, W.C. (2005). Trophic Cascades in a Formerly Cod-Dominated Ecosystem, *Science*, 308, 5728, pp. 1621-1623, ISSN 0036-8075

Froese, R., & Pauly, D., (Eds.). (2005). *Fish Base*, Date of Access 28 July 2011, Available from: http://www.fishbase.org/

Goodwin, R.C. (1980). Demographic Change and the Crab-Shell Dichotomy, *Proceedings of the 8th International Congress for the Study of Pre-Columbian Cultures of the Lesser Antilles*, pp. 45-68

GoogleEarth. (2006). GoogleEarth: A 3D Interface to the Planet. Date of Access 28 July 2011, Available from: www.earth.google.com/

Goudie, A. (1992). *Environmental Change* (3rd edition), Clarendon Press, ISBN 0198741677, Oxford

Government of the Bahamas. (2005). Population in Islands, Census Years 1970-2000, Date of Access 26 April 2005, Available from: www.bahamas.gov.bs/bahamasweb2/home.nsf/vContentW/9C7DE759B 4AEFBEA06256ED1

Granberry, J., & Vescelius, G. (2004). *Languages of the Pre-Columbian Antilles*, University of Alabama Press, ISBN 978-0817351236, Tuscaloosa

Hardy, J.T. (2003). *Climate Change: Causes, Effects, and Solutions*, Wiley, ISBN 978-0470850183, Chichester

Harper, C.L. (2004). *Environment and Society: Human Perspectives on Environmental Issues* (3rd edition), Pearson/Prentice Hall, ISBN 978-0131113411, Upper Saddle River

Hawkins, J.P., & Roberts, C.M. (2004). Effects of Artisanal Fishing on Caribbean Coral Reefs, *Conservation Biology*, 18, pp. 215-226, ISSN 0888-8892

Hill, K. (2001). *Cardisoma guanhumi*, Smithsonian Marine Station at Fort Pierce, Date of Access 28 July 2011, Available from: www. sms.si.edu/irlspec/Cardis_guanhu.htm

Hoffman, C.A. (1967). *Bahama Prehistory: Cultural Adaptation to an Island Environment*, Ph.D. thesis, University of Arizona

Hoffman, C.A. (1987). Archaeological Investigations at the Long Bay Site, San Salvador, Bahamas, In: *Columbus and His World*, D.T. Gerace, (Ed.), pp. 237-245, Bahamian Field Station, ISBN 978-0935909234, San Salvador

Hopkins, J.H., Oetter, D. & Blick, J.P. (2011). Analysis of Prehistoric Settlement Patterns on San Salvador, Bahamas, Using the San Salvador GIS Database. In: *Proceedings of the Thirteenth Symposium on the Natural History of the Bahamas*, J. Baxter and E. Cole, (Eds.), pp. 213-228. Gerace Research Centre, San Salvador, Bahamas

Jackson, J.B.C., Kirby, M.X., Berger, W. H., Bjorndal, K.A., Botsford, L.W., Bourque, B.J., Bradbury, R.H., Cooke, R., Erlandson, J., Estes, J.A., Hughes, T.P., Kidwell, S.,

Lange, C.B., Lenihan, H.S., Pandolfi, J.M., Peterson, C.H., Steneck, R.S., Tegner, M.J., & Warner, R.R. (2001), Historical Overfishing and the Recent Collapse of Coastal Ecosystems, *Science*, 293, pp. 629-638, ISSN 0036-8075

Johnson, K.R. (1980). Results of Tests to Determine Mineralogy of an Artifact from San Salvador, Bahamas, *Bahamas Archaeological Project Reports and Papers 1980*, CCFL Bahamian Field Station, San Salvador

Jones, A.R. (1985). Dietary Change and Human Population at Indian Creek, Antigua, *American Antiquity*, 50, 3, pp. 518-536, ISSN 0002-7316

Keegan, W.F. (1986). The Ecology of Lucayan Arawak Fishing Practices, *American Antiquity*, 51, 4, pp. 816-825, ISSN 0002-7316

Keegan, W.F. (1992). *The People Who Discovered Columbus: The Prehistory of the Bahamas*, University Press of Florida, ISBN 978-0813011370, Gainesville

Keegan, W.F. (1997). *Bahamian Archaeology: Life in the Bahamas and Turks and Caicos Before Columbus*, Media Publishing, ISBN 978-9768170033, Nassau

Keegan, W.F., Portell, R.W., & Slapcinsky, J. (2003). Changes in Invertebrate Taxa at Two Pre-Columbian Sites in Southwestern Jamaica, AD 800-1500, *Journal of Archaeological Science*, 30, pp. 1607-1617, ISSN 0305- 4403

Keigwin, L.D. (1996). The Little Ice Age and Medieval Warm Period in the Sargasso Sea, *Science*, 274, pp. 1504-1508, ISSN 0036-8075

Kelley, J.E. (1992). Epistemology 101 for Landfall Students: An Appreciation of an Important New Book, *Terrae Incognitae*, 24, pp. 101-110, ISSN 2040-8706

Kjellmark, E. (1996). Late Holocene Climate Change and Human Disturbance on Andros Island, Bahamas, *Journal of Paleolimnology*, 15, pp. 133-145, ISSN 0921-2728

Krech, S. (1999). *The Ecological Indian: Myth and History*, W.W. Norton, ISBN 978-0393321005, New York

LaPilusa, T. & J.S. Heilveil. (2011). Genetic Diversity of the Blue Land Crab, *Cardisoma guanhumi* Latreille (Crustacea: Decapoda), on Andros Island: Implications for Sustainable Management, Paper Presented at the 14th Symposium on the Natural History of the Bahamas, Gerace Research Centre, San Salvador, Bahamas, June 2011

Lawson, B. (1993). *Shelling San Sal: An Illustrated Guide to Common Shells of San Salvador Island, Bahamas*, Bahamian Field Station, San Salvador

Morales, A., & Rosenlund, K. (1979). *Fish Bone Measurements: An Attempt to Standardize the Measuring of Fish Bones from Archaeological Sites*, Steenstrupia, Copenhagen

Mumby, P.J., Dahlgren, C.P., Harborne, A.R., Kappel, C.V., Micheli, F., Brumbaugh, D.R., Holmes, K.E., Mendes, J.M., Broad, K., Sanchirico, J.N., Buch, K., Box, S., Stoffle, R.W., & Gill, A.B. (2006). Fishing, Trophic Cascades, and the Process of Grazing on Coral Reefs, *Science*, 311, pp. 98-101, ISSN 0036-8075

Myers, R.A., Barrowman, N.J., Hutchings, J.A., & Rosenberg, A.A. (1995). Population Dynamics of Exploited Fish Stocks at Low Population Levels, *Science*, 269, pp. 1106-1108, ISSN 0036-8075

Nelson, J.S., Cromman, E.J., Espinosa-Pérez, H., Findley, L.T., Gilbert, C.R., Lea, R.N., & Williams, J.D. (2004). *Common and Scientific Names of Fishes from the United States, Canada, and Mexico* (6th edition), American Fisheries Society Special Publication 29, American Fisheries Society, ISBN 978-1888569612 , Bethesda

Newsom, L.A. & Wing, E.S. (2004). *On Land and Sea: Native American Uses of Biological Resources in the West Indies*, University of Alabama Press, ISBN 978-0817313159, Tuscaloosa

Obregón, M. (1987). Columbus' First Landfall: San Salvador, In: *Columbus and His World*, D.T. Gerace, (Ed.), pp. 185-195, Bahamian Field Station, ISBN 978-0935909234, San Salvador

Obregón, M. (1989). Columbus' First Landfall, Paper Presented at the Phileas Conference, Ft. Lauderdale, Date of Access 26 June 2009, Available from: http://muweb.millersville.edu/~columbus/data/soc/OBREGON1.SPK

Osborn, T.J., & Briffa, K.R. (2006). The Spatial Extent of 20th-Century Warmth in the Context of the Past 1200 Years, *Science*, 311, pp. 841-844, ISSN 0036-8075

Ostrander, G.K., & Brocksmith, R. (1997). *The Fishes of San Salvador Island: The Bahamas*, Oklahoma State University, ISBN 978-0935909647, Stillwater

Pandolfi, J.M., Bradbury, R.H., Sala, E., Hughes, T.P., Bjorndal, K.A., Cooke, R.G., McArdle, D., McClenachan, L., M.J.H. Newman, Paredes, G., Warner, R.R., & Jackson, J.B.C. (2003). Global Trajectories of the Long-Term Decline of Coral Reef Ecosystems, *Science*, 301, pp. 955-958, ISSN 0036-8075

Pauly, D., Christensen, V., Dalsgaard, J., Froese, R., & Torres, F. (1998). Fishing Down Marine Food Webs, *Science*, 279, pp. 860-863, ISSN 0036-8075

Pauly, D., Froese, R., & Christensen, V. (1998). Response to J.F. Caddy, et al. *Science*, 282, p. 1383a, ISSN 0036-8075

Pickering, K.A. (1997). The Population of Guanahani, Date of Access 28 July 2011, Available from: http://www.columbusnavigation.com/lclog4.shtml

Perry, L., Dickau, R., Zarrillo, S., Holst, I., Pearsall, D.M, Piperno, D., Berman, M.J., Cooke, R.G., Rademaker, K., Ranere, A.J., Raymond, J.S., Sandweiss, D.H, Scaramelli, F., Tarble, K., & Zeidler, J. (2007). Starch Fossils and the Domestication and Dispersal of Chili Peppers (*Capsicum* spp. L.) in the Americas, *Science*, 315, pp. 986-988, ISSN 0036-8075

Quitmyer, I.R. (2003). Zooarchaeology of Cinnamon Bay, St. John, U.S. Virgin Islands: Pre-Columbian Overexploitation of Animal Resources, *Bulletin of the Florida Museum of Natural History*, 44, 1, pp. 131-158, ISSN 0071-6154

Rainey, F.G. (1940). *Porto Rican Archaeology*. Scientific Survey of Porto Rico and the Virgin Islands, 18, Pt. 1, New York Academy of Sciences, New York

Ramcharan, S. (2004). *Caribbean Prehistoric Domestic Architecture: A Study of Spatio-Temporal Dynamics and Acculturation*, Master's thesis, Florida State University

Redman, C.L. (1999). *Human Impact on Ancient Environments*, University of Arizona Press, ISBN 978-0816519637, Tucson

Reitz, E.J. (1994). Archaeology of Trants, Montserrat, Part 2: Vertebrate Fauna, *Annals of the Carnegie Museum*, 63, 4, pp. 297-317, ISSN 0097-4463

Reitz, E.J., & Wing, E.S. (1999). *Zooarchaeology*, Cambridge University Press, ISBN 978-0521485296, Cambridge

Robertson, R. (2003). The Edible West Indian "Whelk" *Cittarium pica* (Gastropoda: Trochidae): Natural History with New Observations, *Proceedings of the Academy of Natural Sciences of Philadelphia*, 153, 1, pp. 27-47, ISSN 0097-3157

Rose, G.A. (2004). Reconciling Overfishing and Climate Change with Stock Dynamics of Atlantic Cod (*Gadus morhua*) over 500 Years, *Canadian Journal of Fisheries and Aquatic Sciences*, 61, pp. 1553-1557, ISSN 0706-652X

Rouse, I. (1992). *The Taínos: Rise and Decline of the People Who Greeted Columbus*, Yale University Press, ISBN 978-0300056969, New Haven

Russ, G.R. (1991). Coral Reef Fisheries: Side Effects and Yields, In: *The Ecology of Fishes on Coral Reefs*, P.F. Sale, (Ed.), pp. 601-635, Academic Press, ISBN 978-0126151817, San Diego

Sauer, C.O. (1966). *The Early Spanish Main*, University of California Press, ISBN 978-0521088480 (reprint), Berkeley

Scudder, S.J., & Quitmyer, I.R. (1998). Evaluation of Evidence for Pre-Columbian Human Occupation at Great Cave, Cayman Brac, Cayman Islands, *Caribbean Journal of Science*, 34, 1-2, pp. 41-49, ISSN 0008-6452

Serrand, N. (2002). *Exploitation des Invertébrés Marins et Terrestres par les Populations Saladoïdes et post-Saladoïdes du Nord des Petites Antilles (500 B.C.-1200 A.D.): Etude de Cas et Comparaisons*, Ph.D. thesis, Université Paris I Panthéon-Sorbonne

Shaklee, R., Fry, G., & Delvaux, T. (2007). An Archaeological Report on the Storr's Lake Site, San Salvador: 1995-2005, *Bahamas Naturalist and Journal of Science*, 2, 1, pp. 31-39, ISSN 1819-4230

Sibert, J., Hampton, J., Kleiber, P., & Maunder, M. (2006). Biomass, Size, and Trophic Status of Top Predators in the Pacific Ocean, *Science*, 314, pp. 1773-1776, ISSN 0036-8075

Sridhar, V., Loope, D.B., Swinehart, J.B., Mason, J.A., Oglesby, R.J., & Rowe, C.M. (2006). Large Wind Shift on the Great Plains during the Medieval Warm Period, *Science*, 313, pp. 345-347, ISSN 0036-8075

Steadman, D.W. (1995). Prehistoric Extinctions of Pacific Island Birds: Biodiversity Meets Zooarchaeology, *Science*, 267, pp. 1123-1131, ISSN 0036-8075

Steadman, D.W., & Stokes, A.V. (2002). Changing Exploitation of Terrestrial Vertebrates During the Past 3000 Years on Tobago, West Indies, *Human Ecology*, 30, 3, pp. 339-367, ISSN 1402-6902

SPSS, Inc. 1997. SYSTAT Version 7.0 for Windows ®, Statistical Data Analysis Package, SPSS Inc., Chicago

Turgeon, D.D., Quinn, J.F., Jr., Bogan, A.E., Coan, E.V., Hochberg, F.G., Jr., Lyons, W.G., Mikkelsen, P.M., Neves, R.J., Roper, C.F.E., Rosenberg, G., Roth, B., Scheltema, A., Thompson, F.G., Vecchione M., & Williams, J.D. (1998). *Common and Scientific Names of Aquatic Invertebrates from The United States and Canada: Mollusks* (2nd edition), American Fisheries Society Special Publication No. 26, American Fisheries Society, ISBN 1888569018, Bethesda

White, V. (1985). *The Outermost Island: An Oral History of San Salvador, the Bahamas*, Bahamian Field Station, ISBN 978-0935909197, San Salvador

Whyte, T.R., Berman, M.J., & Gnivecki, P.L. (2005). Vertebrate Archaeofaunal Remains from the Pigeon Creek Site, San Salvador Island, the Bahamas, *Proceedings of the 10th Symposium on the Natural History of the Bahamas*, Gerace Research Centre, San Salvador, June 2003, pp. 166-178

Wilson, R.C.L., Drury, S.A., & Chapman, J.L. (2000). *The Great Ice Age: Climate Change and Life*, Routledge, ISBN 978-0415198424, New York

Wilson, S.M. (1999). Cultural Pluralism and the Emergence of Complex Society in the Greater Antilles, Paper Presented at The 18th International Congress for Caribbean Archaeology, St. George's, Granada, Date of Access 28 July 2011, Available from: http://uts.cc.utexas.edu/~swilson/wilson_iaca99.html

Wing, E.S. (1969). Vertebrate Remains Excavated from San Salvador Island, Bahamas, *Caribbean Journal of Science*, 9, pp. 25-29, ISSN 0008-6452

Wing, E.S. (1989). Human Exploitation of Animal Resources in the Caribbean, In: *Biogeography of the West Indies*, C.A. Woods, (Ed.), pp. 137-152, Sandhill Crane Press, ISBN 978-1877743030, Boca Raton

Wing, E.S. (2001). The Sustainability of Resources Used by Native Americans on Four Caribbean Islands, *International Journal of Osteoarchaeology*, 11, pp. 112-126, ISSN 1047-482X

Wing, E.S., & Reitz, E.J. (1982). Prehistoric Fishing Communities of the Caribbean, *Journal of New World Archaeology*, 5, 2, pp. 13-32, ISSN 0147-9024

Wing, E.S., & Scudder, S.J. (1983). Animal Exploitation by People Living on a Tropical Marine Edge, In: *Animals and Archaeology*, C. Grigson & J. Clutton-Brock, (Eds.), pp. 197-210, BAR International Series 183, Oxford, ISBN 0860542955, British Archaeological Reports

Wing, S.R., & Wing, E.S. (2001). Prehistoric Fisheries in the Caribbean, *Coral Reefs*, 20, pp. 1-8, ISSN 0722-4028

Wing, E.S., deFrance, S.D., & Kozuch, L. (2002). Faunal Remains from the Tutu Site, In: *The Tutu Archaeological Village Site: A Multidisciplinary Case Study in Human Adaptation*, E. Reich, (Ed.), pp. 141-165, Routledge, ISBN 978-0415239905, New York

Winter, J. (1980). A Preliminary Archaeological Survey of San Salvador, Bahamas, In: *Bahamas Archaeological Project Reports and Papers 1980*, D. Gerace (Ed.), pp. 1-3, Bahamian Field Station, San Salvador

Winter, J. (1981). 1981 Archaeological Site Reconnaissance: San Salvador, Cat Island, and Rum Cay, In: *Bahamian Archaeology Project Reports and Papers 1981*, D. Gerace, (Ed.), pp. 4-7, Bahamian Field Station, San Salvador

Winter, J. (1982). Reconnaissance of San Salvador, January 1982, In: *Bahamas Archaeology Project Reports and Papers*, D. Gerace, (Ed.), pp. 8-10, Bahamian Field Station, San Salvador

Winter, J. (1997). Ward/Minnis Site, *Proceedings of the 17th International Association for Caribbean Archaeology*, Radisson Cable Beach Resort, Nassau, July 2007, pp. 36-38

Winter, J. & Gilstrap, M. (1991). Preliminary Results of Ceramic Analysis and Movements of Population into the Bahamas, *Proceedings of the 12th International Congress for Caribbean Archaeology*, pp. 371-386

Winter, J., & Wing, E. (1995). A Refuse Midden at the Minnis Ward Site, San Salvador, Bahamas, *Actas de XV Congreso Internacional de Arqueología del Caribe/Proceedings of the 15th International Congress for Caribbean Archaeology*, pp. 423-434

World Commission on Environment and Development. (1987). *Our Common Future*, Oxford University Press, ISBN 978-0192820808, Oxford

Worm, B., Barbier, E.B., Beaumont, N., Duffy, J.E., Folke, C., Halpern, B.S., Jackson, J.B.C., Lotze, H.K., Micheli, F., Palumbi, S.R., Sala, E., Selkoe, K.A., Stachowicz, J.J., & Watson, R. (2006). Impacts of Biodiversity Loss on Ocean Ecosystem Services, *Science*, 314, pp. 787-790, ISSN 0036-8075

Yesner, D.R. (1980). Maritime Hunter-Gatherers: Ecology and Prehistory, *Current Anthropology*, 21, 6, pp. 727-750, ISSN 0011-3204

8

Tidal Wetlands Restoration

Roy C. Messaros[1], Gail S. Woolley[1],
Michael J. Morgan[1] and Patricia S. Rafferty[2]
[1]*US Army Corps of Engineers, New York District*
[2]*US National Park Service, Northeast Region*
USA

1. Introduction

Jamaica Bay (the Bay) is located within the Boroughs of Brooklyn and Queens, New York City, and covers 67.3 square kilometers and opens into the Atlantic Ocean via Rockaway Inlet (Fig. 1). The Bay contains disturbed tidal salt marsh wetlands and upland ecosystems, mud flats, parks, landfills, residential urban communities, commercial and retail facilities, and J. F. Kennedy International Airport. Also present are tidal creeks, navigational channels, and areas of open water. In the early 1900s, Jamaica Bay was an extensive estuarine ecosystem that sustained large expanses of tidal salt marsh. Jamaica Bay was renowned for its abundant and diverse shellfish and ecological importance as a nursery and feeding ground for numerous bird species (JBERRT 2002) and various fish species rely on the Bay for habitat (USFWS 1997). Jamaica Bay is also valuable for various bird species during seasonal migration (NYCDEP 2006). Other wildlife including amphibians and reptiles also rely on these salt marsh wetlands as primary habitat (Tanacredi & Badger 1995).

The Jamaica Bay ecosystem is part of the Gateway National Recreation Area, a unit of the National Park Service, and is connected to the lower bay of New York Harbor by Rockaway Inlet. The Jamaica Bay Federal navigation channel extends from offshore of Rockaway Point, Queens, through Rockaway Inlet and bisects at the southern edge of Floyd Bennett Field (Barren Island), Brooklyn, with one branch extending north into the upper part of Jamaica and a second branch extending east into lower Jamaica Bay (Fig. 1). The dominant littoral drift is to the west along the south shore of Long Island (Kana 1995) and has almost doubled the length of the Rockaway spit since the early 19th century (Englebright 1975). Along the northern New Jersey shoreline, the dominant littoral drift is to the north which causes the elongation of Sandy Hook. These shoreline patterns are attributed to the effect of Long Island in shielding the area from waves from north and northeast. Since the 1930s the Rockaway inlet has been stabilized by jetties. Urbanization of the Rockaway Beach barrier island during the 20th century has effectively halted the delivery of sand to Jamaica Bay via overwash during periods of storm surge. An increase in water depth, such as that caused by dredging of navigation channels, modifies the hydrodynamics and generally leads to an increase in tidal range. Within the Bay alteration to the tidal range also changes the marsh hydroperiod, e.g., plants at the same elevation are inundated for a longer portion of the tidal cycle than at that same elevation prior to dredging of the channels in Jamaica Bay. This may further enhance tidal currents and exacerbate erosion. Mean tidal range within Jamaica Bay varies between 1.5 m and 1.6 m.

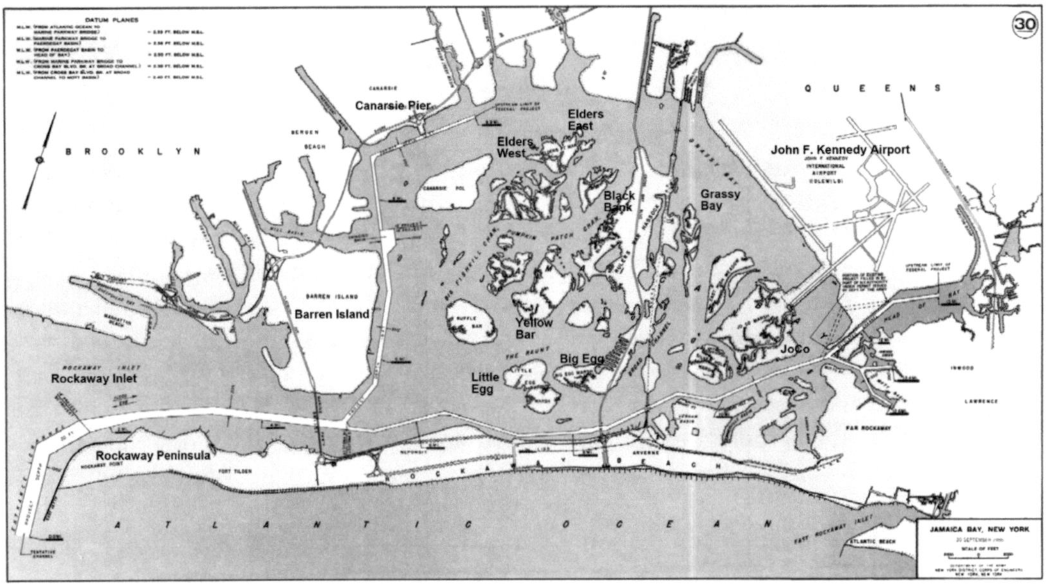

Fig. 1. Jamaica Bay, located within the Boroughs of Brooklyn and Queens, New York City, opens into the Atlantic Ocean via Rockaway Inlet.

The undertaking of tidal salt marsh restoration projects on the order of magnitude presented herein, lead by the U. S. Army Corps of Engineers, will be a unique opportunity to measure and assess the success of such large scale efforts. To prevent continued loss of wetlends decisive action must be taken by placing emphasis on improving compliance and maintaining a true baseline (Race & Fonseca 1996). Race & Fonseca (1996) further noted that past mitigation projects nationwide indicate that the success rate remains low overall. Elders Point East and Elders Point West will establish baselines for future restoration/mitigation efforts in Jamaica Bay and elsewhere. Roberts (1993) reported that the sober reality regarding marsh losses is likely to be that mitigation projects have a high degree of failure. Reporting results of tidal salt marsh restoration projects in Jamaica Bay may be pivotal for establishing baselines and achieving success for future efforts.

2. Historical marsh loss in Jamaica Bay

2.1 Trends

Jamaica Bay was historically more land than open water (Fig. 2), however, vegetated emergent marsh islands in the Bay are being lost at an alarming rate (Hartig *et al.* 2001, GATE & JBWPAC 2007). In 1907, 3430 hectares (ha) of the Bay was shallow water and 6549 ha consisted of marsh islands (Hartig *et al.* 2002). Most of the wetland loss in Jamaica Bay prior to early 1970s can be attributed to human activity such as dredging or filling (Black 1981). When direct anthropogenic sources of loss are removed, a pattern of consistent marsh island loss persists. Excluding areas affected directly by dredging and filling, of the 950 ha of

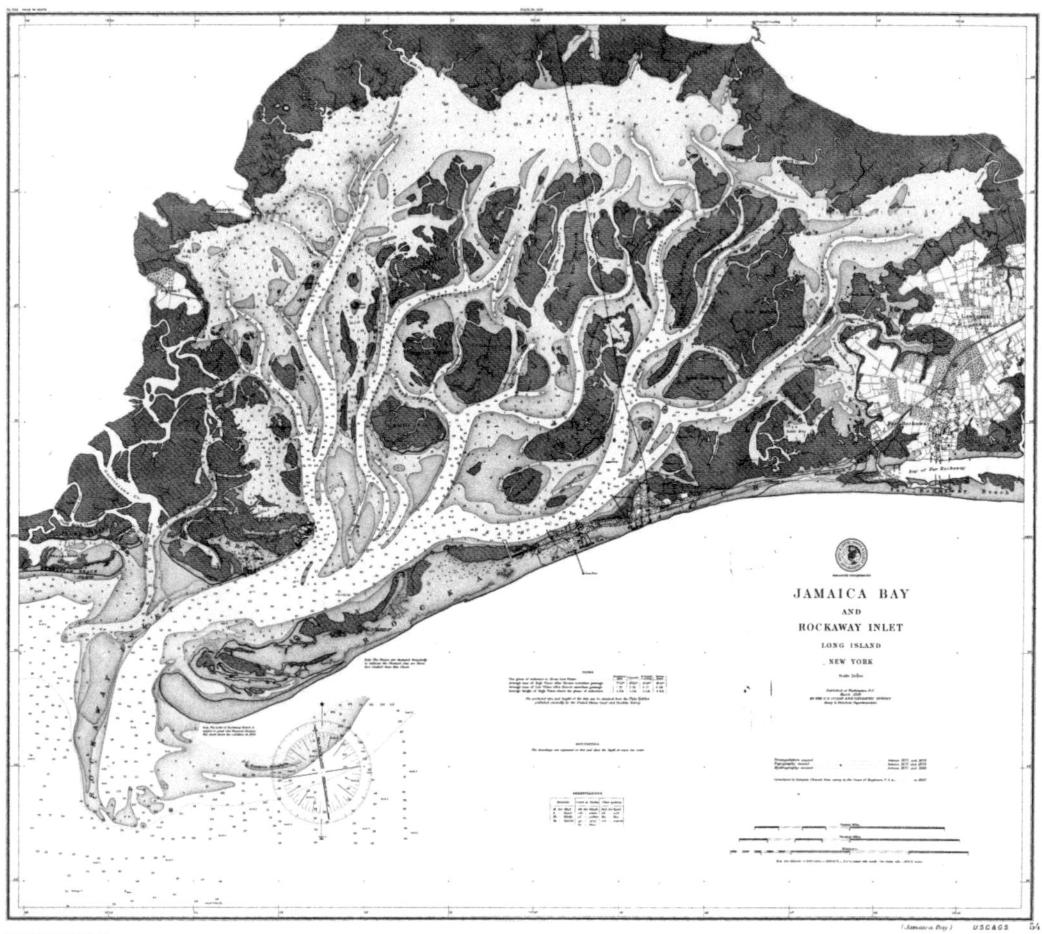

Fig. 2. Jamaica Bay and Rockaway Inlet 1899.

vegetated marsh island in the Bay in 1951, only 355 ha remained as of 2003. During that 49 year period, 63% of the Bay's salt marsh islands were converted from emergent vegetated habitat to submerged and intertidal habitat (Table 1). The calculated average rate of marsh loss increased throughout that time period from 6.9 ha y^{-1} from 1951-1974 to 13.4 ha $y^{-1}$from 1989-2003 (Table 2) (GATE & JBWPAC 2007). Recent analysis (2003-2008) indicates that the rate of loss may be decreasing to 7.7 ha y^{-1}(Christiano 2010).

	Time Period			
	1951*	1974	1989	2003
Vegetated Marsh (hectares)	950	652	539	355

* From 1951 to 1974, 23 ha of marsh island were calculated as lost due to the construction of West Pond and 115 ha lost as a result of the Broad Creek and Goose Pond marsh impoundments. Other factors accounted for the loss of the remaining 161 ha.

Table 1. Total area of vegetated marsh islands in Jamaica Bay (from GATE & JBWPAC 2007).

	Time Period		
	1951-1974	1974-1989	1989-2003
Average Rate of Loss (hectares /year)	6.9	7.3	13.4

Table 2. Rate of marsh loss of vegetated marsh islands in Jamaica Bay (from GATE & JBWPAC 2007).

Based on aerial photography interpretation, the New York State Department of Conservation (NYSDEC 2001) estimated that approximately 567 hectares (ha) of tidal salt marsh island within Jamaica Bay have been lost since 1924, with the rate of loss rapidly increasing in recent years. Elders Point is currently comprised of two separate islands, Elders Point East (Elders East) and Elders Point West (Elders West) that together total about 4.9 ha prior to the restoration project led by the U. S. Army Corps of Engineers (USACE), NY District in 2005 (Fig. 3). Elders Point was historically one island, comprising approximately 53.4 ha but over the last more than 80 years, marsh loss in the center of the island severed the connection creating two distinct islands separated by mud flat. At Elders Point, between 1994 and 1999, an estimated 89 ha of salt marsh was lost at a rate of 17.8 ha per year (USNPS 2001). Hartig *et al.* (2002) reported marsh loss for Elders Point from 1924 through 1999 with an increasing rate of loss from 0.5% (1924 - 1974) to 8.5% (1994 - 1999). It is estimated that if these trends continue, all remaining salt marsh within the Bay will be lost over the next three decades. Steinberg *et al.* (2004) have speculated that by 2024 all of the interior tidal marsh islands will be lost based on the current rate of deterioration.

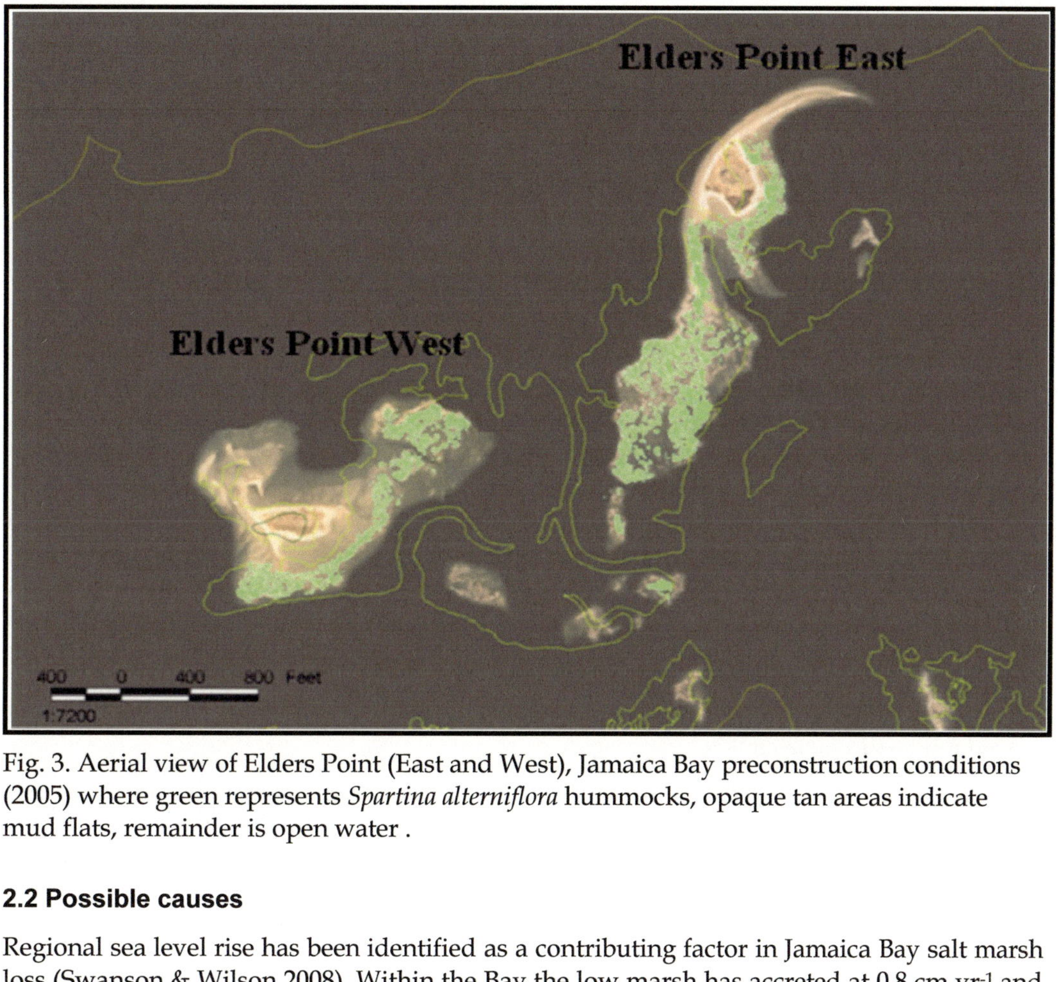

Fig. 3. Aerial view of Elders Point (East and West), Jamaica Bay preconstruction conditions (2005) where green represents *Spartina alterniflora* hummocks, opaque tan areas indicate mud flats, remainder is open water .

2.2 Possible causes

Regional sea level rise has been identified as a contributing factor in Jamaica Bay salt marsh loss (Swanson & Wilson 2008). Within the Bay the low marsh has accreted at 0.8 cm yr^{-1} and high marsh at 0.5 cm yr^{-1}. Prior to 1974 when the New York State Department of Environmental Conservation (NYSDEC) gained regulatory authority over intertidal wetlands, marsh losses have been caused by urbanization and associated land development activities, such as dredging and filling. The historic rate of sea level rise within Jamaica Bay is approximately 2.7 mm yr^{-1} (Gornitz *et al.* 2001) in comparison to the mean eustatic sea level rise of 1.2 mm yr^{-1} (NRC 1987) to 1.5 mm yr^{-1} (Gornitz 1995) during the past century and predicted 1.7 mm yr^{-1} in the next 100 years (IPCC 2007). The difference between the eustatic (global) and the New York (regional) sea level trend has been attributed to local subsidence resulting from crustal readjustments to the removal of ice following the retreat of the last glacial period (Dean *et al.* 1987). The south shore of Long Island, New York is the extent of the leading edge of the Wisconsinan glaciation ice sheet (ca. 20,000 years ago). The area to the south was upwarped while land to the north was depressed beneath the weight of the ice sheet. Much of the Atlantic coast has subsided while the land that was under the ice has rebounded. Marsh losses over the past century do not appear to be related to sea level rise (Kolker 2005); however, sea level rise is likely to be a cause of marsh loss in the future (Hartig *et al.* 2002). Based on ^{210}Pb chronology data Jamaica Bay marsh islands have accreted at rates in excess of the long-term rate of sea level rise at the Battery, NY (0.28 cm y^{-1}) (Kolker 2005). Analysis by Hartig *et al.* (2002) indicates that over the next 80 years, current rates of accretion would only be adequate to maintain Jamaica Bay marsh islands under the most conservative predictions for future sea level rise.

Marsh loss occurs through the undercutting and collapse of peat along the perimeter of marsh islands, widening of tidal creeks, and the development and expansion of pools within the marsh interior (Hartig *et al.* 2002, GATE & JBWPAC 2007). While the causes of marsh loss are poorly understood or not known, Hartig *et al.* (2002) suggest that water logging is contributing to the loss of marsh islands in the Jamaica Bay through the development and growth of interior pools and the subsequent collapse of the root system. Belowground biomass contributes to marsh elevation (Valiela *et al.* 1976, DeLaune *et al.* 1994, Morris & Bradley 1999). Diminished root production can also lead to a loss in marsh elevation. Without sufficient accumulation of belowground organic matter (peat), a marsh that exhibits high aboveground biomass could quickly convert to mudflat or open water if aboveground vegetation dies off (Mendelssohn *et al.* 1981, DeLaune *et al.* 1994). In addition, roots bind sediments and slow sediment compaction (Redfield 1972, DeLaune *et al.* 1994, Rybczyk & Cahoon 2002, Cahoon *et al.* 2002, 2003). Subsidence has been found to be lower in a vegetated marsh compared to adjacent unvegetated pools (Erwin *et al.* 2006). The hydrology of a tidal marsh can be defined as the frequency (how often) and duration (how long) that a marsh is flooded. Mechanisms that will alter hydrology, and thus increase water logging, include increased sea level rise, changes in tidal range, changes in the rate of accretion, and subsidence. Eutrophication and goose grazing have also been identified as possible causes of wetland loss in Jamaica Bay.

Changes in sediment availability, distribution and accumulation may contribute to the loss of marsh islands in Jamaica Bay (Gordon & Houghton 2004, Goodbred *et al.* 2004). Westward growth and stabilization of the Rockaway Inlet, dredging and ocean disposal of sediments from the Rockaway Inlet, development within the watershed, shoreline hardening channelization of runoff through storm sewers and combined sewer overflows, and trapping of sediments within navigation channels and borrow pits (e.g., Grassy Bay) are mechanisms by which sediment availability, distribution, and accumulation may have been altered within the Bay. Inorganic mass of marsh sediments has decreased at Yellow Bar Hassock (Fig. 1) and JoCo marshes (Fig. 1) since European settlement (Peetet *et al.* 2008) while organic matter has increased over the same period (Peetet *et al.* 2004). Kolker (2005), using ^{210}Pb chronology data, determined that accretion rates were higher in the latter half of the twentieth century and lower in the first half. Accretion rates at Big Egg marsh (Fig. 1) were lowest from 1900 to 1920 (0.14 - 0.18 cm y^{-1}) and highest from 1995 to 1999 (0.63 - 0.64 cm y^{-1}). East high marsh accretion rates were lowest from 1900 to 1950 (0.09 – 0.14 cm y^{-1}) and highest from 1950-1980 (0.57 – 0.75 cm y^{-1}). Accretion rates at JoCo were lowest during the 1920s (0.18 cm y^{-1}) and highest during the 1960s (0.59 cm y^{-1}) (Kolker 2005). Accretion rates at the same marshes from 1974 to 2000 were 0.41, 0.35, and 0.46 cm y^{-1}, respectively (Kolker 2005, Cochran *et al.* 2009). Recent short-term accretion rates (2003-2009) measured at Black Bank (Fig. 1) and JoCo marshes are similar (0.48 and 0.44 cm y^{-1}, respectively) (Cahoon 2008). Hydrodynamic modeling indicates that there is little deposition of sediment within navigation channels, however deep pits, such as Grassy Bay (Fig. 1), may serve as sinks for fine sediments (Wilson & Flagg 2008). Additional research is needed to better understand the sources and distribution of sediments within the Bay.

While sediment accretion rates within Jamaica Bay marshes exceed the rate of long-term sea level rise, changes in tidal hydrodynamics have resulted in water level increases within the Bay that exceed regional sea level rise (Swanson & Wilson 2008). Dredging and other development activities have increased the volume of the Bay by 350% (NYDEP 2007) and the mean depth from approximately 1 m to 5 m (Swanson & Wilson 2008). In addition, during the twentieth century, Jamaica Bay experienced an overall increase in tidal range and an amplification of tidal range from west to east as a result of development activities as well as the westward migration and subsequent stabilization of the Rockaway Peninsula. Increases in tidal range have resulted in high tide water levels that are currently 56-78% greater than sea level rise. Prior to 1899, mean tidal range was generally uniform throughout the bay (0.12 m) and tidal height varied from 1.22 m at Plum Beach Channel to 1.28 m at Canarsie (Fig. 1). Combined changes in sea level and increases in sea level rise and tidal range result in tidal height today of 5.0 ft (1.65 m) at Barren Island (modern proxy for Plumb Beach Channel) and 1.58 m at Canarsie Pier (Swanson & Wilson 2008). The rate of marsh loss observed in Jamaica Bay greatly exceeds that which has been observed in other Long Island marshes (Kolker 2005). These marshes have experienced rates of sediment accretion (Kolker 2005) and sea level rise (Kolker 2005, Swanson & Wilson 2008) similar to Jamaica Bay; however, tidal range in these marshes has not changed much (Swanson & Wilson 2008). Thus, increases in the frequency and duration of marsh flooding due to changes in tidal range are likely to contribute to the loss of emergent salt marsh islands within Jamaica Bay (Swanson & Wilson 2008).

Nitrogen loading is frequently indicated to be a factor that may cause or contribute to marsh loss in Jamaica Bay (O'Grady 2001, USNPS 2001, NYCDEP 2007). Nitrogen loading in Jamaica Bay has increased substantially in the past 110 years from an estimated 35.6 kg d^{-1} N, which entered the bay via submarine groundwater discharge, to 15,785 kg d^{-1} that enters the bay via wastewater discharge, subway dewatering, landfill leachate, submarine groundwater discharge, and atmospheric deposition (Benotti *et al.* 2006). High nitrogen levels may result in the reallocation of energy from roots to shoots in *Spartina alterniflora* (Valiela *et al.* 1976, Morris & Bradley 1999, Turner *et al.* 2004). High nitrogen loading may also amplify microbial activity and increase the rate of peat decomposition (Valiela *et al.* 1985). The U.S. Environmental Protection Agency (USEPA) is currently conducting research to evaluate soil respiration, above- and belowground biomass, and root structure at marshes in Jamaica Bay (Wigand *et al.* 2008). In 2009, the USNPS (U.S. National Park Services, P. Rafferty, co-author on this paper) initiated research to evaluate the role of eutrophication on plant function (allocation of resources between above and belowground biomass) at three marshes in Jamaica Bay.

Sulfide toxicity may also contribute to the loss of salt marsh islands in Jamaica Bay. Labile organic carbon resulting from phytoplankton blooms in the eutrophic Bay or direct inputs from water pollution control plants and combined sewer overflows may increase sulfate reduction in marsh sediments and result in elevated pore water sulfide concentrations. Prolonged exposure to high pore water sulfides in greenhouse studies results in stunted growth (>2mM) or death (>4mM) of *S. alterniflora* (Koch & Mendellsohn 1989). *S. alterniflora* seedlings are more sensitive to sulfide exposure than mature plants (Seliskar *et al.* 2004). Reactive iron can serve as a sink for sulfide via the precipitation of iron sulfides (Berner 1980, Goldhaber & Kaplan 1974, as cited in Cochran *et al.* 2009). A 2007 study at JoCo, Elders

East, and Elders West marshes found pore water sulfide concentrations (2-3mM) that are considered to be stressful to *S. alterniflora*. Generally these concentrations were observed at depths of 20 cm or greater and thus are at the lower extent or below the root zone. This study also found that Jamaica Bay marshes have a high degree of pyritization, thus indicating sulfide saturation of the sediment reactive iron pool (Cochran *et al.* 2009). This will result in a decreased capacity of sediment to buffer pore water sulfide levels and may result in the buildup of toxic sulfide levels (Berner 1984, Leventhal & Taylor1990, Raiswell & Canfield 1998, as cited in Cochran *et al.* 2009).

2.3 The sustainability of tidal salt marsh restoration efforts in Jamaica Bay, New York

With the completion of Elders Point East and Elders Point West there is an unprecedented opportunity for monitoring the long term success of such large scale restoration projects. The sustainability of restored/mitigated tidal marsh islands within Jamaica Bay will be critical for incremental environmental improvements in this heavily urbanized ecosystem. The next element for restoration in Jamaica Bay will be Yellow Bar scheduled to begin in 2012 and will be the third large scale restoration project for the Bay. The ongoing multiagency effort to monitor benchmarks for success will be a unique opportunity for lessons learned and improving the sustainability of such restoration efforts in Jamaica Bay in addition to similar tidal salt marsh systems elsewhere.

3. Restoration methods

One of the navigation missions of the U. S. Army Corps of Engineers is to maintain navigable waterways at coastal inlets and may include placement of dredged material to create, restore, or mitigate salt marsh wetlands (Sánchez 2008). The Army Corps of Engineers has been involved with the creation and restoration of wetlands for more than 30 years through the beneficial use of material dredged from navigational channels (Yozzo *et al.* 2004). Restoration at Elders Point East was the first such project led by the Army Corps of Engineers for Jamaica Bay and is part of a multi-phased effort with the first phase successfully constructed on Elders East in 2006. The second phase of these restoration efforts was Elders Point West with construction completed in 2010. Construction of the third restoration element for the Bay is anticipated to begin sometime in 2012.

These current restoration efforts involve a multi-agency group led by the U. S. Army Corps of Engineers, NY District to restore the tidal wetlands of Elders Point East and Elders Point West. Critical to these efforts are the partnering agreements with the Port Authority of NY/NJ (for Elders East) and NYDEC and NYDEP (for Elders West) as well as the land owning agency, U.S. National Park Service. The USNPS wrote the monitoring protocol and their in-kind contributions essentially doubled the monitoring budget annually. For Elders Point East federal, state, and local agencies, with diverse missions, worked collaboratively for more than ten years towards the success of this effort. The cooperative funding among the various agencies has also been a success. Elders Point East was initially a CAP (Continuing Authorities Program) project. When funding for the program changed and less CAP money was available, it was pairing of the need for funding for a Restoration Project with the need for a harbor deepening project (i.e., Mitigation Project) that allowed this effort to come to fruition.

Restoration efforts at Elders Point East and West are being extensively monitored to ensure worthwhile ecological goals have been provided and are a long term sustainable benefit to Jamaica Bay. Monitoring parameters include vegetation, nekton, surface elevation (sediment erosion/accretion) benthic marco-invertebrates, goose/waterfowl grazing impacts, and habitat change. Challenges that need to be better understood, resolved, and overcome include compaction and erosion of placed material, and subsidence. The geomorphic sustainability and plant survival will be an ongoing challenge with the existing physical, chemical, and biological stresses that are present in the Bay. With potential for increase in storm surge via climate change, there are physical benefits to tidal salt marsh wetlands that will act as an energy dissipater against wave action (Dean 1978), but at the same time add further stress to this highly urbanized ecosystem. The objectives of this work are to analyze change in elevation (geomorphology) and to evaluate the initial response for vegetation at the restoration site at Elders East and West. At this time vegetation data is not available for Elders West.

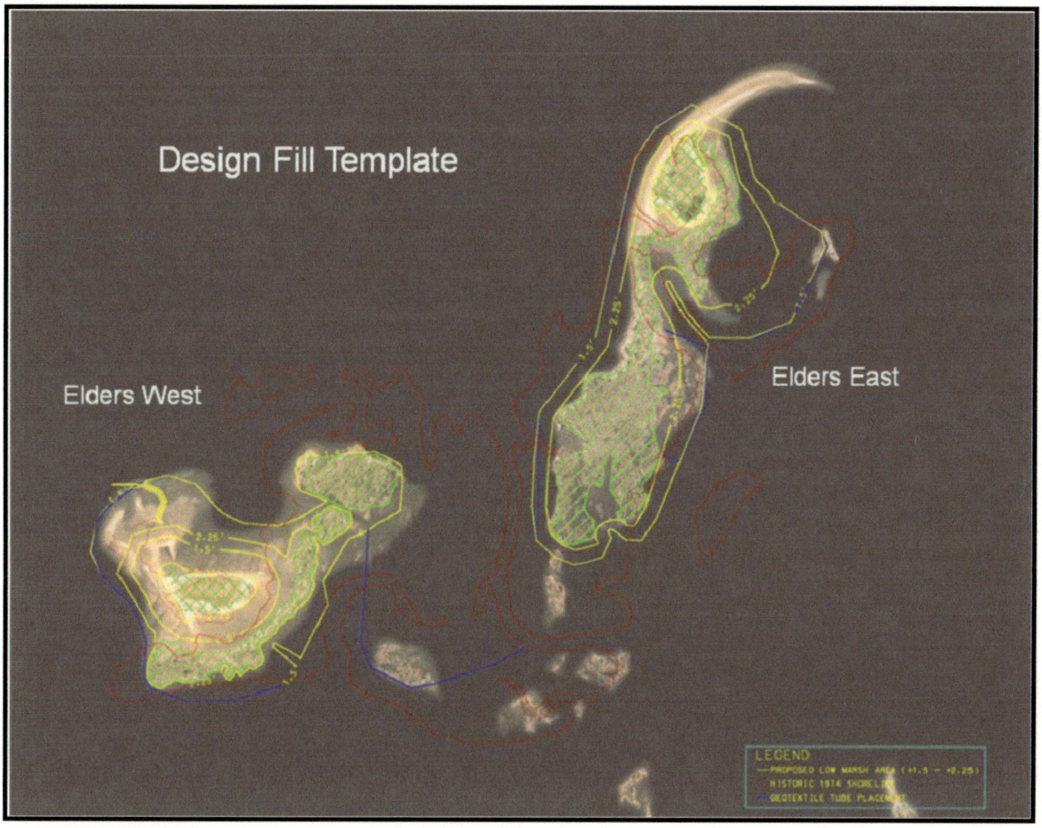

Fig. 4. Design fill template for Elders Point East and Elders Point West based on the 1974 shoreline limits.

3.1 Elders Point East

Elders Point East used material from the maintenance dredging of the Rockaway Inlet navigational channel (120,800m^3/ 158,000 yd^3), Ambrose Channel (35,170 m^3/46,000 yd^3), and purchased from Amboy Aggregate (34,405m^3/45,000 yd^3) totaling 190,374 m^3 (249,000 yd^3) which was composed primarily of sand (98%). The project involved the placement of fill (e.g., sediment from maintenance dredging), regrading the site to appropriate elevations for the target community, and planting with native coastal plant species. This design was based on the approximate extent of the 1974 marsh coverage as reported by NYSDEC. A mixture of *S. alterniflora, Spartina patens,* and *Distichlis spicata* was planted since it is representative vegetation throughout the New York Harbor estuary. Figure 4 is the design fill template for Elders Point East and indicates the planting scheme for *S. alterniflora,* tri-plug plantings (combination of all three species), and existing *S. alterniflora* hummock relocation. Approximately 580,000 *S. alterniflora* plugs, 45,876 *S. alterniflora* pots, 33,640 tri-plugs (*S. alterniflora, Distichlis spicata,* and *S. patens*) were planted over a total of 16.2 ha. Most of the project (i.e., > 95%) was planted with *S. alterniflora.* Figure 5 is an aerial view showing the construction of Elders Point East which illustrates the grids created for plantings. The individual grids (cells) included waterfowl fencing to minimize predation by the Canada Goose population.

Fig. 5. Aerial photograph taken near high tide during construction of Elders Point East October 2006.

3.1.1 Methods for morphological monitoring (survey) Elders Point East

Plantings on Elders Point East were completed in spring 2007, and the baseline monitoring data were collected in July 2007 using real time kinetic global positioning system (RTK-GPS) survey equipment. The first set of follow-on monitoring data was taken in May 2009, again using RTK-GPS technology. The July 2007 data points were collected most densely throughout the southern portion of the island and were regularly spaced approximately 20 feet on-center. Data points were the least dense within the south-central portion of the island, where the muddy conditions can be very difficult to transverse. In the northern half of the island, data were collected along transect lines spaced approximately 100 feet apart. The May 2009 and July 2010 data points were collected along profile lines spaced approximately 100 feet apart, each beginning toward the center of the island and moving radially outward toward the shoreline. The average distance between points along the profile lines was approximately 30 feet and prominent features such as the vegetation line and the centerlines of the tidal creeks were also captured. Data collection was focused along the perimeter of the island where it was anticipated that sediment transport would be the most active. The collected data were post-processed using the National Geodetic Survey's Online Positioning User Service (OPUS) and were plotted with ESRI™ ArcMap® version 9.3. Both data sets were overlaid on top of one another and the areas that did not have a high coincidence of points between the data sets were blocked or "masked," from being used to create a 3-dimensional surface. These areas were blocked because their inclusion would have resulted in a greater degree of interpolation between the two data sets. By only capturing those areas with a higher coincidence of points between the two data sets, the resultant surface is more reliable. A resultant surface between the two data sets was created using ArcMap®'s Spatial Analyst extension. From this surface, areas of sediment erosion, deposition and of no sediment transport were determined.

3.2 Elders Point West

Elders Point West, with construction initiated in late 2009 and completed in mid-2010, involved beneficial reuse of dredge material from Anchorage Channel, the main navigational channel for New York Harbor. The total quantity of placed material was 230,877 m^3 (301,976 yd^3) composed primarily of sand with some silt which created 23.4 ha of new tidal marsh area. The project involved the placement of fill, regrading the site to appropriate elevations for the target community. The planting scheme for Elders West was different from that at Elders East and may allow for a better understanding of species success once that vegetation data becomes available. For this planting scheme no *S. alterniflora* plugs or pots were included. All the low marsh plants were relocated *S. alterniflora* hummocks from the project site which covered approximately 7.0 ha. Of the 23.4 ha of tidal marsh created, 1.62 ha was high marsh transition zone, 0.49 ha upland seeded area, 1.62 ha of no planting acting as a planting control area. A total of 0.61 ha was low marsh hummock relocation, replanted so that they were evenly spaced throughout the designated transplant area. There was also 1.62 ha of no planting zone with a 6.1 m low marsh planted perimeter, 0.65 ha of upland planting and seeding, 7.65 ha of low marsh vegetation, and 0.41 ha of seeded area planted by National Resources Conservation Service. The planted areas included 85,580 high marsh transition plants with a variety of tidal marsh wetland vegetation, 240 shrubs, and 60 wetland trees covering approximately 1.6 ha. Figure 6 is an aerial view of Elders Point West showing the grids that include waterfowl fencing similar to Elders East.

Fig. 6. Construction of Elders Point West July 2010.

3.2.1 Methods for morphological monitoring (survey) Elders Point West

Plantings on Elders Point West were completed in spring 2010, and final as-built survey data were collected in September 2010. A second survey was conducted in May 2011, near the end of the maintenance period of the construction contract for Elders West. On each occasion, survey data were collected with sufficient density to generate 0.5 foot contour lines for the entire project site. An elevation surface for each survey was generated using the ArcGIS Spatial Analyst extension and a third surface was generated by differencing the 2010 and 2011 elevation surfaces. From this surface areas of sediment erosion, deposition, and of no sediment transport were determined.

4. Results and discussion

Perhaps an important outcome of this project is that federal, state, and local agencies, with diverse missions, worked collaboratively for more than nine years towards the success of this effort. The cooperative funding among the various agencies has also been a success. Elders Point East was initially a CAP (Continuing Authorities Program) project and when funding for the program changed and less CAP money was available, it was the pairing of the need for funding for a Restoration Project with the need for a harbor deepening project (i.e., Mitigation Project) that allowed this effort to come to fruition. The construction of Elders East and West did not occur at the same time as a result of funding availability. Figure 7 is an aerial view from 2009 showing the completion of Elders East and preconstruction conditions of Elders West. Figure 7 illustrates the vegetation has become well established through the third growing season (2009).

Fig. 7. Aerial view of Elders Point 2009, inset with arrow indicating location at Elders Point East.

4.1 Elders Point East

Patterns of morphological change (erosion and deposition) from July 2007 to May 2010 at Elders Point East are illustrated in Fig. 8. There is a range in elevation change from approximately - 0.8 m (loss, erosion) to + 0.6 m (gain, deposition) during this time. The GIS methodology used did not permit an overall assessment determining a net change (loss/gain) in volume at Elders Point East. It is apparent however that there is some loss (as much as 0.8 m) along the southwest side of the marsh and an overall gain (up to 0.6 m) on the northeast portion. The elevation in the central portion of Elders East is not included in Figure 8 since this is the interior of the island and not subjected to the same current velocities and sediment transport. Not shown in Figure 8 is the sand spit that extends from the northwest tip of Elders East which has been a well established natural feature of the sediment transport in this area of Jamaica Bay. This spit was not planted with vegetation, remains in relative equilibrium, and is not part of this current tidal marsh restoration effort.

Sediment elevation change from 2007 to 2009 provides a point-in-time measurement of elevation change and is indicative of the tidal energy that is acting upon the restored marsh platform. Anecdotal evidence, particularly in the no planting treatment on the southwest side of the island at Elders Point East, suggested that the variation in plant response may be related to sediment movement. Linear models of individual vegetative metrics for 2009 and the sediment elevation change from 2007 to 2009 were not significant. It is possible that vegetative response is related to other elevation metrics (i.e., initial constructed elevation, 2009 elevation, elevation change from 2008 to 2009). In addition, coordinates for vegetative plots were collected with a Garmin rino 530 GPS unit; thus providing low accuracy for vegetative plot locations. The accuracy of the plot location and the size of the plot buffer both affect pixel selection for GIS computation of a mean elevation change for each plot.

For the first six months after the initial fill occurred (2006) much of the settlement would have taken place and therefore that data would not reflect a meaningful trend for change of elevation. The observed areas of loss (south west side of the marsh) and gain (north east side) are consistent with observed historic sediment transport in this area of Jamaica Bay. There is an existing sand spit that extends from the North West tip of Elders Point East and curves north east for more than 200 m and is visible at mean low water (MLW). This observed pattern of deposition is consistent with observed losses along the south west side of the island that are seen from this data (Fig. 8). Overall there was no appreciable change in elevation of placed material that would have lead to adverse conditions for the vegetation. For example, exposed roots were not observed in areas on Elders East where elevation change was measured to be on the order of 0.7 meters.

Vegetation monitoring at the Elders East and reference marsh (JoCo) seeks to evaluate the response of vegetation to restoration and to determine if vegetation communities at the two marshes are converging. Following two full growing seasons, the vegetation communities in the restored and reference marsh have converged with respect to total canopy cover, *S. alterniflora* stem density, total standing aboveground biomass and annual net belowground production. Total vegetative canopy cover and bare ground canopy cover on the restored and reference marsh are equivalent. Messaros *et al.* (2010) determined that vegetative cover and *S. alterniflora* stem density were greater in the fertilized treatment; no differences were detected between treatments for total standing above- and belowground biomass as well as annual net belowground production. These results suggest that fertilization may affect vegetation form (i.e., the morphology of the aboveground portions of the plant) but not above- or belowground production. The complete discussion for Elders East four year post-construction vegetation monitoring results is available in Messaros *et al.* (2010).

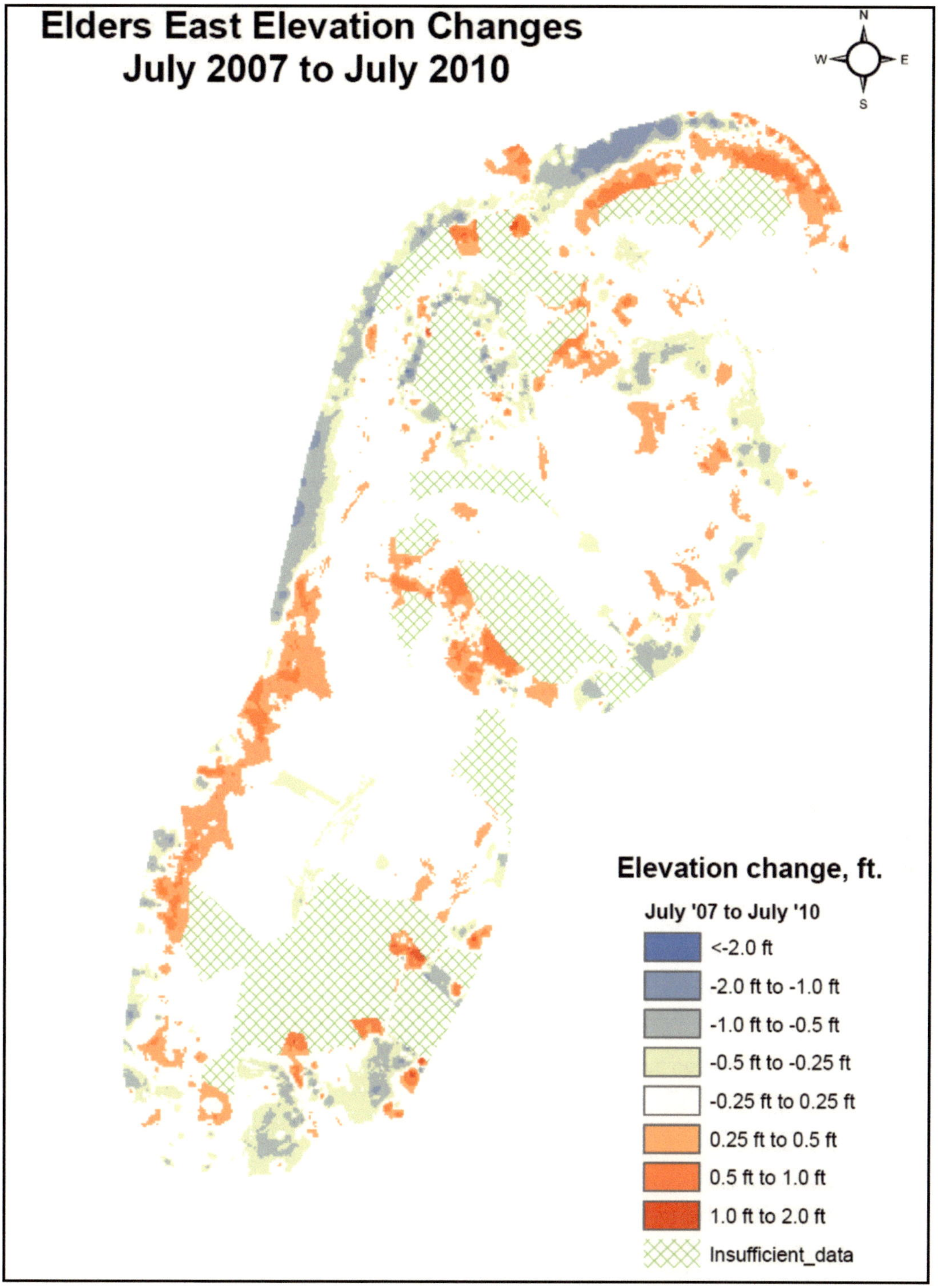

Fig. 8. Elders Point East elevation changes during the monitoring period of July 2007 to July 2010.

Messaros *et al.* (2010) concluded that no significant difference ($p=0.6$) was detected in total vegetative cover between the reference (JoCo) and restored (Elders East) marshes. Average densities (number of stems $0.25m^{-2}$) for vegetation (within $0.25m^2$ plots) indicate no significant difference was detected in stem density of *S. alterniflora* between JoCo and Elders East ($p=0.1$). Stem density of *S. alterniflora* was significantly greater in the fertilizer treatment than in the no fertilizer treatment ($p=0.002$). No difference in stem density of *S. maritima* was detected between the fertilizer treatments ($p=0.3$). Two hummock relocation areas have been monitored. In 2008, stem counts and height of *S. alterniflora* were also evaluated. All vegetative metrics differed significantly between the two hummock relocation areas. Canopy cover, stem density, and height of *S. alterniflora* were all greater within Relocation Area I. The relocation areas differ with respect to when hummocks were transplanted (June vs. October 2006); however, there may be other confounding factors. For example, canopy cover of *S. maritima* is also significantly greater at Relocation Area I. Recruitment of *S. maritima* at the end of the second full growing season would not be related to the timing of hummock transplant. Relocation Area I is noticeably wetter than Relocation Area II and a small tidal creek has developed adjacent to Relocation Area I, thus suggesting differences in hydrology. There are many possible explanations for partial success for restoration projects including poor hydrology (Race & Fonseca 1996). Relocation area II is bordered by a sand ridge that prevents flooding except during the highest tides. Future analysis of elevation surveys should be conducted to determine if differences in elevation are affecting localized hydrology and vegetative response within these treatments. The densities of both *S. alterniflora* ($p<0.001$) and *S. maritima* ($p=0.0008$) stems were significantly greater in hummock Relocation Area I as compared to Relocation Area II.

4.2 Elders Point West

Patterns of morphological change (erosion and deposition) from September 2010 to May 2011 at Elders Point West are illustrated in Fig. 9. There is a range in elevation change from approximately - 0.6 m (loss, erosion) to + 0.6 m (gain, deposition) during this time. The interior portions of the island appear unchanged with some deposition occurring on the southern shore. There are areas of loss on the east and west side of Elders West. It is likely that some of this observed elevation change is the placed material achieving equilibrium given the local currents, tides, and wind.

The first year vegetation monitoring data was not available at the time this publication went to press. Since there is stability in the fill material, the critical substrate for plant growth, it may be speculated that there will be very similar trends in the plant community with Elders West as reported with Elders East. Critical to long term goals is the sustainability of the placed fill and subsequent plant growth. Once vegetation becomes established it will stabilize the fill and thereby lend itself to a successful plant community. While the results for Elders Point West are limited to the one year time point much of the fill placement achieves equilibrium in a relatively short time. This is based on observations noted from the four year monitoring of the fill placement at Elders Point East.

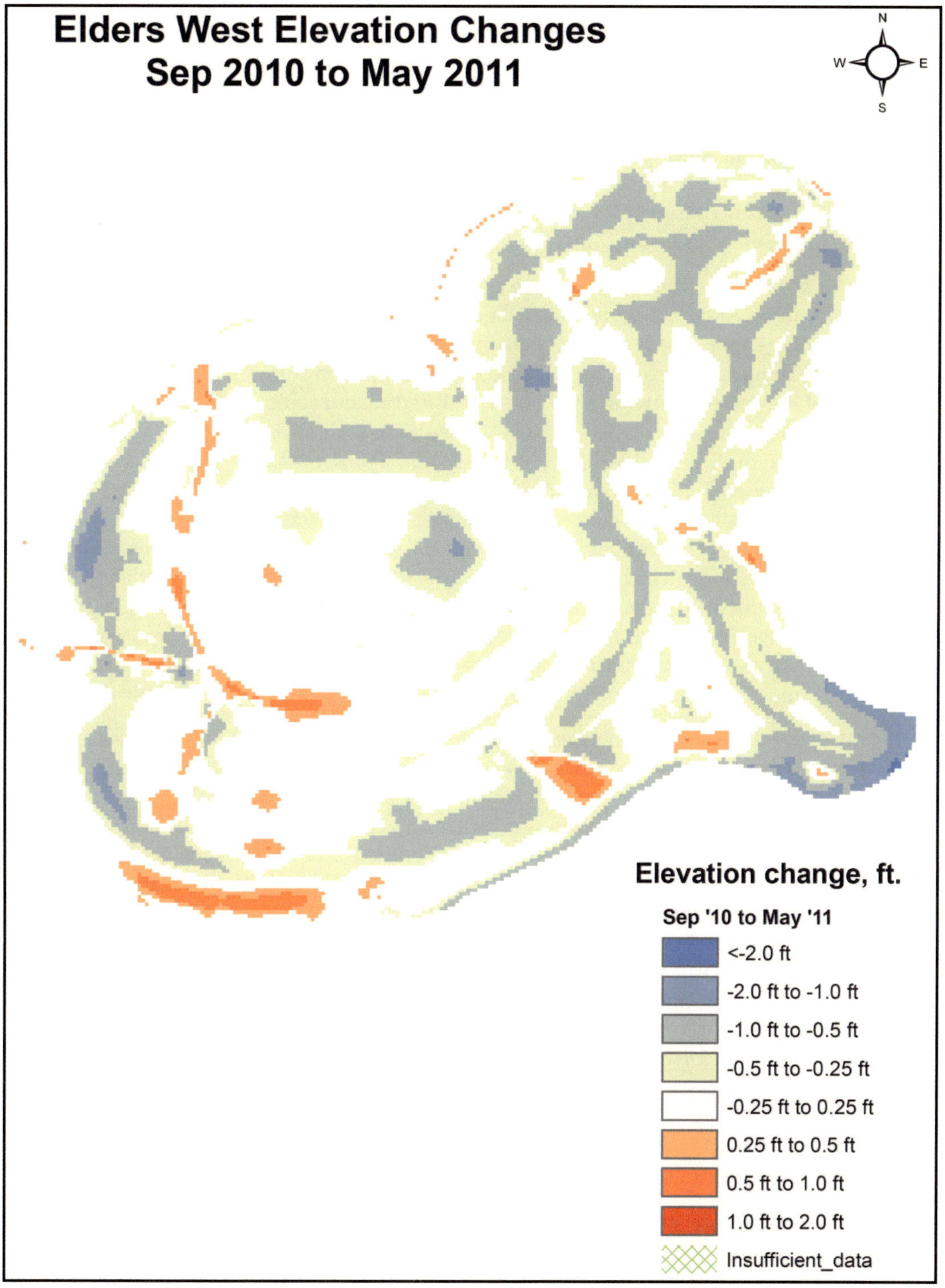

Fig. 9. Elders Point West elevation changes during the monitoring period of September 2010 to May 2011.

5. Conclusion

Based on the survey monitoring to date, no appreciable sediment transport (loss or gain) has occurred and the material has remained in place since the construction of the project in 2006 at Elders Point East and Elders Point West in 2010. On the west side of the Elders East there is some decrease in elevation and likely a result of the historic patterns of sediment transport occurring in a northerly direction of the accreting sand spit at Elders East. Without conducting a sediment transport study the authors reserve a final conclusion but this pattern is consistent with historic observation.

For the *S. alterniflora* plugs (representing >95% of the planted vegetation) the plant communities in the restored (Elders East) and reference marsh (JoCo) have converged with respect to total canopy cover, stem density, total standing aboveground biomass, and annual net belowground production. This occurred after two full growing seasons. In 2008, by the end of the second growing season, the restored marsh achieved 50% vegetative cover, with *S. alterniflora* the dominant species. No significant difference was observed in total vegetative cover and stem density between Elders Point East and JoCo marshes for *S. alterniflora*. It appears that the vegetation for this project is achieving the desired outcome and providing the anticipated ecological benefits. Total vegetative cover was significantly greater in the fertilizer treatment as compared to the no fertilizer treatment. There also appears to be a benefit of increased stem density in *S. alterniflora* when fertilizer is used. However no benefit was detected for total standing above- and belowground biomass as well as annual net belowground production with the use of fertilizer. This project was successful in preventing the return of invasive species such as *Phargmities australis* and also showed no impact from predation as a result of the installation of waterfowl fencing.

Perhaps the most significant conclusion is that there appears to be morphologic stability with the placed material at both Elders Point East and Elders Point West. During the post construction monitoring period the majority of the material has remained unchanged. This may have far reaching implications for future sustainability for restoration/mitigation efforts in Jamaica Bay. Further monitoring is scheduled and subsequent analyses will be essential to develop a more complete understanding of the sediment and plant community dynamics and stability. The need to continue the monitoring programs at Elders Point East and West does not diminish the important observations that have been reported herein as additional efforts are planned in the near future (i.e., restoration of Yellow Bar). Large scale tidal salt marsh restoration efforts in Jamaica Bay may serve as a benchmark for similar projects elsewhere.

6. Acknowledgment

The authors would like to acknowledge the U. S. Army Corps of Engineers, New York District, for the lead role in this multi-agency project in cooperation with U. S. National Parks Services (Gateway) (USNPS), Port Authority of New York and New Jersey (PANY/NJ), New York City Department of Environmental Protection (NYCDEP), New York State Department of Environmental Conservation (NYSDEC), New York State Department of State (NYDOS), National Marine Fisheries Program (NIMFP), NOAA Fisheries: Office of Protected Resources, the National Resources Conservation Service (NRCS), and U. S. Fish and Wildlife Services (USFWS).

7. References

Benotti, M. J., Misut, P. E., Abbene M., & Terracciano, S. A. (2006). "Historic nitrogen loading in Jamaica Bay, Long Island, New York: Predevelopment to 2005." *U. S. Geological Survey Open File Report,* SIR 2007-5051.

Berner, R. A. (1984). "Sedimentary pyrite formation: an update." *Geochimica et Cosmochimica Acta,* 48, 605–615.

Berner, R. A. (1980). "Early diagenesis: a theoretical approach." *Princeton Series in Geochemistry,* Princeton University Press, Princeton, New Jersey, 241 pp.

Black, F. R. (1981). "Jamaica Bay: a history." *Study No. 3,* Division of Cultural Resources, North Atlantic Regional Office, U. S. National Park Service, U.S. Department of the Interior, Washington, DC, USA.

Cahoon, D. R. (2008). Personal communication.

Cahoon, D. R., Hensel, P., Rybczyk, J., McKee, K. L., Proffitt, E., & Perez, B. C. (2003). "Mass tree mortality leads to mangrove peat collapse at Bay Islands, Honduras after Hurricane Mitch." *Journal of Ecology,* 91, 1093–1105.

Cahoon, D. R., Lynch, J. C., Hensel, P., Boumans, R., Perez, B.C., Segura, B., & Day Jr., J. W. (2002). "High-precision measurements of wetland sediment elevation: Recent improvements to the sedimentation-erosion table." *Journal of Sedimentary Research,* 72, 730–733.

Christiano, M. (2010). GIS Specialist, Gateway National Recreation Area. email communication received by P. S. Rafferty on 2/11/2010.

Cochran, H. K., Kolker, A., Hirschberg, D. J., Renfro, A. A., Heilbrun, C., Goodbred, S., Beck, A., & Finiguera, M. (2009). "Sulfur cycling in salt marshes of Jamaica Bay: Possible links to marsh loss." *Draft final report* submitted to: U. S. National Park Service, Narragansett, Rhode Island.

Dean, R. G. (1978). "Effects of vegetation on shoreline erosional processes." Wetland functions and values: The state of our understanding, *Proceedings of the National Symposium on Wetlands,* American Water Resources Association, Minneapolis, Minnesota, 436–456.

Dean, R. G., Dalrymple, A. R., Fairbridge, R. W., Leatherman, S. P., Nummedal, D., O'Brien, M. P., Pilkey, O. H., Sturges III, W., & Wiegel, R. L. (1987). "Responding to changes in sea level: Engineering implications." National Academy Press, Washington, D. C.

DeLaune, R.D., Nyman, J. A., & Patrick, Jr., W.H. (1994). "Peat collapse, ponding and wetland loss in a rapidly submerging coastal marsh." *Journal of Coastal Research,* 10, 1021–1030.

Englebright, S. (1975). "Jamaica Bay: a case study of geo-environmental stresses: a guidebook to field excursions." New York State Geological Association, Hofstra University, Hempstead, New York.

Erwin, R. M., Cahoon, D. R., Prosser, D. J., Sanders, G. M., & Hensel, P. (2006). "Surface elevation dynamics in vegetated *Spartina* marshes versus unvegetated tidal ponds along the Mid-Atlantic Coast, USA, with implications to Waterbirds." *Estuaries and Coasts,* 29, 96–106.

Gateway National Recreation Area (GATE) and Jamaica Bay Watershed Protection Plan Advisory Committee (JBWPPAC). (2007). "An update on the disappearing salt marshes of Jamaica Bay, New York." U. S. National Park Service, U.S. Department of the Interior, Staten Island, NY, 73 pp.

Goldhaber, M. B. & Kaplan, I. R., (1974). "The sulfur cycle." In: Goldberg, E.D., Ahhrenius, G., Dyrssen, D., & Garrels, R. M. (Editors), *The Sea.Volume 5: Marine Chemistry*, Wiley-Interscience, New York, pp. 569–655.

Goodbred, S.L., Cochran, J. K., & Flood, R. D. (2004). "Sedimentation history and budgets for the Jamaica Bay estuary-marsh system: Seasonal to decadal dynamics revealed through radiotracer studies." *In: Proceedings, Jamaica Bay's Disappearing Salt Marshes,* New York: Jamaica Bay Institute, Gateway National Recreational Area, National Park Service, 18 pp.

Gordon, A. L. & Houghton, R. W. (2004). "The waters of Jamaica Bay: impact on sediment budget." *In: Proceedings, Jamaica Bay's Disappearing Salt Marshes,* New York: Jamaica Bay Institute, Gateway National Recreational Area, National Park Service, 18 pp.

Gornitz, V. (1995). "Monitoring sea level changes." *Climate Change,* 31, 515–544.

Gornitz, V., Couch, S., & Hartig, E. K. (2001). "Impacts of sea level rise in New York City metropolitan area." *Global and Planetary Change*: Elsevier Science, Amsterdam, 32, 61-88.

Hartig, E. K., Gornitz, V., Kolker, A., Mushacke, F., & Fallon, D. (2002). "Anthropogenic and climatic-change impacts on salt marshes of Jamaica Bay, New York City." *Wetlands,* 22(1), 71–89.

Hartig, E. K., Kolker, A., Fallon, D., & Mushacke, F. (2001). "Climate change and a global city: the potential consequences of climate variability and change - metro east coast." report for the *U. S. Global Change Research Programs,* National assessment of the potential consequence of climate variability and change for the United States, Columbia Earth Institute, New York.

Intergovernmental Panel on Climate Change (IPCC) (2007). "The physical science basis, contribution of working group I to the fourth assessment report of the Intergovernmental Panel on Climate Change." Cambridge University Press, Cambridge, U. K.

Jamaica Bay Ecosystem Research and Restoration Team (JBERRT) (2002). Final Report 2002. Volumes I, II, and III. Gateway National Recreation Area, Staten Island, New York and Aquatic Research and Environmental Assessment Center, Brooklyn, New York

Kana, T .W. (1995). "A mesoscale sediment budget for Long Island, New York." *Marine Geology,* 126, 87-110.

Koch, M. S. & Mendelssohn, I. A., (1989). "Sulfide as a soil phytotoxin: differential responses in two marsh species." *Journal of Ecology,* 77(2), 565–578.

Kolker, A. (2005). "The impacts of climate variability and anthropogenic activities on salt marsh accretion and loss on Long Island." Stony Brook, New York: Stony Brook University, Ph.D. thesis, 278 pp.

Leventhal, J. & Taylor, C. (1990). "Comparison of methods to determine degree of pyritization." *Geochimica et Cosmochimica Acta,* 54(9), 2621–2625.

Mendelssohn, I. A., McKee, K. L., & Patrick, W.H. (1981). "Oxygen deficiency in *Spartina alterniflora* roots - metabolic adaptation to anoxia." *Science,* 214, 439–441.

Messaros, R., Rafferty, P., & Woolley, G. (2010). "Challenges and successes of tidal wetlands restoration in Jamaica Bay, New York." ASCE Conference Proceedings, Watershed Management 2010, 343-363.

Morris, J. T. & Bradley, P. M. (1999). "Effects of nutrient loading on the carbon balance of coastal wetland sediments." *Limnology and Oceanography*, 44, 699–702.

National Research Council (NRC) (1987). "Responding to changes in sea level, committee on engineering implications of changes in relative mean sea level." National Academy Press, Washington, D.C.

New York City Department of Environmental Protection (NYCDEP) (2007). "Jamaica Bay Watershed Protection Plan." http://home2.nyc.gov/html/dep/html/dep_projects/jamaica_bay.shtml#plan.

New York City Department of Environmental Protection (NYCDEP) (2006). "Planning for Jamaica Bay's future: Preliminary recommendations on the Jamaica Bay watershed protection plan." Jamaica Bay Watershed Protection Advisory Committee, Flushing, New York.

New York State Department of Environmental Conservation (NYSDEC) (2001). "Strategies for addressing loss of intertidal marsh in the Maine District." http://www.dec.state.ny.us/website/dfwmr/marine/twloss.html.

O'Grady, J. (2001). "Why so little marsh grass is waving in the bay." *New York Time,* New York, May 13.

Peeteet, D., Liberman, L., Higgiston, P. (2004). "Paleoecology and marsh compositional changes over the last millennium, Jamaica Bay, New York." *In: Proceedings, Jamaica Bay's Disappearing Salt Marshes.* New York: Jamaica Bay Institute, Gateway National Recreational Area, National Park Service, 18 pp.

Peeteet, D., Liberman, L., Higgiston, P., Sritrairat, S., & Kenna, T. (2008). "Yamekoto JFK – Paleoenvironmental Change from Jamaica Bay Marshes." Presented at: State of the Bay Symposium, New York. http://www.nyc.gov/html/dep/pdf/jamaica_bay/jbaysymp-Dorothy_Peteet_Palaeoecology.pdf. (Oct. 4, 2009).

Race, M. S. & Fonseca, M. S. (1996). "Fixing compensatory mitigation: what will it take?" *Ecological Applications*, 6(1), 94-101.

Raiswell, R. & Canfield, D. (1998). "Sources of iron for pyrite formation in marine sediments." *American Journal of Science*, 298, 219–245.

Redfield, A. C. (1972). "Development of a New England salt marsh." *Ecological Monographs*, 42, 201–237.

Roberts, L. (1993). "Wetlands trading is a loser's game, say ecologists." *Science*, 260(25), 1890-1892.

Rybczyk, J. M. & Cahoon, D. R. (2002). "Estimating the potential for submergence for two wetlands in the Mississippi River Delta." *Estuaries*, 25, 985–998.

Seliskar, D. M., Smart, K. E., Higashikubo, B. T., & Gallagher, J. L. (2004). "Seedling sulfide sensitivity among plant species colonizing *Phragmites* infested wetlands." *Wetlands*, 24(2), 426–433.

Steinberg, N., Suszkowski, D. J., Clark, L., & Way, J. (2004). "Health of the harbor: The first comprehensive look at the state of the estuary." New York: Hudson River Foundation, 82 p.

Swanson, R. L. & Wilson, R. E. (2008). "Increased tidal ranges coinciding with Jamaica Bay development contribute to marsh flooding." *Journal of Coastal Research*, 24(6), 1565–1569.

Tanacredi, J. T. & Badger, C. J. (1995). *Gateway, a visitor's companion*, Stockpile Books, Mechanicsburg, Pennsylvania.

Turner, R. E., Swenson, E. M., Milan, C. S., Lee, J. M., & Oswald, T. A. (2004). "Below-ground biomass in healthy and impaired marshes." *Ecological Research*, 19, 29–35.

U.S. Fish and Wildlife Service (USFWS) (1997). "Significant habitats and habitat complexes of the New York Bight watershed, southern New England - New York Bight, Final Report." Coastal Ecosystem Program, Charleston, Rhode Island.

U.S. National Park Service (USNPS) (2001). "The Jamaica Bay blue ribbon panel on marsh loss and coastal sea level rise, a future agenda for mitigation and pilot investigations, Final Report." National Park Service, Gateway National Recreation Area.

Valiela, I., Teal, J. M., Allen, S. D., Van Etten, R., Goehringer, D., & Volkmann, S. B. (1985). "Decomposition in salt marsh ecosystems - The phases and major factors affecting disappearance of above-ground organic matter." *Journal of Experimental Marine Biology and Ecology*, 89, 29–54.

Valiela, I., Teal, J. M., & Persson, N. (1976). "Production and dynamics of experimentally enriched salt marsh vegetation: Belowground biomass." *Limnology and Oceanography*, 21, 245–252.

Wigand, C., Davey, E., & Johnson, R. (2008). "Nitrogen effects on salt marshes." *Presented at: State of the Bay Symposium*. New York. http://www.nyc.gov/html/dep/pdf/jamaica_bay/jbaysymp-Cathy_Wigand_Nitrogen_Talk.pdf (Oct. 4, 2009).

Wilson, R. E. & Flagg, C. (2008). "Circulation and mixing: Implications for sediment transport." *Presented at: State of the Bay Symposium*. New York, New York. http://www.nyc.gov/html/dep/pdf/jamaica_bay/jbaysymp-Robert_Wilson_Circulation.pdf. (Oct. 4, 2009).

Yozzo, D. J., Wilber, P., & Will, R. J. (2004). "Beneficial use of dredged material for habitat creation, enhancement, and restoration in New York-New Jersey Harbor." *Journal of Environmental Management*, 73, 39–52.

9

Ecological Footprint Applied in Agro-Ecosystems: Methods and Case Studies

Alessandro K. Cerutti[1,3], Gabriele L. Beccaro[1], Marco Bagliani[2,3], Simone Contu[3], Dario Donno[1] and Giancarlo Bounous[1]
[1]*Department of Arboriculture, University of Torino*
[2]*Department Territory and Environment, IRES (Istituto di Ricerche Economico-Sociali),*
[3]*IRIS (Interdisciplinary Research Institute on Sustainability), University of Torino*
Italy

1. Introduction

Covering over a fourth of global land area [1], agriculture has been outlined as one of the most relevant human activities associated with ecosystem degradation. A major cause of environmental impacts from agriculture is due to the high quantity of resource used in terms of water consumption and soil degradation: the Millennium Ecosystem Assessment [1] estimates that 70% of total available water is used in agriculture and the extent of cultivated systems covers 24% of the terrestrial surface. Furthermore, just in the past 100 years, agriculture has been drastically transformed from a resource-based subsistence activity to a highly technological and resource demanding sector. This rapid industrialisation of agriculture, which is based on the use of fossil fuels, produces wastes and residuals that exceed the assimilative capacities of ecosystems and results in alterations to global climate and deterioration of land, air and water in many parts of the world.

This study is motivated by the premise that if agricultural practices degrade ecosystems, they cannot be sustainable, but how can we measure the sustainability of an agricultural system?

In this study some key issues in the evaluation of sustainability of agricultural systems thorough the Ecological Footprint (EF) approach are discussed. More particularly, the first part gives a brief overview of the sustainability concept and the relation between environmental sustainability and agriculture. The second part introduces the Ecological Footprint Analysis (EFA) from a methodological point of view. The third part discusses the application of the EFA in agriculture thorough a review of some case studies in food production systems. The fourth part presents a detailed application of the EFA in an original case study, outlining strength and weakness of the method in the specific application. The fifth part briefly describes how this method can be used as a teaching and outreach tool to make students and stakeholders conscious about the ecosystem demand of our food consumption pattern.

1.1 Understanding sustainability

Sustainability is a broad concept, sometimes also considered ambiguous because it means different things to different people, at different periods of time. This concept (in its broader definition) takes its roots in the beginning of the 70's, conventionally with the publication of the book: "The limits to growth" commissioned by the Club of Rome to the research group of Donatella Meadows [2]. In this book, for the first time, scenarios of infinite economic growth in a finite World were presented with all of the problems and misconception of this way of thinking. Nevertheless, the sustainability concept was almost ignored until the end of 80's, when the Brundtland Report was published by the World Commission on Environment and Development [3]. This report describes sustainable development as "the development that meets the needs of the present without compromising the ability of future generations to meet their own needs" [3, page 43]. However, the definition proposed in the Brundtland report does not articulate the concept of "needs" and the mechanisms for achieving a sustainable society. Indeed, after the publication of this report, numerous definitions of sustainability have been suggested, yet none have been universally accepted, contributing to make this concept more misunderstanding.

One of the most consolidated concepts in sustainability is the needing of the three pillars of economy, society and environment simultaneously (fig 1). As a consequence a lot of definitions start with this concept; e.g. the most accepted definition of sustainable agriculture can be summarized as follows: "To be sustainable, a farm must produce adequate yields of high quality, be profitable, protect the environment, conserve resources and be socially responsible in the long term [4].

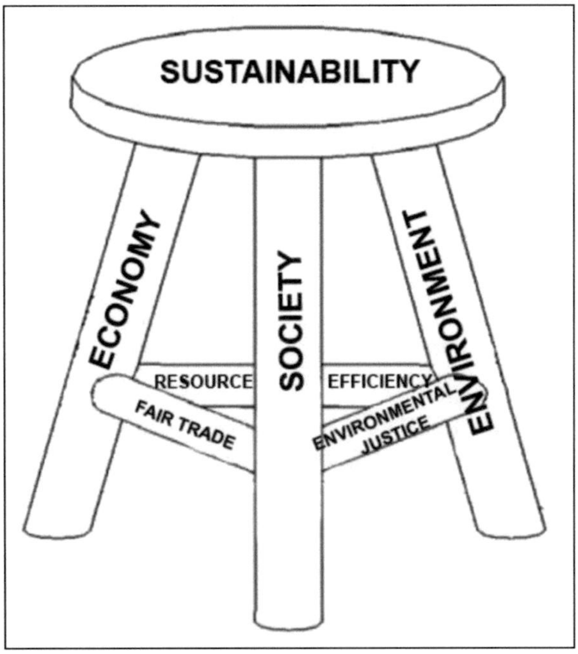

Fig. 1. Graphical representation of the famous metaphor of the "three legged stool" about sustainability. Each of the three legs is a pillar of sustainability: economy, society and environment. Conjunctions of the legs represent aspects than are involved in two pillars simultaneously. For example fair trade considers economy and society, but not environment; resource efficiency considers economy and environment, but not society; environmental justice considers society and environment, but not necessarily economy.

1.2 Ecosystem services and agriculture sustainability

Ecological systems both contribute to and are affected by the production of goods and services that contribute to human wellbeing and are of value to people [5]. These positive functions are commonly defined as ecosystem services. As agriculture is one of the dominant drivers for land use change [1], understanding how agriculture impacts ecosystem services is gaining more and more importance in the scientific research.

A number of works [1] [5] [6] outlined the interrelation of agriculture and ecosystem services. Ecosystems provide several direct benefits to agriculture, such as when pollinators increase agricultural crop yields. Nevertheless, most of benefits to agriculture from natural systems are provided indirectly; e.g. intact forests can minimize flooding by slowing snowmelt and water discharge, moderate regional climate, and remove and store atmospheric carbon dioxide, a greenhouse gas. Another example is the service provided by wetlands, which can reduce the load of nitrogen in surface water originating from agricultural fields and destined for a coastal estuary where eutrophication causes hypoxic conditions and reduced fish productivity.

On the other hand, agricultural practices greatly influence ecosystem services both in positive and negative ways. For example a positive effect is the habitat for bird species provided by specific planting on agricultural lands due to conservation policies. Nevertheless, in most cases, agricultural practices reduce the ability of ecosystems to provide goods and services. Main causes of ecosystem damages from agriculture are caused by the use of fertilizers and pesticides, which can increase nutrients and toxins in groundwater and surface waters, contributing to eutrophication of aquatic habitats and degradation of soils. Moreover, agricultural outputs to ecosystems may change species composition or reduce biodiversity, which may also affect negatively the production of goods and services, because the ability of ecosystems to provide some services depends both on the number and type of species in an ecosystem.

Furthermore, ecosystems degraded by agricultural practices may generate significant disservices back to agricultural systems [5]. For example habitat loss in areas located nearby agricultural fields, may no more provide the conditions for completing the life cycle of insects useful for natural pest control. Thus, pest damages to the production may increase, inducing farm technicians to amplify pesticides doses, which lead to more significant damages to habitats nearby the farms. As a consequence, it can be easily pointed out that negative effects on ecosystems are actually negative effects on the agricultural systems themselves. Thus the obtaining of a conservative agriculture is one of the main aims of modern agronomic research.

One of the most important issues of a conservative management of farms is the obtaining of a sustainable yield; nevertheless it is highly difficult to define and measure what sustainable yield is [7] [8]. Nowadays it is well known that yield is affected by a number of parameters, both natural, such as the pedo-climatic conditions, and artificial such as the agricultural techniques and growing structures. Clearly all of the agricultural practices (such fertilization, irrigation, heating of greenhouses etc...) contribute both the yield and to the environmental burden of the production. The difficulty is to evaluate a trade-off between positive and negative effects, considering also that those agronomic parameters have usually a combinatory effect on the crop, which does not reflect the sum of the effects of single agricultural practice alone.

Considering these general remarks, some authors [9] state that the coming 50 years as the final period of the rapidly expanding human environmental impacts on the global scale, because future agriculture (with all the other human activities) will modify, perhaps irreversibly, ecosystem services obtained from the global bio-geological cycles. Nevertheless, if environmental concern are mostly focused on the global scale, the possible actions for mitigate negative effects can be done on the local scale. Thus better understanding of how to achieve sustainability at the farm level is important because at this level a lot of changes can be made in productive protocols and technologies to reduce the environmental impacts of farming activities. As a result, practical assessment of sustainability at the field level is one of the essential points in order to achieve the sustainability of the whole food and fibre industry and, therefore, to mitigate environmental impacts on the global scale.

2. The ecological footprint in a nutshell

Since the publication of the Limit to growth [2] in the early '70, it is well known that services provided by ecosystems are limited. Even if a limitless growth is still the main objective of public policies in most countries, the scientific debate on the extent to which human development can be maintained in the light of environmental constraints is still vital and prolific.

During the last decades of the XX century several tools and indicators have been developed to measure the impact of production and consumption of goods and services to human systems in order to give indications of behaviours and lifestyles more compatible with the finiteness of the natural resources. One of such method is the Ecological Footprint Analysis (EFA) which estimates the total area of the terrestrial and aquatic ecosystems necessary to supply all resources utilized and to absorb all resultant emissions needed to sustain a human population or in the production of particular products. EFA provides a single value (hectares or global hectares) that comprises of various environmental burdens and which can be disaggregate down to the most detailed level of the single consumption. The aggregation capability of the EFA thus enables easy comparison of results arising from different scenarios.

Results of EFA are generally expressed in global hectares (gha) which represent a productivity-weighted area used to report both the demand on ecosystem services and the biocapacity of the Earth. In order to convert effective land surface into global hectares equivalence factors (EQF) have been introduced [10]. These factors are corrections of the land components based on the different productivities of each land type; therefore the gha unit gives a standardized and productivity-weighted value of the EFA results [11].

The continuous development of the analytical methods by various research groups around the world led to Wackernagel and collaborators to create in 2004, the Global Footprint Network (GFN), a network of research institutions, scientists and users of EFA which aims improve the calculation methods.

2.1 The land components

The original EFA method was introduced by [12] and further developed by [10]. In the original definition just 4 land components were considered:

i. Cropland: is the land needed to grow all crop products, including livestock feeds, fish meals, oil crops and rubber. It is the most bioproductive of the land use types considered by in the GFN standards.
ii. Forest land: is the land needed annually to harvest fuel wood and timber to supply forest products. The yield used in the forest land component in the GFN standards is the net annual increment of merchantable timber per hectare, accounted on the base of the production quantities of 13 primary timber products and three wood fuel products [11].
iii. Built-up land: is the land covered by all type of human infrastructure, such as houses, commercial buildings, industrial structures but also streets and concrete covered lands. This type of land, even if the smaller component on the global scale [11], is quite important because it is assumed that built-up land is a transformed cropland which lost its original biocapacity.
iv. Energy land: is the land needed to produce and to absorb emissions caused by the use of energetic carriers, such as fossil fuels, electricity and renewable energy. In the original calculation the consumption of 80-100 GJ from fossil fuels was equivalent to 1 hectare of energy land [10].

In more than 20 years from the original formulation several improvements have been performed and land components of the EFA have been revised. More particularly two land components have been added:

v. Grazing land (or pasture): measures the area of grassland used in addition to crop feeds to support livestock. It can be considered as a more detailed definition than cropland for the evaluation of the land needed for animal products. More particularly grazing land comprises all grasslands used to provide feed for animals, including cultivated pastures as well as wild grasslands and prairies.
vi. Fishing grounds: is the area surface (as a simplification of volume) required to sustain a harvested aquatic species. Also this land type can be considered as a better qualification for the EFA of food products. It is accounted as the surface need to sustain the mass ratio of harvested fish based on its average trophic level.

Furthermore, the energy land component was deeply revised and updated to the Carbon Footprint component. This land type quantifies the land needed to absorb the carbon emissions from all activities: from production of goods to waste management. As most terrestrial carbon uptake in the biosphere occurs in forests, and to avoid overestimations, carbon uptake land is assumed to be forest land by the EFA method [11].

Recently, a new way of considering EF land components has been proposed [13]. Considering the life cycle of a product, a part of the EF is due to the direct land use, such as the crop growing for food products, and a part of the EF is due to the energy which is embedded in the use of materials and services. The embedded energy can be accounted as the carbon footprint needed to produce and to absorb emission from that source, thus this land can be considered as used indirectly in the life cycle of a product. As consequence a novel formulation of the EF component is the following:

$$EF = EF_{direct} + EF_{CO2} \quad (1)$$

where EF_{direct} represents land occupation over time by cropland, built-up land, pasture and forest. This value is counted as the sum of the area occupied by each land type, multiplied by the specific land-related Equivalence factor (EQF). These factors are land specific coefficients that correct the land components on the base of the different productivities of each land type, taking into consideration the average productivity of all the bioproductive areas on Earth. The second component of the footprint (EF_{CO2}) is the indirect bioproductive land required to sequester, through forestation, the atmospheric fossil CO_2 emissions related to the produced goods. This value is obtained by multiplying the product-specific emission of CO_2 by the EQF of forests and the mass of CO_2 sequestered in biomass with a correction of the CO_2 absorbed by oceans [14].

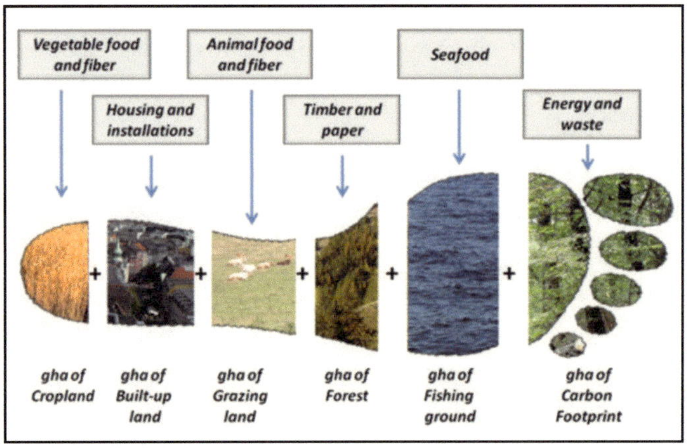

Fig. 2. Schematic representation of the EF land components associated to main goods and services (modified from the GFN).

2.2 Main applications of the EFA

Since the early definition, several changes have been applied in order to being able to perform the EF approach in a number of systems.

The EF method applied to the national scale is also known as the National Footprint Account. This method is standardized by the GFN and, in its last application, it calculates the EF and biocapacity for more than 200 countries and the World [11]. These accounts are published biennially in the WWF's Living Planet Report and are based on the latest complete datasets available, which results usually in a time lag of about three years.

The National Footprint method consists in the quantification of the total production and consumption at the national level. The EF of production, which represents primary demand for biocapacity, is calculated as the amount of a product produced in the Country, divided by the national average yield (or production capability) and then multiplied for the yield factor and the specific EQF. The EF of consumption is obtained starting with the EF of production, summing the EF embodied in imported commodity flows and subtracting EF embodied in exported commodity flows. Even if this method reveals just the apparent consumption, it is considered a sufficient approximation of international trade, because data on stock changes for various goods are generally not available [11].

Furthermore, even if the National Footprint Accounts recognized as one of the best tools to assess the ecosystem demand of countries, data and methodological issues associated with the calculation remain [15]. Main concerns are related to the quality and the country specificity of data source as main data used for the evaluation are obtained from international datasets.

Considering EF applications at the sub-national scale several authors have outlined the potentialities of the EFA in measuring the environmental demand of regions and districts (e.g. [16][17]). For these kinds of application a number of data, which are generally available at a national level, are actually difficult to obtain at the regional and at the local level. Therefore, in the whole calculation method for a sub-national area are generally considered the data available at the commune level [17]. Indeed at this level, it is usually possible to find data regarding different kinds of consumptions such as housing (electricity, gas, and fuel oil), water, waste production and land occupation. Another possibility is to calculate the average consumptions for each commune of a district by weighting the average regional consumptions on the basis of the composition of each commune by employment conditions [16].

Relatively recent applications of the EFA can be found at the company or at the product levels. In the first case some adaptation of the original EF method are proposed (e.g. [18][19]). In these implementations the EFA is usually joined to other economic assessment tools in order to enhance the industrial metabolism of the studied company. The modified EF methods are therefore able to quantify economic flows of a company from an environmental point of view.

On the other hand, methods for the EFA of products are, in most cases, based on the Life Cycle approach used in LCA studies. E.g. [20] proposed to account the EF of a product as the sum of the EF of the process involved in the production. More detailed, the authors suggest to consider impacts related to the materials used during the production; the land occupied, with the account of biodiversity and productivity impacts, labour embedded impacts and waste management impacts. As a consequence some authors [13] considered the EFA of a product, more than a standalone tool, as a specific life cycle impact assessment (LCIA) method, which gives results on a single impact category (land use expressed in gha) instead of different impact categories with specific measure units (such as Global Warming Potential in kg of CO_2 equivalent). Furthermore [13] compared results from EF and Eco-Indicator 99 (single score) and they found that for a large number of products both methods will typically produce the same type of gross ranking results, but not for all products considered in the study. Nevertheless authors suggested that EFA and Eco-Indicator 99 results can get closer if some additional impacts will be taken into account in the EFA, such as tacking in account the complete GHG emission instead of just CO_2.

Another important general issue in the EFA is the evaluation of waste. From an EF perspective, the term waste includes three different categories of materials, thus three different waste management protocols that are differently assessed in the EFA [21]. The first type are biological wastes such as residues of crops and livestock effluents: these materials are already included in the specific land category (cropland for the first and pasture for the second) as the recycling of organic matter in an ecosystem service already accounted in the quantification of such land component. Thus the addition the CO_2 emitted from the decomposition of vegetable and animal residues will lead to a double accounting. The second type are the wastes sent to landfill: these materials are accounted as the infrastructure of the built-up land used for its long term storage as these landfills occupy formerly biologically productive areas. The third waste type are the toxic and pollutant materials that cannot be absorbed in any kind of biological cycle. As the EFA quantify the area needed to the production process of a specific product or to absorb a waste, materials such as plastics which are not part of a biological cycle, have to be accounted just indirectly as the carbon footprint generated by the energy embodied in the production cycle of a specific quantity of material. As a conclusion, the energy embodied within goods (both locally produced and imported) plays a major role in the determination of the total EF in all the EF methods [22], thus robust country-specific embodied energy data to capture the carbon embodied in traded good are needed. These kinds of data have historically been lacking [22], nevertheless with the increasing interest on the carbon footprint methodology to achieve environmental sustainability of consumption and production, more detailed datasets could be available soon.

3. Key-issues in the EFA in agro-ecosystems

As the EFA was developed to assess the ecosystem demand of a given population, the focus of EF of production processes is a relatively new aspect and a complete review of the application of EFA in the agricultural sector has not yet been performed.

For the agricultural sector, three land types are considered sufficient to describe the land composition of the EF of farms [23][24]: cropland, which accounts for the effective land surface where the farm is assessed and for production of animal feeds which were not produced on-farm; forest, which accounts for production of forest resources such as wood for piling; and carbon footprint which accounts for the land required to produce the non renewable energy used on the farm and for the production of the farm inputs.

Nevertheless another land component might be relevant: the built-up land, which accounts for considers the land occupied by infrastructure, e.g. deposit, garages, silos and other structures [25].

Furthermore, some studies [27] consider also the pasture land and the sea area which result when including in the application also the EF of the operators. This approach is still under debate because it is a matter of fixing the system boundaries of the application: those workers would have consumed foods and natural resources regardless their activity in the crop field, thus their environmental impacts are not an externalities of the production. But, on the other hand, when comparing different scenarios (e.g. production protocols such as conventional and organic production) it can be noticed that a scenario might require less labour than another thus the impact of the workers might be decisive in the determination of which scenario has the worst EF. Further investigations are needed to assess definitively this issue.

Typically the EFA of crops is conducted in order to give reference values for the National Footprint Accounts. In this account the EF is evaluated on the basis of the average yield of a species and the national consumption [11][27]or using the conventional global hectare approach [22]. In this case the quantification of the EF of a food item can be simplified in the following equation:

$$EF_{(gha)} = \frac{C_{nation}}{Y_{world}} \times EQF \tag{2}$$

were the EF is obtained dividing the annual tonnage of the food consumption by the annual average global yield and them multiplying by the cropland equivalent factor, which is 2.39 [11]. To this value it has to be added the energy intensity of the production, given by the incorporation of the required energy-land for the supply of cultivation inputs, such as fertilizers, pesticides and machinery. Such parameters are considered as the average requirement for the crop and are usually obtained by international databases such as FAOSTAT and FERTISTAT. This method suffers in the localization of the production, because at different pedo-climatic conditions the EF of the same crop can be very different, thus a place-oriented quantification is suggested [28].

Another example is the National Footprint Accounts' footprint for fruits which is considered to be 0.5–0.6 gha per ton of product on the basis of average global yield, embodied energy of the cultivation and estimation of the impacts of post-harvest management [27]. Thus the EF of imported fruits account simply for the average soil required for the production and the energy for transportation to the imported country.

These rough estimation methods of the EF of crops could be useful for the application of EFA at the large scale, but it is obviously insufficient for detailed applications. Indeed, using general methods, for agricultural products typically 80% of the total EF is clarified by the occupation of land by crops [13], on the contrary, applications with a specific process approach (e.g. [25][26]) reveal that crop land accounts for less than 30% of the total EF and the major role is played by the carbon footprint. As a consequence, different approach may lead to different results and, thus, to different suggestions for improving the studied production system.

3.1 The objective of case studies

The agricultural systems are not, generally, the direct object of an EFA. The environmental pressures generated by human activities in agro-ecosystems are evaluated as a component of the food production systems. More detailed, four main types of case studies for EFA in the agricultural context can be found in the scientific literature:

3.1.1 The EFA of crops

As highlighted in paragraph 1.2, crops are production systems, which utilize ecosystem services to produce commercial goods that are directed to the anthropo-sphere. Further than direct ecosystem services, growers utilize industrial inputs, such as fertilizers and pesticides, and capital goods, such as machineries and infrastructures. All the inputs utilized in the production are involved in the determination of the EF of such production. Thus EFA of crops can be considered a product EFA from a biological based production system instead of a product from an industrial based production system. This difference, which might seem to be small because of the high industrialization of modern agriculture, is however important to be considered in modelling of system for environmental assessment applications. The main difference with industrial systems is that crop production is strongly bounded to specific pedo-climatic conditions which means that crop yields may change dramatically from one year to the other, because different wheatear conditions or because of uncontained pest infections. Being bounded to local conditions also means that in other countries, natural and technical inputs applied in field in order to obtain the same level of production, may vary dramatically in quantity and way of using. As a consequence environmental impacts of crops systems vary greatly from a Country to another and regionalized data are required for precise applications of an environmental impacts assessment tool [29].

A detailed EFA application to crop productions can be found in [30]. Authors applied EFA and other tools on six agricultural crop production systems in Nicaragua: common bean, tomato, cabbage, maize, pineapple and coffee. In this study the EFA included the cropland as well as the machinery and buildings appropriated for the cultivation; resource use was assessed in four land type: (I) the actual cropping area, (II) the built-up area used for cropping purposes and (III) the area for sequestration of carbon dioxide as a consequence of direct and indirect energy use for purchased inputs and in transportation, as well as for labour and other services. Furthermore, (IV) an estimated area appropriated for biological conservation was added.

Applications of the EFA on crops are conducted generally using a mass based functional unit, thus results are commonly presented in EF per tonnes of production and expressed in gha/t (e.g. [24] [25] [26] [31]). Nevertheless other ways to express results are proposed and may lead to different conclusions (see paragraph 3.3).

3.1.2 The EFA of food products

In this case the EFA is mainly conducted as an EFA of product in which the investigated product is a result of biological and industrial production processes. It has to be noted an important distinction between the EFA for primary agricultural products, such as wheat or vegetables, and the EFA for secondary transformed food products, such as pasta or already-prepared soups. In the second case the production reflects an industrial process which is mainly linear and it could be very different to a biologically base production. Detailed study on such kind of production should consider both the field and the industrial phase (e.g. [32]).

The EF of food products is often referred to the land required for the final commercial unit (such as a 1 kg box of pasta or a 250 g jar of jam) instead of a ton of primary production from the field. A detailed application of the EFA to an agricultural product has been performed by [26], in which authors applied a product EFA to two typical Tuscan wines, one conventional and one organic, to determine which type of wine production present the greater environmental burden. They considered four production phases: (I) agricultural, i.e. preparation and planting of the vineyard, grape; (II) production, treatments and harvest; (III) winery, i.e. crushing, fermentation and stabilization; (IV) packing, i.e. bottling, corking and labelling; and the distribution phase. Thus they set the reference flow for the environmental impacts to 1 bottle of wine. Their results showed that higher EF value is obtained for conventional compared to organic production. This was mainly due to the higher demand in the agricultural phase, which contributed most to the EF value of both productions.

Furthermore, the EFA has been used as one environmental assessment method (together with Carbon Footprint and Water Footprint) for the Environmental Product Declaration (EPD) of several cereal-based products from the Barilla Company [29]. In the EPD of pasta the agricultural phase accounts for almost the 49% of the whole product EF, the industrial phase (which comprehends: elaboration, packaging and distribution) accounts for 9%, and the consumer phase (which comprehends the impact of cooking) accounts for 42% of the whole product EF. Beside the application in the agricultural context, this study highlights also the good potential for EFA as an environmental certification tool as the base for EPDs of food products.

3.1.3 The EFA of farms

Several authors consider the farm as a good level of application of an environmental assessment method (e.g. [23] [24] [29]). Results of environmental impact assessments applied to a farm are related exclusively to the farm and cannot be used extrapolated for compare different crops or different production protocols without specific modifications in the method [24]. Nevertheless several studies use the farm level in order to assess the performance of an environmental assessment method (e.g. [23] [24] [33]). In such studies several impact assessment methods are applied to the same farm (or the same farm scenarios) in order both to discuss results and evaluate the method.

Recently the application of EFA on farms was deeply discussed and some corrections have been proposed [34] in order to account negative environmental externalities of farms, such as the decrease of soil fertility, soil erosion and biodiversity loss. Authors decided to account such effects trough an economic quantification of each externality and then the relative valorisation in term of emergetic values (seJ/year) as suggested by [35].

Other authors [36] claim that in order to apply the EFA at the level of farm or company, the economic transactions (such as the cost of capital goods and services) has to be accounted from the environmental point of view. Thus some methods are proposed to incorporating the Input-Output methodology in the EFA (e.g. [36] [37]). Another method for evaluate the economic performance of the company from an environmental point of view is proposed by [19] with the joint implementation of EFA and Cost Accounting. In their approach, the production chain is divided in several activity centres that best reflect both the generation of impacts on ecosystems and the process of economic value production. The environmental pressures of the different activity centres are represented in term of their appropriation of ecological productive land following the EFA accounting system [10] [11]. Results of this method showed the high level of accuracy obtained in EF calculation; more particularly in the application in two livestock farms it was possible to capture small differences in ecological footprint due to little variations of the productive process, depending on the slightly different utilization of mechanised breeding techniques [38].

Although some methodological aspects that are still under debate, according to most studies, the EFA applied on farm is an interesting level of analysis because it allows to visualize the virtual land needs of a farm, including direct and indirect resources use, and may guide technicians and growers to strategy for reduction of the EF of their farms.

3.1.4 The EFA of diets

Another object of application of EFA in the context of food production and agro-ecosystem consumption is the environmental impact evaluation of the diet of a given population. Also in this level of analysis several methods are proposed, but the EFA is one of the most used for didactical and divulgation purposes because of the easy way of calculation via the National Footprint Matrix [11] and because of the easy understanding of the results. Several example of application of EFA to a food consumption patter may be found in scientific reports for NGOs (e.g. [39]) or Research Institutes (e.g. [40]) as well in scientific literature (e.g. [41]). All these studies highlight that the food consumption patterns that are more environmental friendly are also the healthiest. This effect can be observed by the role that meat products play in the evaluated consumption patterns; more particularly [32] stated that, in comparison with the current Scottish diet, a healthier diet can reduce the Scottish EF by between 15 and 20%.

These studies reveal that eating mainly fruit, vegetables and wholegrain foods, and moderate amounts of meat, dairy and eggs is not only good for us but also better for the environment. On this topic [42] defined the double pyramid metaphor which represents that the most consumed foods in a healthy diet (such as seasonal vegetables and fruits) are the items with the lower EF and vice versa the food with the higher EF values (such as beef, cheese and butter) should be consumed very rarely for a healthy diet.

3.2 EFA and scenario evaluation

The main scope of the EFA is to quantify numerically the ecosystem demand of a population or a product and the obtained EF result may be used in several ways. The most important application is the comparison of the EF of a population and the biocapacity of the ecosystems where the population is located in order to assess whether a population is able or not to live with just their own resources (e.g. [10]). Nevertheless a relevant application of EFA results is in scenario comparison of production processes, because its ability to aggregate several environmental impact sources in a single coherent value.

In the agricultural sector the EFA as a tool for scenario evaluation is conducted at two level with two main scopes: at the field level, in order to identify of the best agricultural practice (e.g.[43][44]) and at the farm level in order to identify which production protocol shows the best environmental performance (e.g. [38][45]). Thus the EFA can be considered as one of the available tools to identify best agricultural practice for sustainable food production [7].

The study from [43] was not only one of the first studies in EF in the agricultural sector, but also one of the first EF applications of all time, thus it is considered a very important step in the development of the method. In this study the environmental impacts of tomato production in some hydroponic greenhouses and open fields are compared thorough the EFA. All agricultural inputs, energetic carriers and direct land use for four different scenarios (two greenhouse systems and two open field systems) have been accounted. With the hydroponic systems is it possible to obtain a big amount of product because the higher yield of the system (an average of 465 t/ha compared to an average of 70 t/ha in open field systems). Thus one may think that hydroponic systems are more sustainable because they can produce an higher amount of product for the same occupied land, but considering the EF result, the hydroponic approach required a 14 to 21 times higher EF than conventional farming to produce the same amount of tomatoes. This difference reflects the different resource use intensity of each agricultural technique, thus the requirement of ecosystem services that is embodied in the use of each agricultural input [32].

A recent application of EFA for the evaluation of the best agricultural techniques has been performed by [44]; in this case the effects of different fertilization methods in an Italian nectarine orchard have been assessed. The study represents also one of the first applications of a statistical evaluation of the EFA results in order to assess if different EF values of each scenario are statistically significant or not. It was possible to perform the statistical analysis because of the dataset collected directly on the orchard during an agronomic trial with a randomized block design set up for performing the Analysis of Variance (ANOVA). The randomized block design considered three replications of four treatments has been applied: liquid slurry (LS), covered slurry (CS), solid fraction (SF), mineral nutrition (MN). Each of these treatments has been modelled in a scenario considering also two different alternatives in swine manure disposal (MN1, MN2). All the agricultural inputs for each scenario has been collected directly on field and the total EF of each system has been referred to the unit of production (more detailed, the considered functional unit was gha/t of nectarine produced). The ANOVA results showed that different total EF values reflect statistical significant differences among systems and Tukey's Test indeed highlighted three statistical classes: the first composed by LS, CS, SF systems which total ecological footprint is not significant different among each other (average 0.96 gha/t); the second is composed just by the MN2 system which has the highest ecological footprint (1.14 gha/t) and the third with the MN1 system which has middle-value (1.01 gha/t) but present significant differences with all the other systems. Thus, using the ANOVA it was possible to define mathematically that differences in the EF of the fertilization systems with manure are non significant among themselves but are significantly lower (about 15%) in respect to the other fertilization systems. Without the application of a statistical method it would not be possible to assess such result.

Moreover, [45] performed a detailed application of the EFA to the production of different Brazilian boiler feed production scenarios. The authors designed the scenarios as a small matrix of different feeds and their sources. They choose this method in order to assess the different environmental characteristics existing in the grain produced in different region of Brazil. Applying the EFA to the resulted scenarios they were able to identify the environmentally best combination of production protocol and product source, but also which were the most problematic aspects of the supply chain from the environmental point of view.

Another important issue in the application of EF for scenario analysis is the recurring concept that foods with lower EF can be considered more sustainable (e.g. [39]). Following this theoretical framework, differences in EF values represents the quantification of the sustainability of a scenario in comparison to another one and can be used to quantify precisely the advantage of a specific agro-technique among the others. Thus differences in EF values for each scenario can be used as an objective basis for discussion of policies for promoting techniques or production protocol. E.g. the value of 1.07 gha, found by [26] as the difference in the EF of two wine production scenarios, can be considered the numerical quantification of the sustainability of the organic protocol (1.12 gha/t) in relation to the conventional protocol (2.19 gha/t).

3.3 The choose of the functional unit and the efficiency problem

As highlighted in previous paragraphs EFA results may be expressed in several ways in relation to different functional units. The use of the term functional unit is not proper for the EFA method, but it is originally used in LCA studies. Nevertheless as further applications of EFA to products are more frequently using a life cycle approach [13], the LCA nomenclature can be adapted. Thus the term functional unit in the EFA refers to the measure unit under which are related the environmental impacts of the whole production process. In particular, the most used functional units in the application of environmental impact assessment tools in the agricultural sector are:

i. mass based functional units, in which the environmental impacts are related to a specific amount of products produced. In EFA applications typical mass based functional unit are tones of product, and generally results are expressed in gha/t or gm^2/kg. This category is the more common in EFA because it is immediately understanding and also because being the unit used in the first studies of EFA of products.
ii. Product based functional units, in which the ecosystem demand is related to a single final item of the production process. This unit is used when the final product is made of different materials with different production or transformation processes, such as cardboard and wheat for a box of pasta [32] or glass and grapevine for a bottle of wine [26].
iii. Land based functional units, in which the environmental impacts are related to a specific amount of land directly involved in the production. This category is not commonly used in EFA, partly because land use is not directly a service and does not provide a productive function; nevertheless it can give very interesting results. Results of EFA using such method are expressed in global hectares per cultivated hectares allowing the visualisation of the difference between the area occupied by the crop field and the total area used considering all upstream processes [26][30] expressed in global hectares (gha).
iv. Economic-value based functional units, in which ecosystem demands are related to the economic value of the products produced. Results are expressed in global hectares per unit of income, usually 1000 € or 1000 USD. This functional unit is useful when the economic eco-efficiency of the systems is to be optimised [46], but it is up to now rarely applied in EFA of foods.
v. Nutrient based functional unit, in which the ecosystem burden is related to the amount of a specific nutritional characteristic bound in the final product. This is a typical unit of food products because it allows to compare foods with very different characteristics on a common parameter. For example in feed production, wheat and soy are really different products with different characteristic, thus relating the environmental impacts of their production to a mass may mislead results. On the other hand wheat and soy are both used for feeding livestock mainly for their protein content, thus a protein-based functional unit may allows to better relating the environmental impacts of the two production systems. [30] mixed EM and EFA in an evaluation of environmental burdens related to the production of a calorific unit in fruits, and thus their unit is gha/Gcal.

As a general remark, [24] outlined that the mode of expression of results (for the whole farm, per unit area, or per unit product) strongly affects the rankings obtained. E.g., the mass based functional unit is easy to comprehend and widely used, but carries the problem of over-evaluating efficiency within sustainability research. By simply looking at environmental impacts per unit product, it is possible to evaluate the eco-efficiency of production, but it is not possible to estimate the sustainability of such production because efficiency does not necessarily lead to sustainability. Indeed, in their paper, [31] underline that only using mass based functional units may well lead to a preference for high input-high output systems, which, when concentrated at regional scale, have been shown to cause major pollution problems [24]. Furthermore, when using a mass based functional unit, problems may arise in how to account for different food quality [47]. E.g. in fruit production, the same orchard can usually produce fruit of different quality (e.g. size, colour, firmness or sugar content) that is targeted to different markets (e.g. fresh market or industrial processing). This issue should be considered in the definition of the functional unit and an allocation procedure relating the environmental impacts to the different products which are consistent with the goal of the study should be devised [47].

On the other hand, the use of a land based functional unit allows to quantify the impacts of cultivating a certain area with a specific agricultural technique. This parameter is also called the impact intensity of a farm [46]. Nevertheless, using a land based functional unit alone may well lead to a preference for low input-low output systems, which may decrease impacts at regional level, but may create a need for additional land use elsewhere, giving rise to additional impacts [24].

As a general result the recommended method of application is to present results using a matrix of EFA with multiple functional units [30]. In this study several functional units have been considered in order to evaluate crops from different point of view: due to this method it was possible to highlight that crops may show difference environmental performance with different parameters: e.g., in the revenue based functional unit cabbage was the crop with by far the least area needed to generate an income of 1000 USD (0.7 ha/1000USD), while pineapple, the crop with the largest EF, required more than nine times as large an area as cabbage (6.5 ha/1000USD). When comparing the EF to produce 1 Gcal, maize and beans were the most favourable crops, with the lowest EF (0.17 and 0.23 gha/Gcal, respectively) and tomato the least, with a five times higher value (1.0 gha/Gcal). The index obtained with the use of a land based functional unit shows that growing beans on 1 ha resulted in the smallest additional area (2.4 gha/ ha crop), while on the other hand, tomato and cabbage required the highest (6.5 and 7.5 gha/ha crop, respectively).

4. A focus case study: The orchard

4.1 The environmental demand of fruit production systems

Fruit production is considered an agricultural sector with low environmental impacts in comparison to other food sectors [40] [48]. Some studies (e.g. [41]) highlight that fruit production requires less bioproductive land compared to all animal products because livestock needs both direct land for breeding and land for feed production. [49] shows that some fruit systems may be less impacting from the environmental point of view than some open field crops; nevertheless quantification of the sustainability of fruit production is required to make specific considerations and comparisons.

When evaluating the environmental burdens in fruit production it is very important to remember the differences between the open field crop systems (where assessment tools are mainly applied) and the perennial crops [47]. There are several differences between these two systems that are related mainly to the impossibility of removing the orchard at the end of the year. Indeed, pests and diseases that are host-tree permanent may remain in the orchard for many years and require continuous control. Due to this effect, orchards are among the most intensively sprayed agricultural systems, in order to avoid visible and internal fruit damage and to satisfy international commercial quality standards [50]. The main environmental risks relating to the use of conventional pesticides are the negative effects on the animal and plant communities exposed both in the orchard and in the other terrestrial and aquatic ecosystems to which pesticides are lost [51]. Another important environmental aspect associated with pesticide use that has to be considered is the resource consumption and environmental impacts associated with the production and distribution of synthetic pesticides. [52] found that pesticide use represents up to 20% of total energy consumption in integrated apple production in New Zealand. Furthermore, because of the development of resistant strains of some pests, there is a tendency for increased frequency of treatments [50] and for increased application rates (around 500 L per ha) in order to retain more pesticides on the leaves and the fruits.

Another key difference between annual and perennial crop is soil management. The operations aimed to maintain and improve soil quality (such as fertilization, irrigation and weed management) are obviously different in the two systems, but in both cases they play a major role in the determination of the environmental impacts of the production [25].

4.2 The importance of a systemic view of the orchard

According to systems thinking, the component parts of a system can be better understood through their relationships with each other and with the environment, rather than in isolation (e.g. [53]). An agricultural system can be defined as a group of interacting components, operating together for a common purpose, capable of reaching as a whole to external stimuli; it is unaffected directly by its own outputs and has a specified boundary based on the inclusion of all significant feedbacks [54].

Various authors are proposing that system thinking is necessary in order to achieve the understanding of the system dynamics which is necessary for proper assessment of the sustainability of the production in agricultural systems (e.g. [55] [56]). The principal reason for this is that agricultural systems are complex, and therefore, it is not possible to modify a single component without generating effects on the other components of the systems. A second important reason is that sustainability is an interdisciplinary issue which requires complex models and multidisciplinary analytical tools.

Most of the papers on the application of an environmental impact assessment method in fruit production describe the orchard from a thermodynamic point of view. They consider the orchard system from interactions that it has with the natural environment in which it is embedded. Thus, orchards are considered systems, which constantly interact with the environment through the use of energy and materials across the boundary (fig 3).

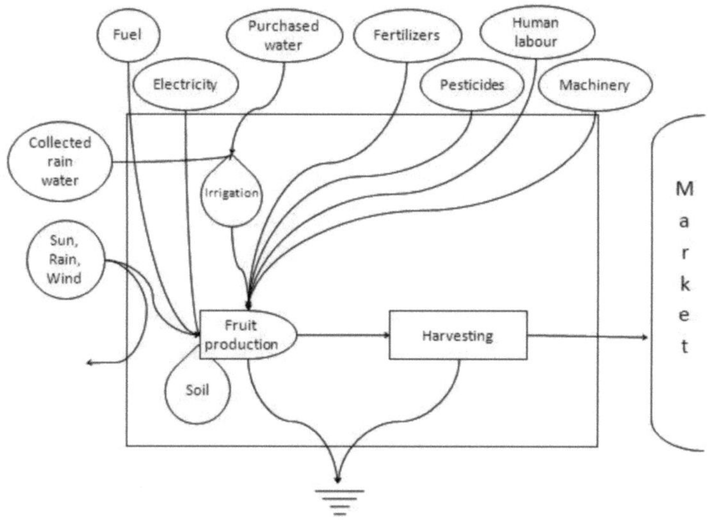

Fig. 3. General energy system diagram of an orchard. This diagram is commonly used for Emergy Analysis and represents the fruit production system from a thermodynamic point of view.

Another way of modelling orchards is by considering interactions instead of energy flows. From a systemic point of view, fruit production can be seen as the integration of various systems (fig 4).

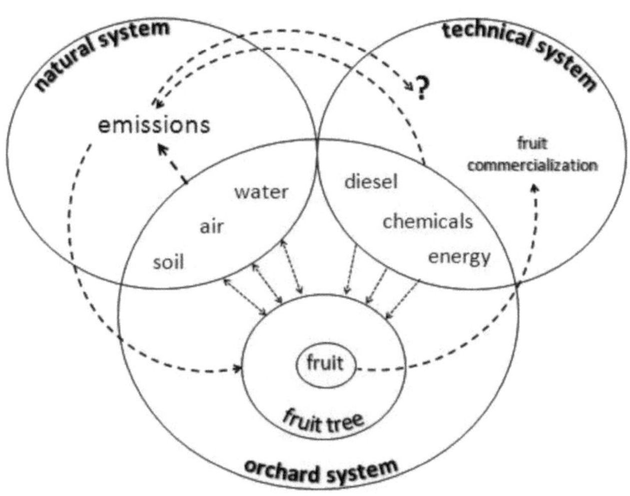

Fig. 4. One of the possibly representations of the orchard from a systemic point of view modified from Page (2009). Orchards are at the same time part of natural system and part of technical system, but they are also a system themselves with subsystems (e.g. fruit trees). Interfaces components of the three systems are represent with common areas of the three cycles and arrows are interactions between components. "Emissions" comprehends all kind of emissions, in water, soil and atmosphere.

In order to model the fruit production, three systems have to be considered: the natural, the technical and the orchard itself. The natural system can be simplified as the biotic and abiotic components of the environment in which the orchard is embedded. The interfaces between natural and orchard systems are mainly soil, air and water (as a local part of pedosphere, atmosphere and hydrosphere). Furthermore, the orchard system is dependent on several ecosystem processes provided by biotic components (see paragraph 1.2).

As system theory suggests, all systems are in fact components of still larger systems and all components of systems are in fact systems made up of still smaller components [56]. For example, fruit trees are at the same time components of the orchard system and systems themselves. More precisely fruit tree sub-system is the core of the orchard system, because effects of natural and technical systems can be seen mostly at the plant level [55].

It is important to underline that when using a systems thinking approach, almost all the processes transferring matter or energy form one system to another produce effects (expected or unexpected) in the other systems. The expected feedbacks of the inputs from technical system into the orchard are the production of fruits and income, but there are also effects on the natural system with the production of different emissions. Studying the chain of effects within the natural system may be very complicated (e.g. [57]), but in the end it is well known that changes in the natural system will affect the orchard and technical systems.

As a consequence, applying system thinking to orchard system allows highlighting connections with related systems and becoming conscious of environmental boundaries that have to be considered for a sustainability assessment.

Experts of system theory underlined that from one system to another there are flows (or cycles) of various forms of energy, matter and information. Looking at orchard systems, different flows of information can be seen. Probably the most important information flow is the genetic, carried with the choice of the cultivar, but from an environmental point of view a key information flow is represented by agricultural practices. Farmers constantly choose the type of orchard management among their knowledge, thus productive protocol and orchard design can be considered as expression of information flows. This information flow may play an important role in determining the environmental burdens of the orchard. For example [58] identified a significant correlation between orchard design and input management. Particularly they found that low density apple orchards required more water but less P-fertilizer per hectare.

Another example of the importance of considering the whole orchard system for sustainability is the shift to high-density plantings of dwarf rootstock trees [49] in integrated fruit production. Low canopy height decrease the use of machinery and a more open canopy helps reduce some diseases and improves pesticide coverage and efficacy.

Furthermore, apart from rare studies in the past decade (e.g. [51]), the application of system thinking to orchards is just at the beginning and new research paradigms have to be investigated. A recent study highlighted the inefficiency of plot based studies in orchards for assessing fruit quality, landscape patterns and sustainability [58]. The authors suggested that splitting orchards in plots was necessary to understand the bases of fruit production in the past decades, but the plot concept will no longer be useful to meet the future challenges, because the plot concept is unable to catch the complexity of the orchard system.

4.3 Example of application in apple production: methods

As described in the previous paragraph, orchards are complex biological productive systems. In order to obtain reliable environmental assessments in orchards, instead of considered only the one-year field operations, all the impacts related to the entire lifetime of the orchard have to be accounted. One of the direct consequences is that some resources are used annually whilst others are present during the whole lifetime of the orchard thus a multiple-stage model of the orchard has to be considered. [47] suggested a 6 stages model that has been tested and validated by [25]. The model comprehends 6 different stages during the overall fruit production (nursery through to orchard removal) and point out that all the stages contribute to the environmental burden of the system even if just the main orchard stages actually produce fruit to the market.

Therefore, in this EFA application, we distinguish and evaluate the impacts of each of the following stages:

Stage 1 (ST1). Nursery stage (accounted for 2 years). This stage was evaluated as the average processes and resources needed to obtain rootstocks, scions and finally young plants.

Stage 2 (ST2). The establishment stage (occurs just one time). This stage was evaluated as the common practice of removing previous installation and preparing the field for the orchard. Plastic, steel, wood resources and energy for the orchard installation have been added in proportion to the lifetime of the orchard.

Stage 3 (ST3). Low yield production due to young plants (accounted for 3 years). This stage includes all the one-year field operation (see ST4) but all impacts and resource use are proportioned to a reduced production due the youth of the plants.

Stage 4 (ST4). Full production (accounted for 13 years). This stage includes all the one-year field operation, particularly:

- tree management: this category comprises of operations aimed to improve orchard productivity, facilitate harvest and prevent disease proliferation [47].
- pest and diseases management: pesticide applications, usually by air-blast spraying several times per season
- understorey management: the management of the soil between the rows seeks to prevent competition for water or nutrients with the trees and erosion [52].
- irrigation: trees received water through drip pipe irrigation directly under the tree canopy. This system requires pumping systems that consumes electricity.
- weather damage prevention: hail prevention nets were installed, opened and closed once per season, with two field crossings by hydra-ladder.

Stage 5 (ST5). Low yield production due to declining plants (accounted for 2 years). This stage includes all the one-year field operation (see ST4) but all impacts and resource use are proportioned to a reduced due the old age of the plants.

Stage 6 (ST6). The destruction of the orchard (occurs just one time). This stage was mainly accounted for machinery and fuel.

Thus, the system boundary of the model includes production of differentiated apple farming inputs and their transport to the field, fuel and electricity use during nectarine farming, nursery, orchard installation and destruction (fig 5).

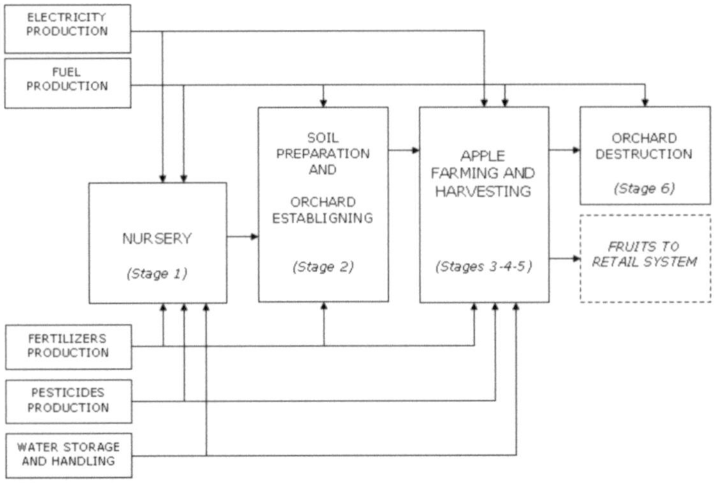

Fig. 5. System boundary and modelling of the apple production phase. Dotted box refers to processes are not included in the application.

The data collection was made in a commercial apple (Malus domestica Borkh var. Golden Deliciuos) orchard in Cuneo province, Northern Italy, managed according to the Italian Integrated Fruit Production (IFP) protocol. Impacts and resources use for all of the farming operations were obtained directly on field during years 2009-2010. All other information required (e.g. nursery impacts and resources use) were collected through questionnaire to the related growers and specialized field technicians.

Following information provided from the owner of the orchard in the field registry, the yield of each year has been collected, but instead of the real value we decided to consider an average yield as well an average resource use of agricultural inputs. The average yield has been assessed as 29.48 t/ha (fig 6). Considering local pedoclimatic conditions, agrotechniques and cultivar the average yield in the low-yield years (ST3, ST5) has been modelled as 20 t/ha (fig 6).

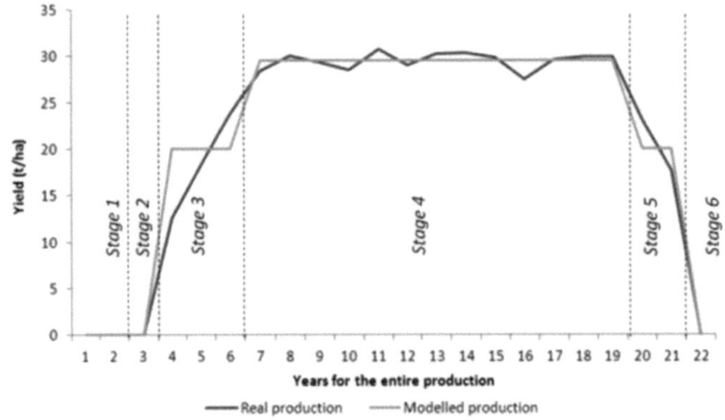

Fig. 6. Real and modelled production of the investigated apple orchard thorough its whole lifetime.

Regarding the EFA method, a product EFA was applied, with the following formula:

$$EF_{product}\left(\frac{gha}{t}\right) = \frac{(\sum_{st1-st6} EF_{direct}) + (\sum_{st1-st6} EF_{CO_2})}{\sum_{st1-st2} Yield} \quad (3)$$

in which the sum of the gha required for each production stage (counted once for each year of incidence in the production) gives the total land required and the sum of the yield of each stage for each year of incidence is the total yield of the orchard in its lifetime. Total land required (gha) on total yield gives the footprint of 1 t of nectarine produced.

As highlighted in previous paragraphs, the use of different functional units may be useful to compare different production from several points of view. In this application just one production is considered and the use of different functional units is necessary, nevertheless, a further EF method was applied in order to highlight some potentialities of the evaluation. The second functional unit used was a land-based unit (1 hectare of orchard) applying this formula:

$$EF_{land}\left(\frac{gha}{gha_{farm}}\right) = \frac{(\sum_{st1-st6} EF_{direct}) + (\sum_{st1-st6} EF_{CO_2})}{Area_{farm} \times EQF_{cropland}} \quad (4)$$

in which the whole EF of the productive system is divided by the biological carrying capacity of the farm, given by the multiplication of the real hectares of the farm by the Equivalence Factor of cropland. Using this method all the environmental impacts are related to the use of 1 ha of orchard and it will be possible to evaluate the extra land needed per each ha of the farm needed to maintain the production.

4.4 Example of application in apple production: Results and discussion

The $EF_{product}$ of the investigated system was 1.59 gha/t apple produced. The footprint land-components were distributed as follows: cropland 7.5%, built-up land 0.01%, forest 0.18%, energy land 92.55%. The built-up land and the forest land percentages are very low (less than 0.2%), and they are confirmed as irrelevant in the EFA of fruit, at least in the production stages. Indeed further application of the method to the whole fruit commercial system (production, transformation, packaging and distribution) is required to extend this remark to the whole supply-chain.

Considering the incidence of the direct and indirect land use in the whole production system EFdirect accounts for 11.27% and EF_{CO2} accounts for 88.73%. The latter indicator states the indirect land use as a proxy of the energy applied in the production system. This energy is applied in various forms: not only electricity and diesel, but also the embodied energy in chemical material (e.g. fertilizers and pesticides) and all the other resources. Without using such additional energy the productive system would have a lower yield (e.g. product lost from pest attacks or from lower fertilization). Thus the EF_{CO2} can be considered as an indicator of the energy intensity production protocol applied to the system in order to amplify the productivity.

Specific incidence of the EF_{direct} and the EF_{CO2} per each stage to the whole production system is presented in fig 7, in terms of total gha of that stage divided by the total tonnage of apple produced from the orchard across all years.

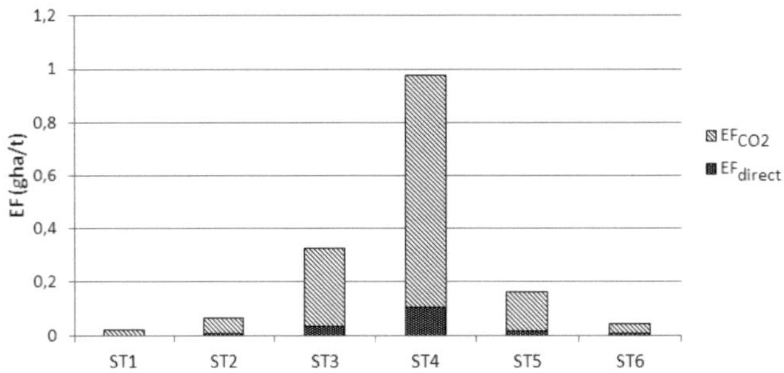

Fig. 7. Comparison of the direct and indirect contribution in the EF results accounted by the mass based functional unit: tonnes of apple produced at the farm gate.

Among the stages involved in apple production, ST4 (operations and resources for production high yield years) has, as expected, the highest footprint value: 61.17% of the overall footprint. The other stages make substantially lower contributions to the overall impact, specifically: ST1=1.34%, ST2=4.02%, ST3=20.52%, ST5=10.26%, ST6=2.68%.

This study confirms that the gaps suggested by other authors [47] and evaluated in previous works [25] can be significant. In the case study, the ST4 (high yield field operations and resources use) contributes to almost 62% of the footprint of the whole system; therefore the other stages in total contribute to about 38% of the footprint. The impact from the stages that occur just one time (ST1, ST2, ST6) have a relative small contribution to the total EF (about 8%), nevertheless, we have to consider that the impact of such stages is spread over the lifetime of the orchard, thus in absolute term is actually relevant.

These results suggest that applying the EFA only to the high yield production, as presently proposed (e.g. [26] [30]) will probably underestimating the real footprint in some situations, depending on the production protocol. More studies are required to verify the average gap for each fruit species; when these data are available, consideration of all stages in the application of EFA (and other environmental or sustainability indicators) is strongly advised.

Further interesting remark may arise from the use of the second method: the EF_{land} of the entire production system was 13.61 gha/gha$_{farm}$, and the proportion of the footprint land-component was the same as the EF obtained with the mass based functional unit. Assuming that all land needed for the production of agricultural inputs may be accounted as cropland, it is possible to transform the EF_{land} to real hectares, using the EQF for cropland given by [11]. The result is 5.15 ha/ha$_{farm}$. This value means that for each real hectare of the farm, 5.15 hectares are needed (directly and indirectly) to support the production. Thus it is possible to assess that the specific production protocol weight globally on the ecosystem services for 5 times the land in which it is applied.

5. EFA for education

Beside the use of the scientific results obtained with EFA for improving agro-techniques and developing policies, one of the most important aspects of this tool is the educational outreach. EFA results are easy to be understood and relatively easy to be translated in lifestyle patterns.

Recently [59] conducted a study in order to assess whether tracking the EF of a family in its daily life consumption pattern may encourage a more sustainable behaviour at the household level. Their research started from the statement that inability to measurably personalize the link between global unsustainable consumption and individual lifestyles is a main factor of stopping the changing towards a more sustainable. Thus they assessed the EF of 18 families in the UK, throughout a number of questionnaires and EF results were given back to the family members in order to increase their environmental awareness of their own lifestyle. Then, intended and actual changes in EF, as a reflex of changes in consumption patterns, were evaluated. Authors assessed a significant reduction of the household EF once families become aware of their own ecosystem burdens, thus they stated that EFA can be used successfully to inspire reductions in the environmental impacts of individuals.

The EF tool, in various forms, can also be applied in schools; indeed the WWF Switzerland describes several didactic units with a simplified EFA application [60]. In these units students are brought to think about their consumption pattern in terms of ecological burden of their behaviour. For this type of application a precise quantification of the EF of items or services used is not really necessary, what is very important is to highlight clearly that some consumption behaviours required more ecosystem services that others. E.g. in the EF board game [60], eating apples costs 1 footprint-point and eating meat everyday costs 6 footprint-points; probably more detail on the evaluation of such points should be given, but the important thing is that students are able to understand the difference in the environmental burden. Another interesting case study of EFA in school is the research performed by [61]. Authors developed a didactic research letting students apply EFA to their own school in Israel during years 2008-2009. The educational program consisted of two components: (1) a theoretical, interdisciplinary component based on content material from the fields of the social sciences and ecology and (2) a practical part that included the calculation of the EF of the school and exploring ways to reduce it. Looking at the results of their study, food and electricity consumption are the major components of the school's EF and transportation is the smallest component because most students either walk to school or use public transportation. From the educational point of view, authors highlight that the EFA can be an important educational tool in order to help students understand the linkage between behavioural choices and their impact on the ecological systems. Furthermore, EFA also enables students to think critically about the choices they make and the environmental consequences of those choices and to take the opportunities and responsibilities they have as members of a larger community for active participation and collaboration in moving toward sustainability [61].

Another EFA's ability is the efficient communication of the notion that any society depends on ecosystem services that might be beyond its local bio-capacity and that natural goods and services needed to support human systems might be located virtually everywhere in the World. While part of any population's ecological footprint is placed on domestic sources, uneven global distribution of population and natural resources combined with economic growth, processes of economic integration (i.e., globalization) and increasing international trade imply that the ecological footprint of any society is increasingly placed on ecosystems beyond its own boundaries [28]. Thus the EFA helps to act locally, in a specific system, but also to think the environment on a global scale. In order to highlight the dependence of human system on ecosystem services on the global scale several metaphors and tools have been developed. One of the most significant tools is the Earth Overshot Day, developed by the GFN. This calculation allows to identify the day on which the global EF (measured in global hectares) is equal to the Earth carrying capacity (also measured in global hectares) that nature can regenerate in that year (for a detailed description of the method see the [62]). For the rest of the year, we are accumulating debt by depleting our natural capital and letting waste accumulate. Thus the identification of the day of the overshoot of ecosystem services helps conceptualize the degree to which we are over-budget in our use of nature, but also it allows to visualize the size of the gap between a sustainable level of ecological demand and how much is currently required to support human activities globally.

6. Conclusion

As can be easily seen thorough proposed literature and researches, sustainability of agricultural systems is becoming more important and sustainability assessment methods continue to evolve. Indeed sustainability is considered one of the most pressing and urgent challenges that agriculture is facing today. The need for research and education to meet this challenge has been identified in almost every recent study on agricultural research needs. More specifically, it is often outlined that nowadays we need a new agriculture that more nearly mimics the structure and function of natural ecosystems (e.g. [63]).

In order to assess the sustainability of an agricultural system in a quantitative manner, indicators which are consistent with the criteria for sustainability are required. There are several studies of direct and indirect environmental assessed methods in food production, in which a multitude of assessment methods has been applied to a large variety of systems, nevertheless just with the use of few indicators the relation to sustainability can be assessed.

The most common methods applied for evaluating environmental impacts of agriculture may be different concerning various aspects, for example: the global objective, the set of environmental issues considered the definition of the system boundaries, the considered functional unit and the calculation algorithms. Therefore, standardization of research methods and protocols when applying environmental assessment methods in food production is needed, otherwise result may be impossible to compare from system to system. Being able to compare the results from different studies would be important also in order to identify sustainability threshold, as suggested by several authors (e.g. [64]).

As a general result of the research and taking into account that any discussion of sustainable agriculture depends on the context [49], EFA can easily discriminate the environmental burdens of each component of the system of the specific case study. Therefore, in food production, the EFA could help to improve one of the three aspects of sustainability (ecology, economic and social). However, although the ecological footprint is an indicator easy to understand, it is not an indicator easily applied by non-experts; thus the application of EFA directly by field technicians should only be considered when utilizing a pre-constructed and standardized sheet of calculation.

Thus, like various authors (e.g. [24]), we recommend that environmental evaluation methods should be used with great caution, considering which method is most appropriate given the specific study case.

7. References

[1] Millennium Ecosystem Assessment (MA), 2005 Synthesis, Island Press, Washington DC (2005) http://www.maweb.org (last access 22.11.2010).

[2] D.H. Meadows, D.L. Meadows, W.W. Behrens, J. Randers, E.K.O. Zahn, *Limits to Growth,* Universe Books, New York, 1972.

[3] WCED (World Commission on Environment and Development): *Our Common Future* ("Brundtland-Report"), Oxford University Press, 1987.

[4] J.P. Reganold, P.K. Glover, P.K. Andrews, H.R. Hinman, "Sustainability of three apple production systems". *Nature,* vol 410, pp. 926–929, 2001.

[5] V.H. Dale, S. Polasky, "Measures of the effects of agricultural practices on ecosystem services" *Ecological Economics,* vol. 64, pp. 286–296, 2007.

[6] W. Zhang, T.H. Ricketts, C. Kremen, K. Carneyd, S.M. Swintona, "Ecosystem services and dis-services to agriculture" *Ecological Economics,* vol. 64, pp.253-260, 2007.

[7] J. Ferng, "Local sustainable yield and embodied resources in ecological footprint analysis—a case study on the required paddy field in Taiwan" *Ecological Economics* vol.53, pp.415- 430, 2005.

[8] Z. Mózner, A. Tabi, M. Csutora, "Integrating the environmental impacts of agriculture into the Ecological Footprint methodology" in *9th International Conference of the European Society for Ecological Economics,* Boğaziçi University, Istanbul 2011.

[9] D. Tilman, K.G. Cassman, P.A. Matson, R. Naylor, S. Polasky, "Agricultural sustainability and intensive production practices", *Nature,* vol. 418, pp. 671-677, 2002.

[10] M. Wackernagel, W. Rees, *Our ecological footprint: reducing human impact on the earth.* New Society Publishers, Gabriola Island, BC, 1996.

[11] B. Ewing, W. Reed, A. Galli, J. Kitzes, M. Wackernagel, *Calculation Methodology for the National Footprint Accounts, 2010 Edition,* Oakland: Global Footprint Network, 2010.

[12] W.E. Rees, "Ecological footprints and appropriated carrying capacity: what urban economics leaves out" *Environment and Urbanization,* vol. 4, pp.121–130, 1992.

[13] M.A.J. Huijbregts, S. Hellweg, R. Frischknecht, K. Hungerbuhler, A.J. Hendriks, "Ecological footprint accounting in the life cycle assessment of products" *Ecological Economics,* vol. 64, pp. 798-807, 2008.

[14] C. Monfreda, M. Wackernagel, D. Deumling, "Establishing national natural capital accounts based on detailed Ecological Footprint and biological capacity assessments" *Land Use Policy,* vol. 21, pp. 231-246, 2004.

[15] Risk & Policy Analysts Ltd., *A review of recent developments in, and the practical use of, ecological footprint methodologies: A report to the Department for Environmental Food and Rural Affairs,* Defra, London, 2007.

[16] L. Lewan, C. Simmons, *L'impronta ecologica e l'analisi della biocapacità come indicatori di sostenibilità per le aree geografiche subnazionali,* Rapporto Finale Progetto Indicatori Comuni Europei EUROCITIES, 2001.

[17] M. Bagliani, A. Galli, V. Niccolucci, N. Marchettini "Ecological footprint analysis applied to a sub-national area: The case of the Province of Siena (Italy)", *Journal of environmental management,* vol. 86, pp. 354-364, 2008.

[18] M. Lenzen, S. Lundie, G. Bransgrove, L. Charet, R.F. Sack, *A novel ecological footprint and an example application.* ISA research paper 02-02, University of Sidney, 2002 http://www.isa.org.usyd.edu.au/publications/documents/SydneyWaterFootprint.pdf (accessed 20.07.11).

[19] M. Bagliani, F. Martini, "A joint implementation of ecological footprint methodology and cost accounting techniques for measuring environmental pressures at the company level", *Ecological Indicator,* doi:10.1016/j.ecolind.2011.09.001

[20] E.A.M. Limnios, A. Ghadouani, S.G. Schilizzi, T. Mazzarol, "Giving the consumer the choice: A methodology for Product Ecological Footprint calculation" *Ecological Economics*, vol. 68, pp. 2525–2534, 2009.

[21] J. Kitzes, M. Wackernagel, "Answers to common questions in Ecological Footprint accounting", *Ecological Indicators,* vol. 9, pp. 812-817, 2009.

[22] J. Kitzes, A. Galli, M. Bagliani, J. Barrett, G. Dige, S. Ede, K. Erb, S. Giljum, H. Haberl, C. Hails, S. Jungwirth, M. Lenzen, K. Lewis, J. Loh, N. Marchettini, H. Messinger, K. Milne, R. Moles, C. Monfreda, D. Moran, K. Nakano, A. Pyhälä, W. Rees, C. Simmons, M. Wackernagel, Y. Wada, C. Walsh, T. Wiedmann, "A Research Agenda for Improving National Ecological Footprint Accounts", in *the International Ecological Footprint Conference – Stepping up the Pace: New Developments in Ecological Footprint Methodology, Policy & Practice,* Cardiff, Wales, United Kingdom, 2007.

[23] M.A. Thomassen, I.J.M. de Boer, "Evaluation of indicator to assess the environmental impact of dairy production systems" *Agriculture, Ecosystems and Environment,* vol. 111, pp. 185–199, 2005.

[24] H.M.G. Van der Werf, J. Tzilivakis, K. Lewis, C. Basset-Mens, "Environmental impacts of farm scenarios according to five assessment methods", *Agriculture, Ecosystems and Environment,* vol. 118, pp. 327–338, 2007.

[25] A.K. Cerutti, M. Bagliani, G.L. Beccaro, G. Bounous, "Application of Ecological Footprint Analysis on nectarine production: methodological issues and results from a case study in Italy" *Journal of Cleaner Production,* vol. 18, pp. 771-776, 2010.

[26] V. Niccolucci, A. Galli, J. Kitzes, M.R. Pulselli, S. Borsa, N. Marchettini, "Ecological footprint analysis applied to the production of two Italian wines" *Agriculture, Ecosystems and Environment,* vol. 128, pp. 162–16, 2008.

[27] N. Chambers, C. Simmons, M. Wackernagel, *Sharing nature's interest: ecological footprint as an indicator of sustainability.* Earthscan Publications Ltd., London and Sterling, VA, 2000.

[28] M. Kissinger, D. Gottieb, "Place oriented ecologial footprint analysis - The case of Israely grain supply" *Ecological Economics,* vol. 69, pp.1639-1645, 2010.

[29] L. Milà i Canals, A. Azapagic, G. Doka, D. Jefferies, H. King, C. Mutel, T. Nemecek, A. Roches, S.Sim, H. Stichnothe, G. Thoma, A. Williams, "Approaches for Addressing Life Cycle Assessment Data Gaps for Bio-based Products" *Journal of Industrial Ecology,* vol. 15, pp. 707–725, 2011.

[30] M. Cuadra, J. Bjorklund, "Assessment of economic and ecological carrying capacity of agricultural crops in Nicaragua" *Ecological Indicators,* vol. 7, pp.133–149, 2007.

[31] M. Wackernagel, W. Rees "Perceptual and structural barriers to investing in natural capital: Economics from an ecological footprint perspective" *Ecological Economics,* vol. 20, pp. 3-24, 1997.

[32] L. Ruini, L. Campra, M. Marino, A. Filareto, M. Bagliani, S. Contu, "Ecological Footprint as large-scale pasta producer Key Performance Indicators (KPI). The State of the Art in Ecological Footprint Theory and Applications" in *Proceeding of the Footprint Forum* 2010, Siena (Italy), pp 133-134.

[33] A.K. Cerutti, M. Bagliani, G.L. Beccaro, C. Peano, G. Bounous "Comparison of LCA and EFA for the environmental account of fruit production systems: a case study in Northern Italy" in *Proceedings of the VII International conference of life cycle assessment in the agri-food sector*,2010, vol 2, pp. 99-104.
[34] F. Agostinho,J.R. Siche, E. Ortega, "True ecological footprints for small farms in Brazil" in *Proceedings of the International Ecological Footprint Conference*, Cardiff, Wales, UK, 2007.
[35] S. Zhao, Z. Li, W. Li "A modified method of ecological footprint calculation and its application" *Ecological Modelling*, vol. 185, pp. 65–75, 2005.
[36] T. Wiedmann, J. Barrett, "A review of the ecological footprint indicator - perceptions and methods" *Sustainability*, vol. 2, pp. 1645–1693, 2010.
[37] J. Ferng, "Using composition of land multiplier to estimate ecological footprints associated with production activity" *Ecological Economics*, vol. 37, pp. 159-172, 2001.
[38] M. Bagliani, M. Carechino, F. Martini, *La contabilità ambientale applicata alla produzione zootecnica. L'Impronta Ecologica dell'allevamento di bovini di razza piemontese.* Collana Ambiente, Regione Piemonte, Torino, 2009.
[39] D. Deumling, M. Wackernagel, C. Monfreda, *Eating up the Earth: How Sustainable Food Systems Shrink our Ecological Footprint.* Redefining Progress, 2003, Available at: http://www.rprogress.org (27.03.09).
[40] S. Frey, J. Barrett, *The Footprint of Scotland's diet: the environmental burden of what we eat. A report for Scotland's Global Footprint Project.* Stockholm Environment Institute, 2006.
[41] P. W. Gerbens-Leenes, S. Nonhebel, "Consumption patterns and their effects on land required for food", *Ecological Economics*, vol. 42, pp. 185–199, 2002.
[42] Barilla CFN *The double pyramid: Healthy food for people, sustainable for the planet.* Supporting technical paper, 2011, http://www.barillacfn.com/ (15.09.2011).
[43] Y. Wada, *The appropriated carrying capacity of tomato production: comparing the ecological footprints of hydroponic greenhouse and mechanized field operations,* Ph.D. Thesis. University of British Columbia, 1993.
[44] A.K. Cerutti, M. Bagliani, G.L. Beccaro, F. Gioelli, P. Balsari, G. Bounous, "Evaluation of the sustainability of swine manure fertilization in orchard through Ecological Footprint Analysis: results from a case study in Italy", *Journal of Cleaner Production,* vol. 19, pp. 318-324, 2011.
[45] F. de Alvarenga, V.P. da Silva, R. Soares, "Comparison of the Ecological Footprint and a life cycle impact assessment method for a case study on Brazilian broiler feed production" *Journal of Cleaner Production*, doi: 10.1016/j.jclepro.2011.06.023
[46] P. Mouron, T. Nemecek, R.W. Scholz, O. Weber, "Management influence on environmental impacts in a apple production system on Swiss fruit farms: combining life cycle assessment with risk assessment" *Agriculture, ecosystems and environment*, vol. 114, pp. 311-322, 2006.
[47] L. Milà i Canals, G. Clemente Polo, "Life cycle assessment of fruit production" in: Mattsson B., Sonesson U., *Environmentallyfriendly food processing*. Woodhead Publishing Limited and CRC Press LLC, Cambridge and Boca Raton, 2003, pp. 29–53.
[48] A. Carlsson-Kanyama, M.P. Ekstrom, H. Shanahan, "Food and life cycle energy inputs: consequences of diet and ways to increase efficiency" *Ecological economics*, vol. 44, pp. 293-307, 2003.

[49] D. Granatstein, E. Kupferman, "Sustainable horticulture in fruit production", *Acta Horticulturae (ISHS)*, vol. 767, pp. 295-308, 2006.

[50] S. Simon, J. Bouvier, J. Debras, B. Sauphanor, "Biodiversity and pest management in orchard systems. A review", *Agronomy for sustainable development*, vol. 30, pp. 139–152, 2009.

[51] S.R. Gliessman, *Agroecology. The ecology of sustainable food system*, CRC press, Taylor and Francis Group, 2006.

[52] L. Mila i Canals, G.M. Burnip, S.J. Cowell, "Evaluation of the environmental impacts of apple production using Life Cycle Assessment (LCA): Case study in New Zealand" *Agriculture, Ecosystems and Environment*, vol. 114, pp. 226-238, 2006.

[53] L. Von Bertalanffy, *General System Theory: Foundations, Development, Applications*, Eds George Braziller Inc, New York, 1976.

[54] C. Spedding, *An introduction to agricultural systems*, London Elsevier Science Publications, 1988.

[55] G. Page, *An environmentally-based systems approach to sustainability analyses of organic fruit production systems in New Zealand*. PhD Thesis. Massey University, Palmerston North, New Zealand, 2009.

[56] J.E. Ikerd, "The need for a system approach to sustainable agriculture" *Agriculture, ecosystems and environment*, vol. 46, pp. 147-160, 1993.

[57] M. Hauschild, "Spatial Differentiation in Life Cycle Impact Assessment: A decade of method development to increase the environmental realism of LCIA" *The International Journal of Life Cycle Assessment*, vol. 11, pp.11-13, 2006.

[58] T. Nesme, F. Lescourret, S. Bellon, D. Plenet, R. Habib, "Relevance of orchard design issuing from growers' planting choices to study fruit tree cropping systems" *Agronomie*, vol. 23, pp. 651–660, 2003.

[59] M. Sutcliffe, P. Hooper, R. Howell, "Can Eco-Footprinting Analysis Be Used Successfully to Encourage More Sustainable Behaviour at the Household Level?" *Sustainable Development*, vol. 16, pp. 1–16, 2008.

[60] WWF Switzerland, didactic materials. http://www.wwf.ch/it/fare/essereattivi/formazione/scuola/materialedidattico/impronta_scuola.cfm (15.09.2011)

[61] D. Gottlieb, E. Vigoda-Gadot, A. Haim, M. Kissinger, "The ecological footprint as an educational tool for sustainability: A case study analysis in an Israeli public high school" *International Journal of Educational Development*, vol. 32, pp.193-200, 2012.

[62] Global Footprint Network, overshot day description www.footprintnetwork.org/earthovershootday/ (15.09.2011)

[63] W. Rees "The eco-footprint of agriculture: A far from (thermodynamic) equilibrium interpretation" in *Agricultural Biotechnology: Finding Common International Goals*", National Agric. Biotech. Council, Ithaca, NY 2004, pp. 87–110.

[64] H.M.G. Van der Werf, J. Petit. "Evaluation of the environmental impact of agriculture at the farm level: a comparison and analysis of 12 indicator-based methods," *Agriculture, ecosystems and environment*, vol. 93, pp. 131-145, 2002

10

Techniques for Assessment of Heavy Metal Toxicity Using *Acanthamoeba* sp, a Small, Naked and Free-Living Amoeba

Nakisah Mat Amin[1,2]
[1]*Department of Biological Sciences, Faculty of Science and Technology,*
[2]*Institute of Oceanography, Universiti Malaysia Terengganu,*
Kuala Terengganu, Terengganu
Malaysia

1. Introduction

Heavy metals are natural component of earth crust that cannot be degraded and destroyed but accumulate through food chain and produce potential human health risks and ecological disturbances (Loka et al., 1990). Other than occur naturally, their quantity in the environment increased due to anthropogenic activities (Greger, 1999). Heavy metals are toxic to most organisms at specific concentrations; and often cause serious upset both at cellular and genomic levels of cells. Various organisms have been employed as indicator of the metal contamination; they are mainly multicellular organisms such as fishes and bivalves (Shazili et al., 2004) based on their suitability to accumulate the metals from the environment. Some of these organisms have ability to balance between bioaccumulation and depuration of contaminants in the their body (Chin, 2005), therefore the measurement of the metals in these organisms becomes inaccurate. This chapter describes the use of a single cell organism, *Acanthamoeba,* a small, naked and free-living amoeba to assess the toxicity of heavy metals that are commonly present in our aquatic ecosystem.

2. Why *Acanthamoeba*?

Acanthamoeba is an amoeba genus, a type of small, naked and free-living amoebae. This amoeba genus occupies in most habitats such as in soil, air, dust and water. In their natural environment, these free-living organisms feed on bacteria and any particulate matters; some of them, however can infect brain, eyes, skin, and lungs of a variety of mammals, including humans (Cabral & Cabral, 2003).

The interest arose to investigate the toxicity effects of heavy metals using free-living amoebae is due to their ubiquitous in the environment as well as the amoeba cells are considered as the closest cell towards the human diploid cells, since both of them are eukaryotes. The advantage is free-living amoebae are single cell organisms, so any changes in the environment can directly affect the amoebae. Moreover, the trophozoites of amoebae are not covered by a cell wall so they are more sensitive to these environmental changes, hence they are suitable candidates for hazard prediction and environmental risk assessment due to heavy metal pollution. Heavy metals that were investigated in the present study were commonly occur in our environment and potentially can impose hazards to human's health such as lead, zinc, cadmium and mercury.

Acanthamoeba sp used in this study was a local species, isolated from water at Setiu Wetlands, Terengganu, Malaysia and labeled as *Acanthamoeba* sp. (SW isolate). The amoeba was maintained axenically in Polypeptone medium at 30°C. The cytotoxicity and genotoxicity potential of heavy metals on the amoeba involved determination of the amoeba viability, IC_{50} of the metals (the concentrations that could inhibit the 50% growth of *Acanthamoeba*), observation on the changes on their ultra-cellular structure, the amoeba membrane permeability, mode of cell death and DNA damage due to exposure of the amoeba to the metals.

3. Experimental approach

Assessements for the metal's toxicity on *Acanthamoeba* are summarized in two categories; the cytotoxicity and Genotoxcity studies. The cytotoxicity study involved determination of amoeba viability and IC_{50} values, observation on changes in the amoeba morphology, membrane permeability and determination of mode of cell death. In the genotocity study, the main focus was observation on the DNA damage after amoeba exposure to the metals.

3.1 Cytotoxicity study

Mercury (mercury chlorides), cadmium (cadmium dichloride), lead (lead nitrate) and zinc (zinc sulphate) were first prepared as stock solutions by dissolving appropriate amount of these metals (in powder form) in polypeptone media to give 100 ppm for mercury and cadmium, 500 ppm for zinc and 50 ppm for lead.

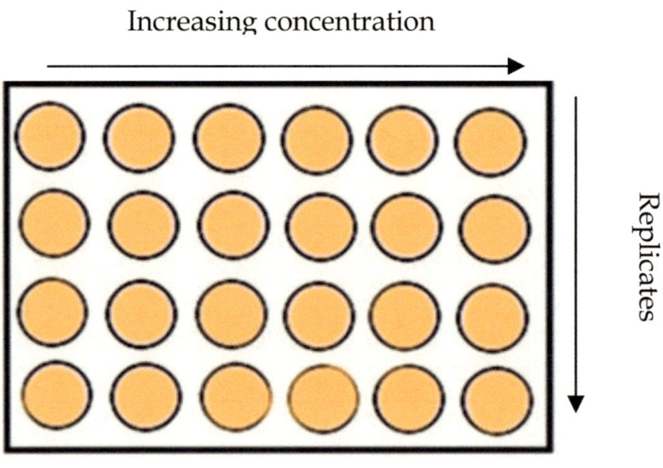

Fig. 1. Experimental design in 24-well plate for treatment of heavy metals on *Acanthamoeba*.

The cytotoxicity test on amoebae was conducted in 24-well plates. The amoebae used were taken from confluent growth cultures to ensure only healthy amoebae were used in the experiment. To get various concentration of metals, an appropriate volume of each metal stock solution was transferred into 24-wells containing 10^4 amoebae/mL following $M_1V_1 = M_2V_2$ formula. For cadmium and mercury, six different concentrations with three replicates of metal solutions (0 ppm, 4 ppm, 8 ppm, 12 ppm, 16 ppm, 20 ppm) were tested on *Acanthamoeba* sp. The final volume in each well of 24-well plates was 1000 µL. Then, the plates were sealed with parafilm and incubated for 72 h at 30°C. Similar techniques were applied for getting various concentrations of zinc and lead in toxicity test on *Acanthamoeba*. Concentrations of lead used were 0 ppm, 6 ppm, 12 ppm, 18 ppm, 24 ppm and 30 ppm whereas concentrations for zinc were 0 ppm, 30 ppm, 60 ppm, 90 ppm, 120 ppm and 180 ppm.

3.1.1 Determination of amoeba viability and IC_{50} of metals against *Acanthamoeba*

The cytotoxicity of metals on the amoebae was investigated using the eosin dye assay to evaluate the cell viability. The technique was described by Wright et al., (1988) for determination of viable cells for *Entamoeba histolytica*. Eosin is a synthetic and acidic dye attracted to the tissue elements of a basic nature such as cytoplasm of cells (Sims, 1998). The principle of this assay was based on the release of eosin dye by viable trophozoites cytoplasm after being incubated with 0.1 M NaOH solution. In this assay, the floating trophozoites were considered as dead cells and thus were discarded leaving only the viable trophozoites attached to the 24 well plate surfaces.

Briefly, after incubation, the culture medium of the amoeba cultures in 24-well plates was discarded using a micropipette. The plates were then immediately washed with sodium chloride solution (0.9%) at 37°C. This procedure was completed quickly and the plates were not allowed to cool in order to prevent the detachment of amoebae. Then, the plates were allowed to dry at room temperature and the amoebae were fixed with methanol and allowed to dry. Next, the plates were stained with aqueous eosin (0.5%) for 15 min. Stained plates were washed once with tap water and then twice with distilled water and allowed to dry again. Later 200 µL of 0.1 M sodium hydroxide solution was added into each well to dissolve the protein and release the dye. The resulting solution was later transferred into 96-well plate. The optical density of the solution in each well was determined at 490 nm using a microplate reader. The percentage of inhibition of amoeba growth was calculated from the optical density of the treated wells relative to control wells multiply by 100. A dose-response curve representing percentage of inhibition and metal concentration was plotted. The IC_{50} values of the metals against *Acanthamoeba* were derived from this curve.

3.1.2 Visualization of changes in the amoeba morphology

To visualize the morphology of amoebae in detail after treatment with heavy metals, the amoebae were harvested and processed for scanning electron microscopy following techniques by Nakisah et al. (2008).

The amoeba was cultured on a glass slide in 6 well plate containing heavy metals for 72 h at 30°C. Then, the culture was fixed with warm glutaraldehyde in PBS (2.5% Glutaraldehyde v/v) for 2 h after the removal of all medium. After that, the amoebae on the cover slide was washed 3 times for 10 min each with PBS buffer. The amoebae were later post-fixed with 1% osmium tetraoxide at room temperature for 1 h. The amoebae then were subjected to dehydration process using a graded series of ethanol from 30% 50%, 60%, 70%, 80%, 90%, 95% (for 10 min each) and 100% (twice for 10 min each). The amoebae were later critical point dried before the cover slip containing amoebae was attached to aluminum stub and coated with gold dust. The sample was then viewed under Scanning Electron Microscopy (JEOL JSM-6360LA, analytical SEM).

3.1.3 Membrane permeability study by AOPI staining

In membrane permeability study, fluorescence dyes of acridine orange (AO) and propidium iodide (PI) staining were used to examine the integrity of *Acanthamoeba* membrane after the amoebae (10^4 cells/mL) were treated with the metals (at their IC_{50} concentrations). The experiments including controls were done in 15 mL culture flasks for 72 h at 30°C. After that, the amoebae were harvested by centrifugation at 3000 rpm for 15 min. The supernatant then was discarded and the pellets from each treatment were washed once with PBS and re-centrifuged at the same speed for 5 min. The final pellets obtained were re-suspended in 100 µL AO/PI solution (The AOPI solution was prepared by adding 2 µL of acridine orange (1 mg/mL) and 2 µL propidium iodide (1 mg/mL) in 996 µL PBS). The cell suspensions were incubated for 10 min in the dark since both dyes were light sensitive. The *Acanthamoeba* cell suspension was placed onto a slide and covered with a cover slip. The slide was then viewed under fluorescence microscopy in the dark condition

3.1.4 Determination of apoptosis type of cell death by DNA laddering assay

DNA laddering assay was employed to determine the mode of cell death after the amoebae were exposed to IC_{50} concentrations of the metals. Amoebae (10^4 cells/mL) were incubated in polypeptone media at 30°C in the presence and absence of heavy metals for 72 h. Amoebae were collected by centrifugation at 3000 rpm for 15 min. The amoeba DNA was extracted using Dneasy Kit (Qiagen) according to the manufacturer's protocol. The concentration of DNA was first determined from the OD_{260} using a spectrophotometer (Beckman, USA). The integrity of DNA then was inspected by running the samples on a 1.2 % of agarose gel at 20 volts and visualized using Ethidium bromide.

$$\text{DNA concentration} = A_{260} \times 0.05\ \mu g/\mu L \times D,$$

Where A_{260} is the absorbance measured at 260 wavelength and D is the dilution factor.

3.2 Genotoxicity study by alkaline comet assay

The analysis of DNA damage in heavy metal-treated *Acanthamoeba* was performed by Alkaline Comet Assay (Collins, 2004). This assay is also known as the alkaline version of the single cell electrophoresis that is used for a wide range of applications including DNA damage and repair studies, genetic toxicology, radiation biology, and environmental monitoring (Cotelle and Ferard, 1999). In this study, the analysis of the DNA damage in *Acanthamoeba* sp was performed after 2 h treatment with heavy metals (at their IC_{25} concentrations) in 15 mL culture flasks. Lower concentration of metals is essential in comet assay in order to avoid false positive results on the DNA damage. After treatment, the alkaline Comet assay described by Collins (2004) and employed by Nakisah et al. (2009) for *Acanthamoeba* was followed.

First, the amoebae were harvested and centrifuged at 1500 rpm for 7 minutes. The supernatant was discarded and the pellet was washed with Ca^{2+} and Mg^{2+} free PBS and was centrifuged again. After that, the pellet was mixed thoroughly with 80 µL 0.7% low melting agarose (LMA) and placed above the hardened first layer of 0.6% normal melting agarose (NMA) that was prepared earlier. Following that, the cover slip was placed on the agar and the slides were left on ice to solidify the LMA for 5 min. The cover slip then was removed and the gel was again coated with 200 µL of 0.5% LMA as the third layer to prevent the nuclear DNA from escaping during the cell lysing process.

The cells were incubated in alkaline lysis buffer for one hour at 4°C. Next, the slides were placed in horizontal electrophoresis tank containing electrophoresis buffer (pH>13). The electrophoresis was carried out at 25 V/cm and 300 mA for 5 min. After that, the slides were neutralized with neutralizing buffer of Tris-HCl (pH 7.5) three times, 5 min each. The slides were then stained with 20µg/mL ethidium bromide and were left overnight at 4°C before viewed under fluorescence using 590 nm filter.

3.2.1 Comet scoring

Three slides were used for each treatment and 50 cells were counted from each slide. Comets were scored from 0-4 as described by Collins (2004) for detecting DNA damage in individual cells. To compare the significance among groups in alkaline comet assay, Kruskal Wallis test was done. A p value of <0.05 was considered statistically significant.

4. Results & discussion

Evidences on the toxic effect of the four metals (cadmium, mercury, lead and zinc) on *Acanthamoeba* sp are described and highlighted in this chapter.

4.1 Determination of amoeba viability and IC_{50}

The cytotoxic effects of the four metals against *A.polyphaga* after 72 h of treatment are shown in Fig. 2. Based on eosin dye assay, there were significant increases of inhibition of amoeba population after exposure to increased concentration of metals ($p < 0.05$). The IC_{50} values of the metals against *Acanthamoeba* sp with comparison with data for a reference amoeba (*A. polyphaga* CCAP 1501/3A) are shown in Table 1. Based on IC_{50} value obtained in the present study, *Acanthamoeba* spsp (SW) is more tolerate to these metals then *A. polyphaga* (CCAP 1501/3A). This observation indicates the former *Acanthamoeba* which was a local species, and was isolated from the environment can adapt with higher concentration of the metal presents in the environment.

Metal	*Acanthamoeba* spp	IC_{50} values(ppm)
Mercury	*A. polyphaga* (CCAP1501/3A)	2.2
	Acanthamoeba sp (SW isolate)	4.0
Cadmium	*A.polyphaga* (CCAP 1501/3A	3.6
	Acanthamoeba sp (SW isolate)	7.0
Lead	*A. polyphaga* (CCAP1501/3A)	8.6
	Acanthamoeba sp (SW isolate)	9.0
Zinc	*A. polyphaga* (CCAP1501/3A	42
	Acanthamoeba sp(SW isolate),	36

Table 1. The IC_{50} values of heavy metals against *Acanthamoeba* sp (SW isolate) in comparison with *A. polyphaga* (CCAP1501/3A).

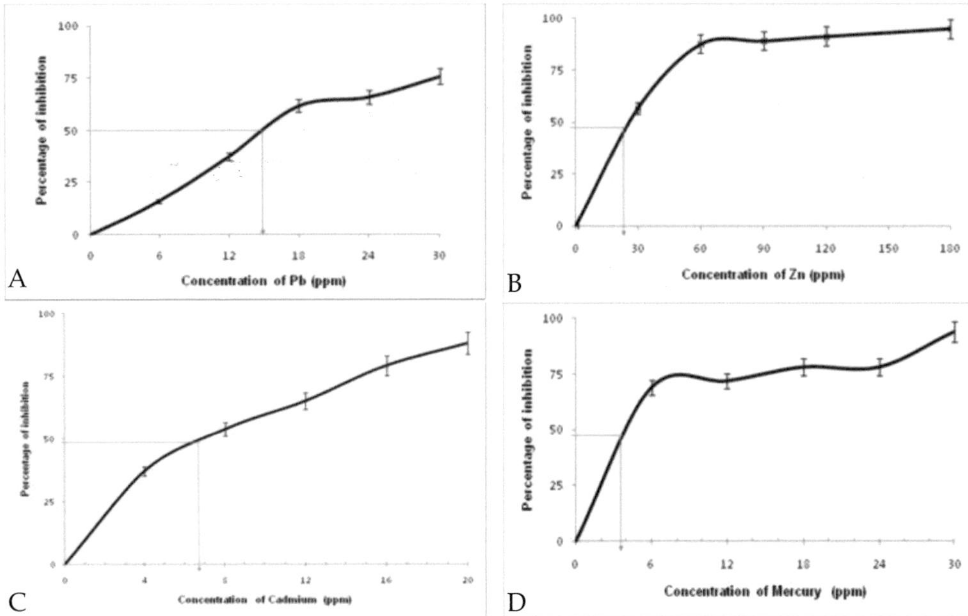

Fig. 2. Percentage of inhibition of *Acanthamoeba* sp. (SW isolate) exposed to (A) Lead, (B) zinc, (C) Cadmium and (D) Mercury. Data presented are means of three replicates. The IC_{50} values are derived from the dose-response curve for each metal.

4.2 Visualization of changes in the amoeba morphology

Details toxicity of the four metals on the amoeba can be visualised under scanning electron microscopy. Severe damage on the amoeba morphology as compared with controls is apparent (Fig. 3). Tiny projections on the surface of *Acanthamoeba* (called acanthopodia), the prominent characteristics of the genus, were almost completely lost. This structure is crucial for the amoeba to move or adhere to the substratum. Therefore, the presence of the metal ions in the amoeba media or in the environment will affect the biology and movement of this amoeba.

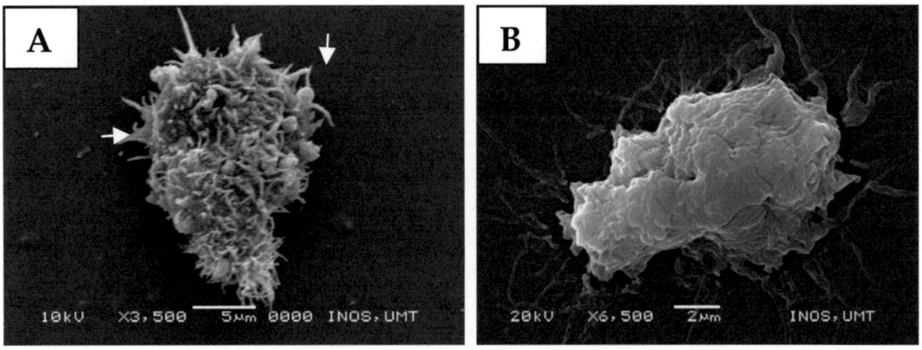

Fig. 3. Scanning electron microscopy of trophozoite of *Acanthamoeba* sp (SW isolate). (A) In control and (B) after treatment with IC_{50} of lead. Acanthopodia (arrows) are prominent on the surface of control *Acanthamoeba* but these structures are lost in lead-treated amoeba.

4.3 Membrane permeability study by AOPI staining

The integrity of *Acanthamoeba* membrane was affected after being exposed to the metals (Fig.4). Under fluorescence microscopy, the heavy metals-treated *Acanthamoeba* were stained orange or green with orange granules observed in their cells, in contrast with control *Acanthamoeba* which were stained green. Chromatin condensation observed as green to red fluorescence precipitates in the cytoplasm of treated *Acanthamoeba* resulted from disintegration of nuclear envelope followed by nuclear fragmentation. The live *Acanthamoeba* only allows acridine orange dye when at low concentration to enter the cells, but not for propidium iodide, thus they stained green. By this technique, apoptotic and necrotic cells can also be visualised.

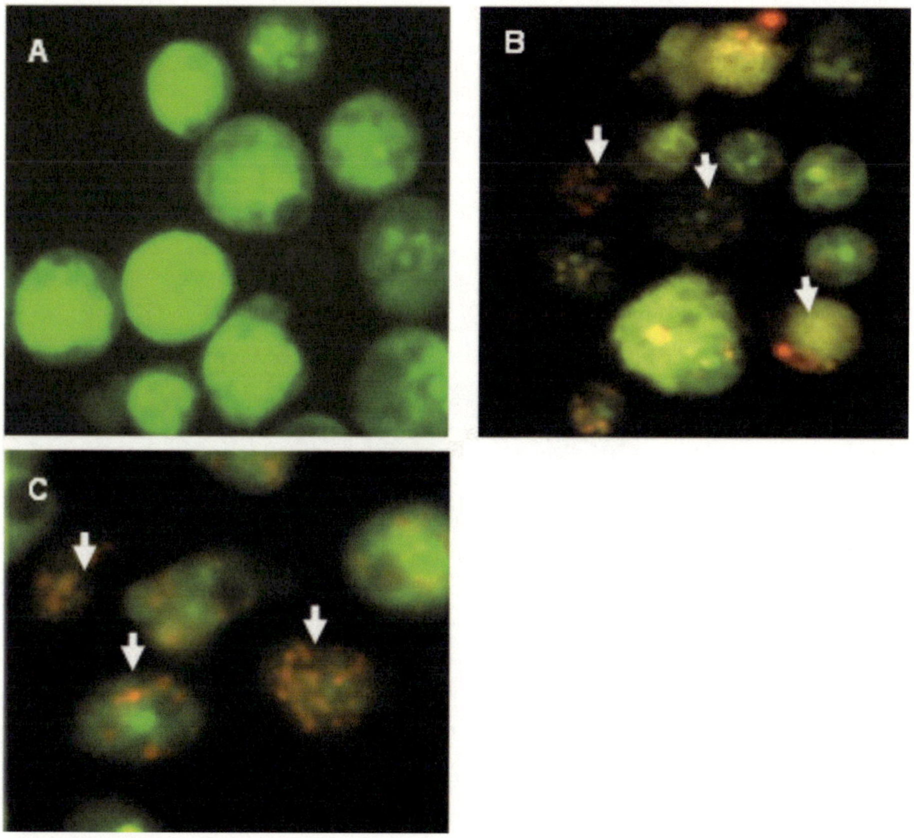

Fig. 4. Fluorescence microscopy of *Acanthamoeba* sp. (SW isolate). (A) Untreated amoebae stained with AOPI dyes display green fluorescence (B) Amoeba stained with AOPI after 72h exposure to cadmium. (C) Amoeba cells stained with AOPI after exposure to mercury. Arrowheads indicate damage on the amoeba membrane. Orange granules inside the cells in B and C are damaged lysosomes. (Magnification X400).

4.4 Determination of apoptosis type of cell death by DNA laddering assay

DNA laddering is one of the more reliable techniques presently employed to study apoptosis both *in vitro* and *in vivo*. Results showed that all metals cause inter-nucleosomal

DNA cleavage in *Acanthamoeba* as shown in the DNA laddering assay to suggest apoptotic type of cell death was induced in this amoeba when exposed to the metals (Fig 5). In contrast, there was no evidence of internucleosomal cleavage or ladder pattern observed from control amoeba. Interestingly, there are two DNA fragments were observed in amoeba treated with Pb and Zn but only one fragment observed in Hg and Cd-treated amoeba. Different mechanism probably involved for Pb and Zn while inducing apoptosis in this amoeba which deserves further analysis and study. Apoptosis occurs in cells injured by certain levels of toxic agent. It is also crucial process for eliminating cancer cells (Guchelaar et al., 1997). From this study, IC_{50} value for the metals used was a sufficient amount to create a toxic environment to the amoeba cells. Morphological estimation for apoptosis is based on cell characteristics such as chromatin condensation, formation of apoptotic bodies from one cell (each having a fragmented piece of nucleus surrounded by a viable cell membrane), shrinkage of cytoplasm, and blabbing of plasma membrane with an irregular outline (Searler et al., 1982).

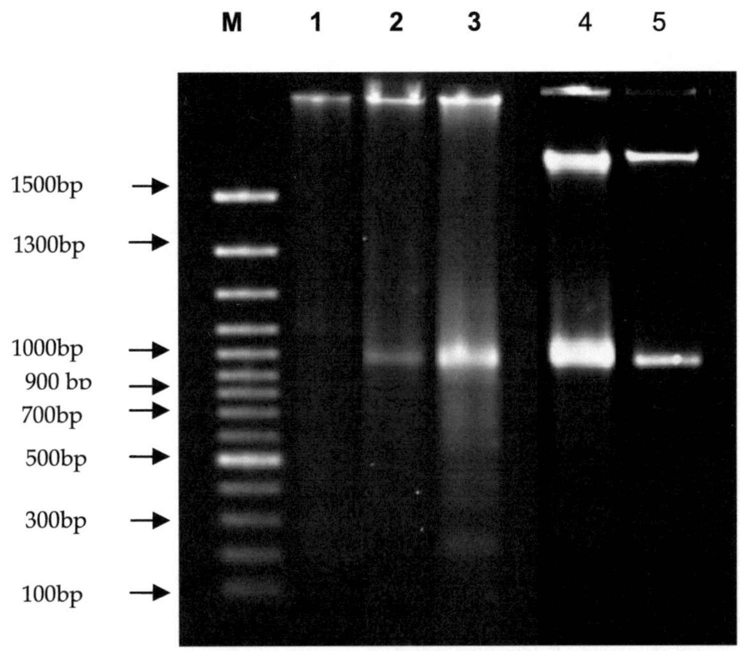

Fig. 5. Agarose gel electrophoresis of genomic DNA of *Acanthamoeba* sp (SW isolate). Explanation for lanes: M-marker, 1-*Acanthamoeba* (control) shows no ladder pattern, 2-amoeba exposed to IC_{50} of cadmium, 3-amoeba exposed to IC_{50} of mercury, 4-amoeba exposed to IC_{50} of lead and 5- amoeba exposed to IC_{50} of zinc. DNA fragmentation at 1000 bp (white arrows) is observed in all metal-treated amoebae, and fragmentation at >1500 bp is observed only in Pb and Zinc-treated amoebae (arrowheads).

DNA fragmentation observed was also reported in other study for humans exposed to lead (Danadevi et al., 2003). Such evidence may be due to a direct effect of lead on the DNA structure, oxidative mechanisms (Stohs & Bagchi, 1995) or indirectly due to another mechanism involving the activation of caspases in the process of cell death (Saleh et al., 2003). Although most reports on apoptosis are for multicellular organisms, apoptosis to occur in unicellular organisms is also possible (Gordeeva et al., 2004).

Zinc has been reported to prevent apoptosis, an effect assumed to be due to the ability of the metal ion to inhibit a Ca^{2+}/Mg^{2+}-dependent endonuclease (Cohen & Duke, 1984; Waring et al., 1990). However, it is also reported that zinc failed to inhibit the formation of apoptotic cells (Cohen et al., 1992). At higher concentration of zinc, when methallothionien is induced, the level of oxidant decreases and therefore causes oxidative damage to DNA by some unknown mechanism.

4.5 Genotoxicity of metals on the amoeba genome

The toxicity of the metals used in this study not only affects the amoebae at the cellular level, but also affect the amoeba's DNA. The scoring of DNA damage was done manually based on the comet length and intensity of the tail and categorizing cells into five classes based on extent of migration of DNA with strands breakage during electrophoresis (Collins, 2004). In this way, the percentage of DNA damage in the comet tail and their score of the tail can be determined. DNA damage was categorized into five classes (as score 0, 1, 2, 3 and 4) during electrophoresis (Fig. 6). The damage in the DNA with scores 4 and 3 are considered very severe and irreversible, compared with scores 2, 1 and 0 (intact DNA). The data obtained were graphically expressed in histograms by plotting frequency of comets and the corresponding DNA damage measurement for each metal.

Assessments of genotoxic effect of mercury, cadmium, zinc and lead on *Acanthamoeba* are illustrated in Figs. 7 and 8. The DNA damage with scores 4 and 3 were only observed in mercury and cadmium-treated *Acanthamoeba* to indicate the degree of toxicity of the two metals imposed on this amoeba. On the other hand, in lead and zinc-treated amoebae, only the DNA damage at score 1 and 2 were observed.

DNA damage of the amoeba cells is due to inadequacy in tolerance capacity of the phytochelatin to arrest the metal ions and forming extensive double-strand breaks. Several investigations have concluded that based on the characteristic appearance of the comets, apoptotic cells can be distinguished from necrotic cells in the alkaline comet assay (Olive et al., 1993: Fairbairn et al., 1996: Kizilian et al., 1999). Apoptotic cells were concluded to form comets with large fan-like tails and small heads while necrotic cells were concluded to form comets with relatively large heads and narrow tails of varying length.

Alkaline comet assay DNA electrophoresis is sensitive in detecting DNA single strand breaks and alkali-labile damage in individual cells (Gedik et al., 1992; Singh et al., 1988). In addition, the visual classification system should be regarded as a reliable way of measuring DNA migration in the comet assay because it is a common practice that only one investigator analyze the samples within one investigation (Forchhammer, 2008). As comet assay is a very sensitive method for detection of DNA damage, a low dose level would be expected to give an acceptable effect compared to high dose that might produce false cytotoxic effect and to avoid only comet with only score 4 found (Hartmann et al., 2003).

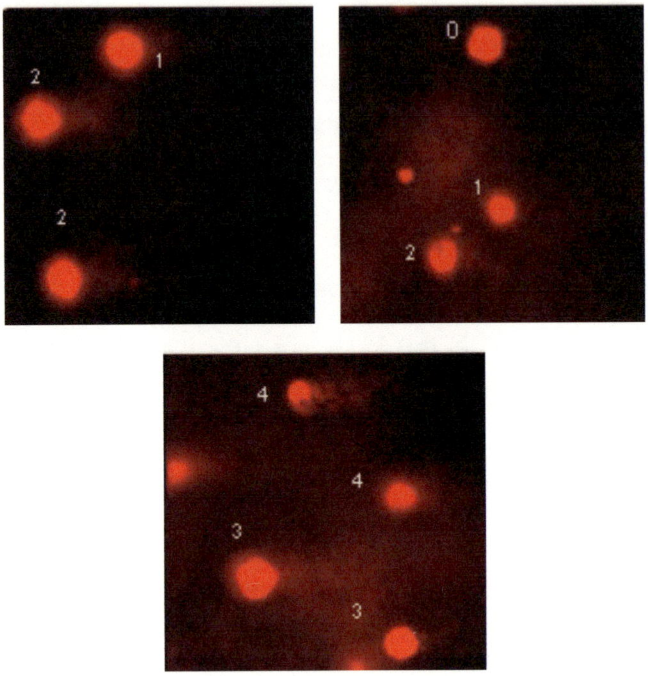

Fig. 6. Cells of *Acanthamoeba* sp. with different scores of the DNA damage following Collins's descriptions after 2 h treatment with the metals. Two hundred of cells on a slide were counted and data obtained are presented in Fig. 7 and Fig. 8.

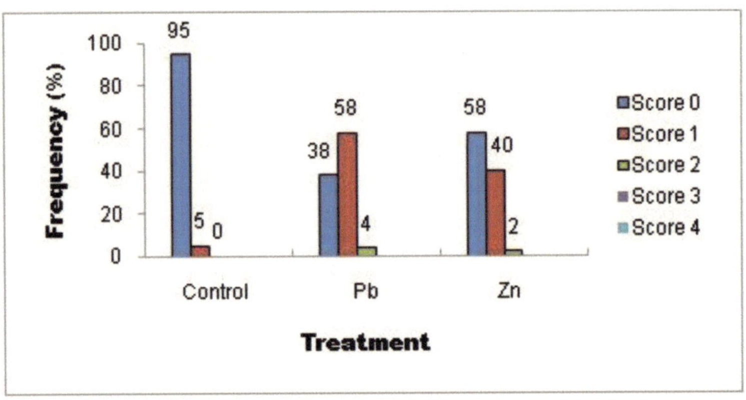

Fig. 7. Frequency of DNA damage scoring of the *Acanthamoeba* sp. (SW isolate) after treated with IC_{25} of lead and zinc. Each data are significantly different among group ($p < 0.05$)(Kruskal-Wallis test).

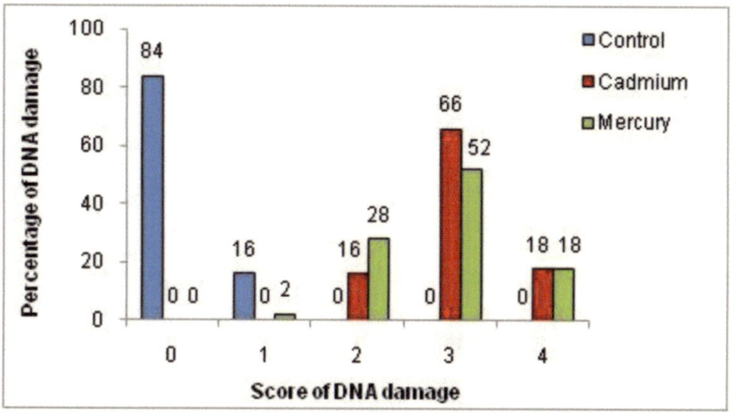

Fig. 8. DNA damage scoring following 2 h treatment with IC_{25} of cadmium and mercury on *Acanthamoeba* sp. (SW isolate) assessed by alkaline comet assay. Each data are significantly different among group ($p < 0.05$)(Kruskal-Wallis test).

In the present study, cadmium is observed to be the most toxic to metals to the amoeba followed by mercury, lead and zinc. The results obtained in a part reflect the chemical and physical properties of the elements and their positions in the Periodic Table. The increase in atomic number (number of protons) from the upper right corner of the periodic table to the lower left corner is reflected in the formulas of the oxygen acids of the elements in their highest states of oxidation. The smallest atoms group only three oxygen atoms about themselves while the larger atoms coordinate a tetrahedron of four oxygen atoms (Holden, 2001).

This study was aimed to investigate the effect on the excess of the metals in the environment to living organisms such as amoeba. Maximum permissible concentration in the environment for all metals used in the present study is shown in Table 2.

Heavy metal	Max conc. in air (mg/m3)	Max. conc. in sludge (soil) (mg/ Kg or ppm)	Max. conc. in drinking water (mg/l)	Max conc. in H_20 supporting aquatic life (mg/l or ppm)
Cd	0.1-0.2	85	0.005	0.008
Pb	-	420	0.01	0.0058
Zn	1.5	7500	5.0	0.0766
Hg	-	< 1	0.002	0.05
Ca	5.00	Tolerable	50	Tolerable >50
Ag	0.01	-	0.0	0.1

Table 2. United State Environmental Protection Agency (USEPA) maximum contamination levels for heavy metal concentration in air, soil and water with World Health Organization (WHO) guidelines (Duruibe *et al.*, 2007).

Heavy metal cations enter the amoeba cells thus gave a physiological or toxic effect to the amoebae therefore affect the growth of the amoeba population. The presence of heavy metals in high concentration in the culture medium of amoebae in this study creates unfavorable or harsh condition to the amoebae to live, therefore promotes the conversion of amoeba trophozoites into cysts stage. Cysts are non-motile inactive stage in the life cycle of protozoa. There is no any form of reproduction occurs in cyst stage of *Acanthamoeba* and as a result no growth or increasing in number of cells in the population was observed.

In higher organisms, lead toxicity can affect their organ system as well as at molecular level. Proposed mechanisms for toxicity involve fundamental biochemical processes. These include lead's ability to inhibit or mimic the actions of calcium which can affect calcium-dependent or related processed to interact with proteins including sulfhydryl, amine, phosphate and carboxyl groups which result in denaturation and inactivation of enzymes and disruption of cell organelle membrane integrity and cell division (Ochiai, 1987). Similar disturbance might occur to the amoebae when exposed to the four metals used but such study was not conducted in the present study.

Heavy metals are known to disturb biochemistry process and also inactivated the enzyme systems in humans (Wiley et al., 2008), as well as in free living amoebae. Heavy metals ions combine with proteins, often with the sulfhydryl groups and inactivated the protein. They may also precipitate the cell proteins. Heavy metals such as cadmium and mercury ions being more hydrophobic than amoebae cell membranes, thus they can cross the cell membranes easily (Wiley et al., 2008). In eukaryote cell membrane, the major lipids are phosphoglycerides, sphingolipids and cholesterols. These lipids appear to participate in a variety of cellular processes, including allowing transportation through the membrane.

Zinc plays a vital role in the biosynthesis of nucleic acids, RNA polymerase and DNA polymerase thus involved in the healing processes of tissues in mammals. However, zinc has been shown to exert adverse reproductive, biochemical, physiological and behavioral effects on a variety aquatic organism (Sirover & Loeb, 1976). The toxicity of zinc to such organisms is influenced by many factors such as the temperature, hardness and pH of the water. Previous study reported that zinc inhibited the activity of DNA polymerase-1 activity in *E. coli* (Miyaki et al., 1977).

All experiments in this study were done using the trophozoites, active stage of amoebae. Trophozoites stage is more sensitive to heavy metals treatments due to absence of cell wall (Diaz et al., 2006). Since amoebae are eukaryotic cells, therefore in many aspects, their morphology and physiology are similar to the cells of multicellular animals, including human. Hence, the cytotoxicity effects of cadmium, mercury, lead and zinc on amoebae can be used for estimating the cytotoxic effects for humans as well.

5. Conclusion

Toxicity of heavy metals (such as zinc, lead, cadmium and mercury) in the environment can be assessed using *Acanthamoeba* sp, a type of free-living amoebae. Morphological observation, cell viability, IC_{50} values, type of cell death and the DNA damage in *Acanthamoeba* are tools to reveal the hazard properties of the metals to living organisms.

6. Acknowledgment

The author would like to express gratitude to her students, Dr Fatimah Hashim, Ms Norhasimah Hassan, Ms Siti Nur Diana Abu Bakar, and Mr Manesh Kumar s/o Niithinathan for their contributions in this work and to Ms Nuraini Nawi for her technical assistance.

7. References

Cabral, F. M. & Cabral, G. 2003. *Acanthamoeba* spp. as agents of disease in humans. *Clinical Microbial Reviews*. 16,273–307.

Chin K.Y. 2005. Bioaccumulation and depuration of pyrene in oyster *Crassostrea iredalei*. A Master Thesis. Universiti Malaysia Terengganu.

Cohen, J. J. & Duke, R. C. 1984. Glucocorticoid activation of a calciumdependent endonuclease in thymocyte nuclei leads to cell death. *Journal of Immunology* 132, 38-42.

Cohen, G. M., Sun, X. M., Snowden, R. T., Dinsdale, D. & Skilleter, D. N. 1992. Key morphological features of apoptosis may occur in the absence of internucleosomal DNA fragmentation. *Biochemical Journal* 286, 331-334.

Collins, A.R. 2004. The comet assay for DNA damage & repair. *Molecular Biotechnology*. 26,249-261.

Cotelle, S. & Ferard, J. F. 1999. Comet assay in genetic ecotoxicology. *Environmental Molecular Mutagenesis*. 34, 246–255.

Danadevi, K., Rozati, R., Banu, B.S., Rao, H.P. & Grover, P. 2003. DNA damage in workers exposed to lead using comet assay. *Toxicology* 187,183-193.

Diaz, S., Martin-Gonzalez, A. & Gutierrez, J. C. 2006. Evaluation of heavy metal acute toxicity and bioaccumulation in soil ciliated protozoa. Environmental International. 32,711-717.

Duruibe, J. O., Ogwuegbu, M. O. C. & Egwurugwu, J. N. 2007. Heavy metal pollution and human biotoxic effects. *International Journal of Physical Sciences* 2(5): 112-118.

Fairbain, D.W., Olive, P.L. & O'Neill, K.L. 1996. The comet assay: A comprehensive review. *Mutation Research* 339,37-59.

Forchhammer, L., Elvira, V.B., Janne, K.F., Pernille, H.D., Claus, N., Jensen, A., Steffen, L., Gitte, F. & Peter, M. 2008. Variation in assessment of oxidative damaged DNA in mononuclear blood cells by the comet assay with visual scoring. *Mutagenesis* 23(3), 223-231.

Gedik, C.M., Ewen, S.W.B. & Collins, A.R. 1992. Single-cell gel electrophoresis applied to the analysis of UV-C damage and its repair in human cells. *International Journal of Radiation Biology* 62, 313–320.

Greger, M.C. Ed (2nd). (1999).*Metal availability uptake transport and accumulation in plant, In: Heavy metal stress in plant*. From biomolecules to ecosystem. Prasad, M.N.V (edt). Narosa Publishing House India.

Gordeeva, A.V., Labas, Y.A. & Zvyagilskaya, R.A. 2004. Apoptosis in unicellular organisms: mechanisms and Evolution. *Biochemistry (Moscow)*, 69,1301-1313.

Guchelaar, H. J., Vermes, A., Vermes, I., & Haanen, C. 1997. Apoptosis: Molecular mechanisms and implications for cancer chemotherapy. *Pharmalogical World Sciences* 19, 119–125.

Hartmann, A., Agurell, E., Beevers, C., Brendler-Schwaab, S., Burlinson, B., Clay, P., Collins, A., Smith, A.,, Speit, G., Thybaud, V & Tice, R.R. 2003. Recommendations for conducting in vivo alkaline comet assay. *Mutagenesis* 11(1), 45-51.

Holden, N.E. 2001. *History of the Origin of the Chemical Elements and Their Discoverers.* Brookhaven National Laboratory Brisbane, Australia.

Kizilian, N., Wilkins, R.C., Reinhardt, P., Ferrarotto, C., McLean, J.R.N. & Mc- Namee, J.P. 1999. Silver-stained comet assay for detection of apoptosis. *Biotechniques* 27, 926 -930.

Loka, B-P.A., Sathe, V. & Chandramohan, D. 1990. Effect of lead, mercury & cadmium on sulphate reducing bacteria. *Environment pollution* 46, 24-28.

Miyaki, M., Murata, I., Osabe, M., & Ohno, T. 1977. Effect of metal cations on mis-incorporation by *E. coli* DNA polymerases. *Biochemical & Biophysical Research Community* 77, 854- 860.

Nakisah M.A, Fatimah. H, & Ali, A.M. 2008. Observation on the cytotoxicity of a plant compound labeled as MK2 on morphology of Acanthamoeba *by* Scanning Electron Microscopy. *Annals of microscopy* 8, 72-75.

Nakisah, M.A., Syafaz-syazwani, S., & Antonina A. 2009. Genotoxic effects of mercury & zinc on *Acanthamoeba* sp., a free living amoebae from Setiu Wetland Water: A laboratory study. *Journal of Sustainability Science & Management* 4, 72-77.

Ochiai, E. I. 1987. *General principles of biochemistry of elements.* Plenum Press, New York, N.Y.

Olive PL, Frazer G, Banath JP. 1993. Radiation-induced apoptosis measured in TK6 human B lymphoblast cells using the Comet assay. *Radiation Research* 136,130-136.

Saleh, A.M., Vijayasarathy, C., Masoud, L., Kumar, L., Shahin, A. & Kambal, A. 2003. Paraoxon induces apoptosis in EL4 cells via activation of mitochondrial pathways.*Toxicology Applied Pharmacology* 190, 47-57.

Searle, J., Kerr, J. F. R., & Bishop, C. J.1982. Necrosis and apoptosis: Distinct modes of cell death with fundamentally different significance. *Pathology Annals* 17, 229–259.

Shazili, N.A.M, Nakisah, M.A. & Tan, Y.L. 2004. *Perna viridis* as indicator of copper & zinc pollution in coastal waters of Johor. *Biomonitoring in Tropical Ecosystems*, Phang & Brown (eds). Pp 151-157

Sims, B. 1998. Basic Staining Mechanism. In The Science of Laboratory Diagnosis. Eds: Crocker, J and Burnett, D. 603pg. ISIS Medical Media, Oxford, UK.

Singh, N.P., McCoy, M.T., Tice, R.R. & Schneider, E.L. 1988. A simple technique for quantitation of low levels of DNA damage in individual cells. *Experimental Cell Research* 175,184-191.

Sirover, M.A. & Loeb, L.A. 1976. Infidelity of DNA Synthesis in vitro: Screening for potential metal mutagens or carcinogens. *Science* 194, 1434–1436.

Stohs, S.J.& Bagchi, D. 1995. Oxidative mechanisms in the toxicity of metal ions. *Biology Medical*.18(2),321-36.

Waring,P., Egan, M., Braithwaite, A., Mullbacher, A. & Sjaarda, A. 1990. Apoptosis induced in macrophages & T blasts by the mycotoxin sporidesmin and protection by Zn^{2+} salts. *International Journal Immunopharmaceutical,* 12;445-457.

Wiley, J.M., Sherwood, L.M. & Woolverton C.J. (Ed (7th)). (2008). *Prescott, Harley & Klein's microbiology*.Singapore: McGraw Hill International Edition.

Wright, C.W, Philipson, M.J and Warhurst, D. 1988. Use of microdilution to access antiamoebic activities of Brucea javanica fruits, *Simarouba amara* stem and number of quasinoids. *Antimicrobial Agents and Chemotheraphy* 32, 1725-1729.

11

Commercial Exploitation of Zooplankton in the Norwegian Sea

Eduardo Grimaldo and Svein Helge Gjøsund
SINTEF Fisheries and Aquaculture
Norway

1. Introduction

Since 1990s there has been increased interest in the exploitation of marine zooplankton like copepods and krill. This has been motivated by the increasing demand for marine bio-resources for human consumption in general, and in particular the growing demand for feed in aquaculture. In Nordic Seas, zooplankton is a key component in the energy transfer from primary producers to higher trophic levels such as herring, capelin, salmon, cod larvae and juveniles, and other species (Skjoldal, 2005). Roughly 70-80% of the zooplankton production in these waters is made up by copepods of the genus *Calanus* (Tande and Miller, 2000). According to general ecological theory about 10% of this production is available to the next trophic level (Lalli and Parsons, 1997). Estimates of the total annual production of *Calanus sp.* vary between 75 million tons y^{-1} for the Nordic Seas (Aksnes and Blindheim 1996) and 300 million tons *Calanus sp.* (mainly *Calanus finmarchicus*) y^{-1} for the Norwegian Sea only (Skjoldal et al. 2004).

This vast resource has great economic potential because it is rich in marine lipids, proteins, amino acids, and pigments. Further, by nature of being low on the food chain it has far lower bioaccumulation of heavy metals, organo-chlorides, dioxins, and other pollutants than higher trophic species now in use (Mizukawaa et al., 2009). Therefore, copepod fisheries have a potential to support the growth of new ventures in markets for functional food, food ingredients, and nutrition products. However, development of a copepod fishery must be pursued wisely (Nicol and Endo, 1999) using the best technology at hand, and implemented within a solid ecosystem based management regime, particularly given the importance of copepods to the marine ecosystem. In open Norwegian waters, *Calanus finmarchicus* is widely found in the upper 50 m during the productive period from April to August (Falkenhaug et al., 1997; Dahle and Kaartvedt, 2000). For practical reasons, harvesting concentrates on adults (CVI) and the life stages CIV and CV (the two last copepodite stages before becoming adults), because in these stages copepods have achieved sufficient body mass - body lengths are from 2.3 to 5.0 mm - depending upon species (Unstad and Tande, 1991). In addition, lipid content of copepods increases with increasing stage, with the two oldest stages being the most lipid rich (Kattner and Krause 1987).

Currently, *Calanus finmarchicus* harvesting uses fine-meshed trawls (~500 µm bar length) with mouth openings that range from 40 to 100 m², depending on the vessel size (Snorre Angell, Calanus AS, Sortland-Norway, Pers. Comm., 2009). However these trawls may be unsuitable for large-scale zooplankton harvesting because their very high towing resistance translates to high fuel consumption and CO_2 emissions.

2. Zooplankton trawls

In Norway, the development of commercial harvesting of zooplankton (*Calanus finmarchicus*) started in the fjords in the late 1950's, with relatively small nets operated from smaller boats (Wiborg and Hansen, 1974). The fishery developed from annual catches of a few tons to more than 50 metric tons by the mid 1970's. By then one had also developed larger trawl-like structures, e.g. beam trawls with rectangular front openings up to 5 m wide and 4 m high, and with coarser jellyfish nets at the front and structural cover nets outside the filtering net (Wiborg and Hansen, 1974). The fishery then evolved only slowly, but through the 1990's there was an increasing interest for *Calanus finmarchicus* as a potential raw material for feed in the rapidly growing aquaculture industry. For precautionary reasons, a general prohibition against harvesting of zooplankton in Norwegian waters was introduced in 2006. In order to support the further development of a sustainable fishery the authorities instead granted a limited annual trial quota for *Calanus finmarchicus.* The trial fishery has been conducted by the Norwegian biomarine company Calanus AS, who developed and patented a harvesting system for *Calanus finmarchicus.* Their trawls have ~ 500 µm meshes and front areas of up to 100 m^2 (~12 m wide, ~8 m high), with reported catch rates up to 2 tons dried weight per hour (Snorre Angell, Calanus AS, pers. comm.) (Fig 1).

Fig. 1. Full scale test of a commercial zooplankton trawl (left), and 1200 kg of copepods (*Calanus* sp.) caught after one hour tow (right). Photo: Snore Angell, Calanus AS.

By 2011 the Norwegian Ministry of Fishery and Coastal Affairs are preparing a public hearing to prepare for a limited, but more open commercial fishery for *Calanus finmarchicus.* In addition, the company Plantonic AS has being granted a special permit to harvest also smaller phyto- and microzooplankton, e.g. as an alternative to cultivated feed for the larval and juvenile stages of cultured fish species. For the latter large, anchored (stationary) nets are used. Mesh sizes can be an order of magnitude or smaller than those used for adult *Calanus finmarchicus*, posing considerably greater challenges with respect to filtration and clogging.

2.1 Flow through plankton nets and trawls

A theoretical model for the flow through fine-meshed nets and trawls is presented in Gjøsund and Enerhaug (2010). They derive basic relations for the flow through and forces on inclined net sections, based on pressure drop and streamline deflection through porous screens, and present parametric expressions for the filtration efficiency and drag on conical nets, cf. Eqs. 1 and 2. The model allows easy assessment of the effect of varying mesh opening, twine thickness, porosity, taper angle and flow (towing) velocity. The filtration efficiency F is defined as the ratio between the average velocity across the net mouth and the velocity of the net through the water (e.g. the towing velocity through quiescent water), see Figure 2. Further in Eqs. 1 and 2, C_D is the overall drag coefficient, F_D is the overall drag force, A_0 is the mouth area, C_N is a normal force coefficient, K is the pressure drop coefficient, K_0 is $K(a = 90°)$, Re_d is the effective Reynolds number, d is twine diameter, ν is kinematic viscosity, a is taper angle, γ is an angle describing streamline deflection, β is the screen porosity, T_γ is a tangential stress coefficient due to streamline deflection, and T_f is a tangential stress coefficient due to friction. The expression for K_0 in Eq. 4 is from Brundrett (1993), C_N in Eq. 6 is from Løland (1991), γ in Eq. 7 is from Gibbings (1973) and the tangential stress component in Eq. 8 is from Taylor and Batchelor (1949). Schubauer et al. (1950) found that Eq. 5 describes the pressure drop well for $a \geq 45°$, and Gjøsund and Enerhaug (2010) argue that it applies to lower taper angles also if the Reynolds number dependency in K_0 is properly accounted for. A suitable model for T_f is not available, and Gjøsund and Enerhaug use a constant value T_f = 0.02. A number of other models for K_0 and C_N exist, see Gjøsund (2006) and Gjøsund and Enerhaug (2010), and may yield equally good or better predictions is some cases. Gjøsund and Enerhaug (2010) conclude that the filtration efficiency is well predicted by Eq. 1 invoking Eqs. 3-6, while there is more uncertainty related to C_D in Eq. 2, presumably due to Eq. 7 being less accurate at smaller taper angles. Note that Eq. 1 is a simplification of the corresponding expression in Gjøsund and Enerhaug (2010, their Eq. 16), and that it must be solved in an iterative manner because K is also a function of F.

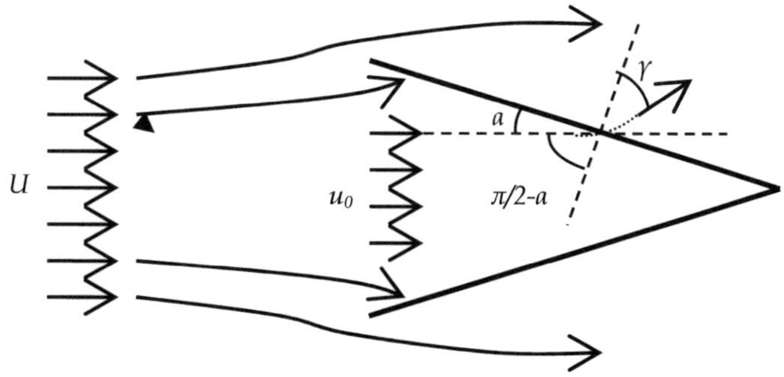

Fig. 2. Sketch of flow through a conical net with taper angle α and filtration efficiency F = u0/U, the angle γ describes streamline deflection through the net wall ("cf." from Gjøsund and Enerhaug, 2010).

$$F = \frac{u_0}{U} = \sqrt{\frac{C_N(K)}{K}} \tag{1}$$

$$C_D = \frac{F_D}{1/2\rho U^2 A_0} = C_N(K) + \frac{F^2\left(T_\gamma + T_f \sin^2(\alpha+\gamma)\cos^2\left(\frac{\pi}{2}-\gamma\right)\right)}{\tan\alpha} \tag{2}$$

$$Re_d = \frac{(u_0 \sin \alpha)d}{v} \tag{3}$$

$$K_0 = \left[\frac{7.0}{Re_d} + \frac{0.9}{\log(Re_d + 1.25)} + 0.05\log(Re_d + 1.25)\right]\frac{1-\beta^2}{\beta^2}, 10^{-4} < Re_d < 10^4 \tag{4}$$

$$K = K_0 \sin^2 \alpha \tag{5}$$

$$C_N(K) = \frac{2}{K}\left[K + 1 - \sqrt{2K+1}\right] \tag{6}$$

$$\gamma = \tan^{-1}\left[\Delta \tan\left(\frac{\pi}{2} - \alpha\right)\right], \Delta = \left[\left(\frac{K_0}{4}\right)^2 + 1\right]^{\frac{1}{2}} - \frac{K_0}{4}, 0 < \gamma < \frac{\pi}{2} - \alpha \tag{7}$$

$$T_\gamma = \frac{\tau}{1/2\rho u_0^2} = 2\sin^2 \alpha \left[\tan\left(\frac{\pi}{2} - \alpha\right) - \tan \gamma\right], 0 < \gamma < \frac{\pi}{2} - \alpha \tag{8}$$

A key element in Gjøsund and Enerhaug (2010) is the importance of viscous effects at very low Reynolds numbers, i.e. as local dimensions and velocities become very small. This Reynolds number effect manifests itself as a dramatic increase in the pressure drop coefficient as mesh dimensions and towing velocities decrease; potentially leading to a strong decrease in filtration efficiency, see Eq. 1 and Figure 2. Hence the filtration efficiency of plankton nets depends strongly on the net parameters and towing velocity. Gjøsund and Enerhaug (2010) compare the theoretical model with flume tank measurements with fine-meshed net cones, and demonstrate among other things how the filtration efficiency for typical plankton nets increases with increasing towing velocity, and decreases with decreasing velocity (Figure 3). This is contrary to common belief, as noted also by Tranter and Heron (1967); there is a widespread and persistent, but incorrect perception that filtration efficiency generally decreases as towing velocity increases. In plankton sampling the towing velocity is therefore often recommended to be low, and it is also assumed that a low towing velocity reduces clogging (Sournia, 1978). However, Tranter and Heron (1967) found that so-called flared samplers clogged more readily in field experiments than unflared samplers. A flared sampler implies reduced velocity inside the net and towards and through the net wall, hence hydrodynamically it is equivalent to an unflared sampler with lower filtration efficiency. Also, as the velocity decreases and the pressure drop coefficient increases, the flow (and thus the plankton) deflects more perpendicularly towards the net wall (Reynolds, 1969), i.e. the angle γ in Figure 3 decreases. This suggests that the clogging rate may actually increase with decreasing velocity in some cases. Here it is crucial to consider clogging with respect to filtered volume and not with respect to tow time or tow distance (McQueen and Yan, 1993); if one measures reduced clogging at low velocities compared to higher velocities, this may simply be due to the low velocity case filtering less water and thus less plankton than assumed. For high towing velocities, e.g. 10 knots and more as sometimes used in so-called high speed sampling, other effects may be important and it is less clear how velocity variations affect filtration. For instance, the global wake field behind the sampler can influence filtration to a greater extent.

Usually only the open area ratio R is considered when designing plankton nets, i.e. the ratio between the open mesh area and the mouth area. A general recommendation is that R should be greater than 3 to have high initial (i.e. before any clogging occurs) filtration efficiency (Tranter and Heron, 1967), and greater than 6 to have an additional buffer against clogging (Harris et al., 2000). However, this ratio involves only the porosity and taper angle of the net; it does not account for the Reynolds number effect described above. Figure 3 shows that recommending low towing velocities and a fixed value of R can be highly misleading in some cases, and may result in low filtration efficiency and potentially also in increasing clogging. This may represent a significant and largely unknown source of error in plankton samples and abundance estimates, in particular for the very smallest plankton and mesh sizes. There are nevertheless other reasons for limiting the towing velocity. If the pressure becomes too high, very fine meshed netting may break (Sournia, 1978), and plankton may be extruded through the meshes and thereby lost or contributing to clogging. Also, the towing resistance basically increases with the towing velocity squared. This is seldom an issue for small sampling nets, but it can be crucial for the fuel efficiency of larger commercial plankton trawls. For commercial plankton trawls, key issues are catch quality, catch- and fuel efficiency and structural reliability, requiring that filtration efficiency, clogging, towing resistance and more are properly balanced in the design process (Larsen, 2009).

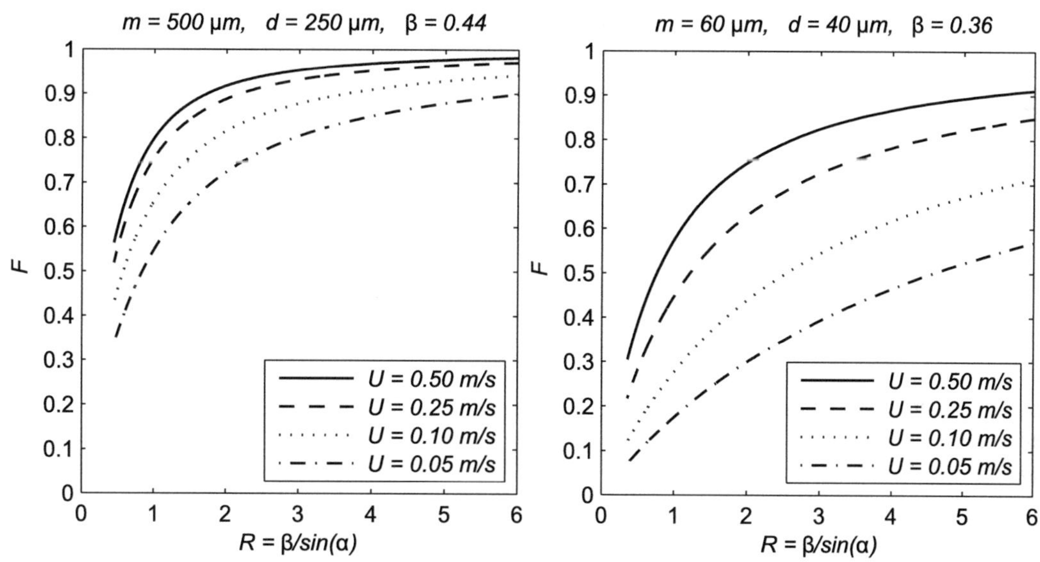

Fig. 3. Predicted initial filtration efficiency F as a function of open area ratio R for two nets with different mesh size m, twine thickness d and porosity β, at four different velocities 5, 10, 25 and 50 cm/s (from Eq. 1, cf. Gjøsund and Enerhaug, 2010).

3. Bubble-enhanced zooplankton harvesting

Bubble-enhanced zooplankton harvesting is a novel environmental friendly platform for commercial harvest of zooplankton at sea, targeting lower energy consumption during towing and less by-catch than existing catching equipment. The working principle of this harvesting platform is based on releasing air bubbles at a depth of 20-40 meter to vertically displace copepods towards the sea surface. Key components are a submerged, towed air bubble diffuser (sparging elements), an air delivery system, and a collector net or surface skimmer (Fig 4).

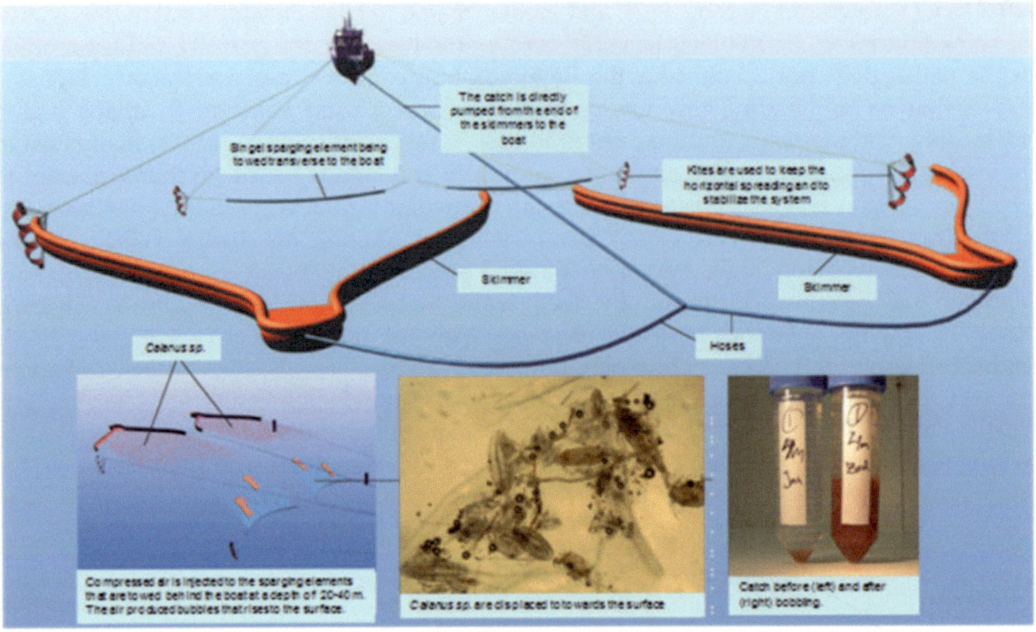

Fig. 4. Initial prototype of a skimmer trawl for commercial harvesting of *Calanus sp*. Air is released from a single perforated hose at 20-40 m depth to produce small bubbles. These bubbles while rising displace *Calanus sp* towards the surface. Here they are collected by skimmer and directly pumped to the boat. Animation: Mats Heide, SINTEF.

Two primary mechanisms associated with micro-bubble technology are upwelling (Leifer et al., 2009) and attachment/flotation. As bubbles rise, they transfer momentum to the surrounding fluid, creating the upwelling flow, which transports deeper water upwards, including entrained zooplankton. For bubble attachment, small bubbles attach and add positive buoyancy to the zooplankton, lifting them upwards. Real-world applications involve both of these synergistic processes.

3.1 Bubble attachment

Bubbles effectively accumulate surfactants through the process termed sparging or flotation. Surfactants are surface active substances including surfactant-bacteria and particles with hydrophilic and hydrophobic sites that energetically prefer to be at air-water interfaces. This bubble collection and transport process is central to wastewater treatment microflotation (Persechini et al., 2000), mining airlift separators (Mao and Yoon, 1997), bioreactors (Wu, 1995), and marine aggregate formation (Mari, 1999). Surfactants affect bubble properties, decreasing gas exchange and rise velocity and thus decreasing dissolution (Leifer and Patro, 2002). Surfactants also stabilize bubbles against breakup (Johnson and Cooke, 1980). Attachment requires several steps. First, the bubble trajectory must intersect the zooplankton close enough for the two to touch. Then, the bubble and zooplankton must attach, rather than "bounce." Finally, the bubble must remain attached long enough to lead to significant vertical advection. Smaller (< 300 µm radius) bubbles are more likely to attach to *Calanus* because of their slower rise velocities (< 6 cm s^{-1}), and because their size is comparable to key *Calanus* dimensions, such as thorax, legs, and antennae. However, because small bubble buoyancy is minimal, they provide little lift buoyancy force. Given that *Calanus* are slightly negatively buoyant; this can lead to minimal or negligible upwards motion unless several small bubbles attach to the copepod.

Laboratory studies showed that the highest zooplankton attachment was for bubbles in the range $50 < r < 300$ µm (own data, unpublished), where r is the equivalent spherical radius (Fig. 5). Although larger bubbles have greater buoyancy, their attachment probability is lower (own data, unpublished). Bubbles comparable in size or larger than *Calanus* - circa 1000-µm radius - have well developed turbulent wakes and boundary layers, and rise fast (25 - 30 cm s^{-1}) compared to small bubbles (Leifer and Patro, 2002). These large bubbles tend to displace the *Calanus* along streamlines around the bubble as they pass, leading to negligible attachment probability. After attachment, the bubble-copepod aggregate rises with a velocity, V_{AG}, determined by the drag resistance of the *Calanus*-bubble aggregate and the buoyancy force, until bubble detachment (or surfacing). In laboratory studies, a linear relationship was found between r and V_{AG}, from 2.5 to 9.0 cm s^{-1} with the highest rise velocity for a 341-µm radius bubble (own data, unpublished). Because *Calanus* are mobile, bubble detachment by body motions can be significant and reduces the *Calanus* vertical advection distance. Then, a combination of *Calanus*'s negative buoyancy and active swimming towards its original depth likely will cause sinking. Thus, successful flotation requires the attachment time scale to be sufficiently shorter than the detachment time scale. Due to detachment, flotation of more active *Calanus* is less efficient, mimicking natural selection.

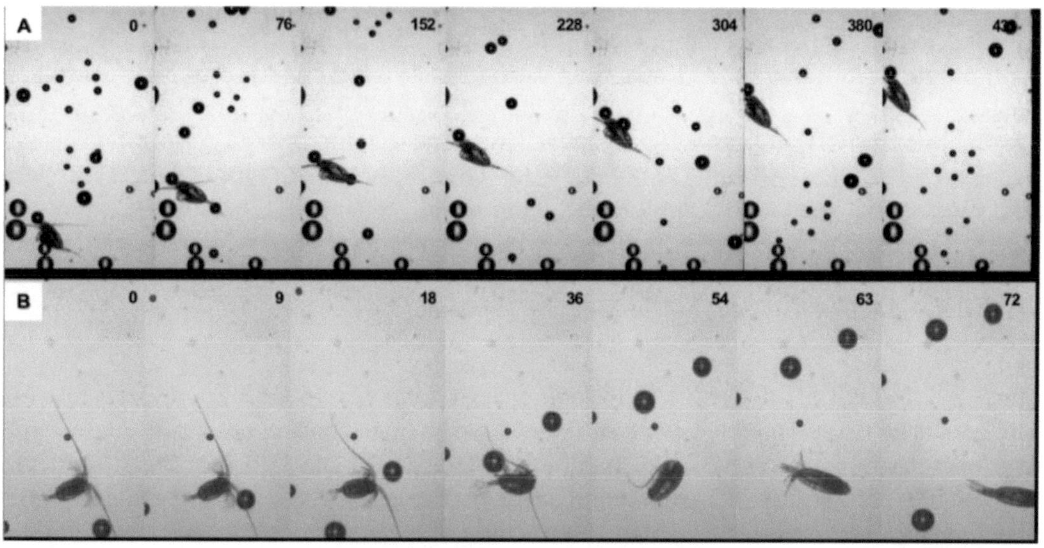

Fig. 5. *Calanus* response to bubbles (timestamp in images is given in milliseconds). A) One r~150-µm bubble attaches to a *Calanus* head, causing upwards motion. B) Detachment of a small bubble (r~75 µm) from the *Calanus* antenna during an escape jump, possibly triggered by a passing larger bubble (r~350 µm).

3.2 Bubble-driven upwelling flow

For sufficiently high bubble concentration, synergistic interactions in bubble plumes create fluids with distinct properties from the surrounding fluid (Leifer et al., 2006). Bubble plumes transfer momentum to the surrounding fluid; creating an upwelling flow, see review in Leifer et al. (2009). The upwelling flow decreases bubble gas exchange with the surrounding fluid because of the reduced transit time across the water column, enhancing bubble survival against dissolution (Leifer et al., 2006).

For point-source bubble plumes, larger fluid velocities are at a peak along the centerline decreasing radially with a Gaussian profile (Milgram, 1983). The fluid velocity increases with height above the source in non-stratified fluids due in part to the increase in buoyancy flux from decreasing hydrostatic pressure. At the surface, the upwelled fluid spreads out in a horizontal intrusion, the outwelling flow. Thermal and haline stratification are common in the marine environment, with cooler and/or more saline (denser) water at greater depth. Thus, marine upwelling flows lift water with increasingly negative buoyancy. Upon encountering a steep density gradient, the bubble plume can significantly (or completely) detrain plume fluid into a horizontal intrusion; however, the bubbles continue rising, entraining new water, unlike for a continuum (single phase) plume such as a sewage outfall (McDougall, 1978). Such horizontal intrusions deposit any transported zooplankton, marine particles, and dissolved gases in a layer, and have been identified in the field (Solomon et al., 2009; Leifer et al., 2009; Leifer and Judd, 2002). Sufficiently strong bubble plumes can support the upwelled fluid through the density stratification to the sea surface.

Laboratory studies showed that upwelling advection has "100%" efficiency (the analogue of bubble attachment) for copepods entrained in the flow, and created faster vertical motions than flotation (own data, unpublished). Moreover, where the upwelling flow is wide copepod jumps cannot exit the upwelling flow (analogue of detachment). Bubble plumes were produced for flow rates, Q, spanning $0.48 < Q < 76.5$ L h^{-1} which produced upwelling flows, V_{up}, from $4 < V_{up} < 37$ cm s^{-1}, with the highest V_{up} for flows with large bubbles or large Qs. For this study, $V_{up}(Q)$, showed a power law increase with Q as $V_{up}(Q) \sim Q^{0.20}$ for small bubbles ($r \sim$ 50-100 μm) and $V_{up}(Q) \sim Q^{0.33}$ for large bubbles ($r \sim$ 400-600 μm), in close agreement with the large-scale plume findings of Leifer et al. (2009), who found $V_{up}(Q) \sim Q^{0.23}$, and also for seep bubble plumes in the open ocean (Leifer, 2009) where $V_{up}(Q) \sim Q^{0.3}$. These parameterizations are in agreement with the finding of the calculations of Lemckert and Imberger (1993) on the Milgram (1983) data set.

3.3 Stationary vs towed bubble plumes

Most published field bubble plume studies are for stationary bubble plumes in static water (e.g., lake destratification studies) (Schadlow, 1992; Lemckert and Imberger, 1993; Singleton et al., 2007), natural marine hydrocarbon seeps (Leifer et al., 2000; Leifer and Boles, 2005; Leifer et al., 2009), or gas blowouts (Topham, 1975; Milgram, 1983). However, typical fisheries applications involve a towed bubble plume (Grimaldo et al., 2010). Potentially, there is a significant difference between a stationary (i.e., fixed) source bubble plume in a uniform horizontal current (Fig. 6A) and a towed source bubble plume through quiescent water (Fig. 6B), even though both geometries appear similar.

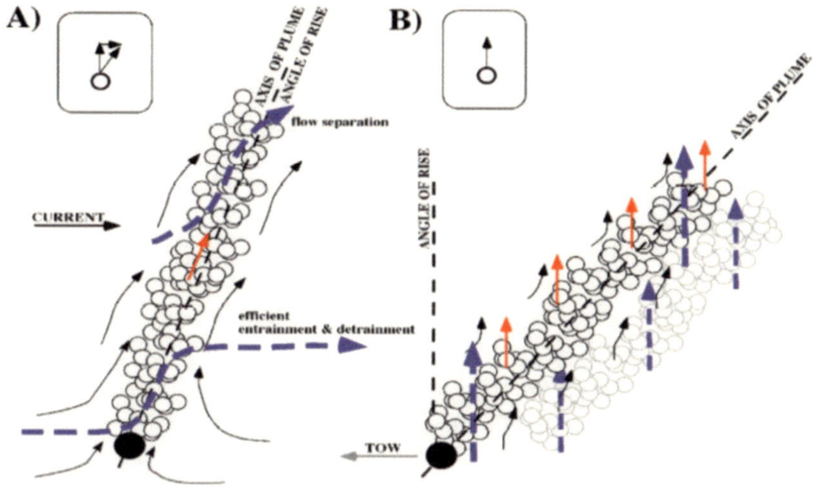

Fig. 6. Two dimensional schematic of a bubble plume for A) a fixed bubble source in a uniform horizontal current and B) towed bubble source through quiescent water. Inset shows details of individual bubble motions. Red and black arrows pertain to bubbles and entrained water, respectively. Grayed bubbles represent plume several seconds earlier, and remnant, persistent fluid motions.

For a stationary (i.e., fixed) source bubble plume in a uniform horizontal current, each bubble rises and is advected by the horizontal currents. As a result, the fluid motions and bubble plume motions are aligned along the plume axis. This allows the bubbles to accelerate the fluid throughout the entire water column. In contrast, for quiescent water (Fig. 6B). Bubbles rise vertically. Thus, for a towed bubble source, the apparent plume angle is vertical as are the fluid motions. However, they are not aligned with the angle of the plume. As a result, a parcel of water experiences vertical advection as a short pulse from the passing bubble sheet, rather than a sustained force. Thus, a towed bubble plume is more analogous to a bubble plume pulse in a horizontal current. Locally the two are identical; however, the boundary conditions are different. For a stationary bubble plume in a horizontal current (Fig 6A), the bubble plume is surrounded by water with no vertical motion. In contrast, in a towed plume the “local” bubble pulse is bounded on the down-tow side by persistent upwelling flows driven by the pulse that already passed. The primary bubble processes underlying the bubble plume trawl are bubble flotation and plume upwelling.

3.4 Towed submerged bubble rafts

Two highly distinct approaches were used to generate bubble plumes during the two field test series in 2008 and 2009 (Grimaldo et al, 2010). One, a bubble raft with tow parallel sparging elements, used a flushed sparger that tended to produce very small bubbles (which a video camera mounted on the raft imaged as milky in appearance) (Fig. 7A-B). The second, a bubble raft with tow transverse sparging elements, used a porous rubber hose that produced larger bubbles (~1-2 mm diameter) (Fig. 7C-D). The tow parallel bubble raft sought to maximize attachment flotation, while the tow transverse bubble raft was designed for using upwelling flotation.

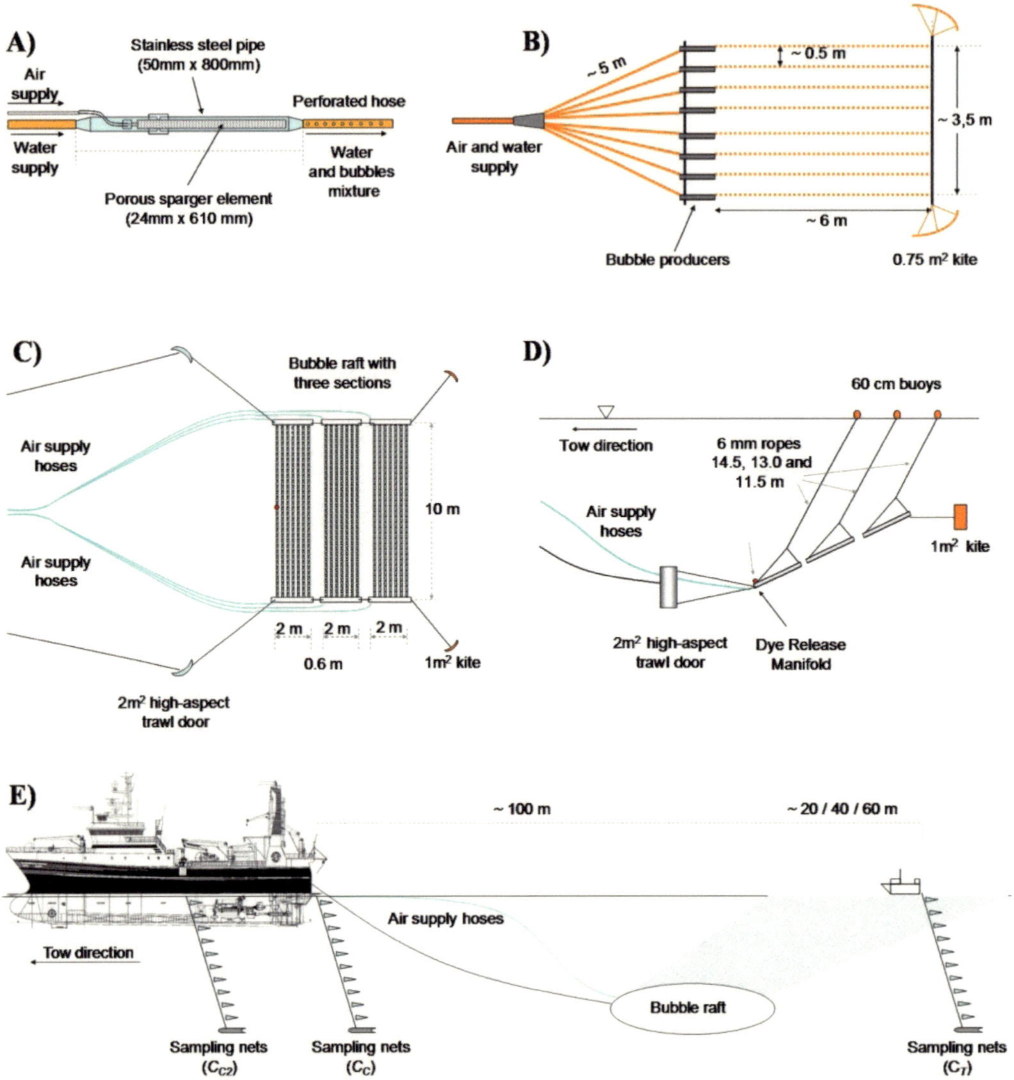

Fig. 7. Schematics for the bubble rafts: (A) Single bubble sparger element. (B) Top view of the raft with tow-parallel sparging elements. (C) Top view of the raft with tow-transverse sparging elements comprised of 3 sub-raft elements and (D) side view. (E) Bubble raft and plankton net deployment (Taken from Grimaldo et al., 2010).

Size distributions were not measured in the field (i.e., at depth, and under tow, and pressure) and significant deviations from laboratory bubbles are likely. With regards to bubble processes, the smaller flushed sparger bubbles are less efficient at creating upwelling flows (Patro *et al.*, 2002), but more efficient at attachment (own data, unpublished) than the larger bubbles from the porous rubber hose. Significantly, bubbles from the flushed sparger from 25-m depth did not always reach the surface, strongly suggesting that the bubbles produced were dissolving during rise. Bubble dissolution reduces the buoyancy flux, decreasing the plume's ability to transport fluid against stratification and maintain coherency against current and wave disruption.

Plume upwelling velocities, V_{up}, were measured for the raft with tow-parallel sparging elements in the experiments of 2008 by injecting dye and measuring the transit time, t, for the dye to reach the sea surface. Values of t were determined with a stopwatch based on the first arrival time at the sea surface, when the boil exhibited green colour (Fig. 8C); and generally showed a high degree of repeatability. Measurements were made for a range of airflows, Q (2100 to 7200 L min^{-1} at STP), and release depths, z_0 (2.5, 5.0, and 7.5 m). Each combination had between 3 and 20 repetitions, depending on variability (more repetitions for higher variability data sets). For the raft with tow transverse sparging elements, the upwelling flow was measured for z_0 = 15 m and Q = 8432 L min^{-1} (STP). However, dye surfacing was difficult to observe and V_{up} was measured only for this combination. V_{up} values were highly variable, but were mainly dependent on the air flow (Q). V_{up} ranged between ~10 and 25 cm s^{-1} and could be described by a power law fit, $V_{up} = Q^b$, where b varied between 0.246 and 0.323, depending on release depth (z_o). V_{up} for z_o>7.5 m were unsuccessful at advecting dye to the sea surface. For the raft with tow-transverse sparging elements, V_{up} was measured (in 2009) for the maximum airflow rate (Q=8432 L min^{-1}) only and was 17.3±2.4 cm s^{-1} for z_o=15 m.

Fig. 8. Photos of the raft with tow-parallel sparging elements: A) During deployment 20 June 2008, B) Surface expression of the bubble plume for 4800 L min^{-1} air from 5-m depth. C) Dye arriving at sea surface. Buoys are 60-cm diameter. *R/V Jan Mayen* stern is 120 m distant. D) Surface bubble plume for 25 m deployment. Photos of raft with tow-transverse sparging elements: E) During deployment on 29 April 2009. F) The trawl doors spreading the bubble raft laterally. G) *R/V Hyas* in position in bubble plume for sampling. H) Tow sampling nets, note high visibility.

The towed submerged bubble rafts were highly successful at elevating *Calanus* concentrations in a thin surface layer. Surface enhancements, ε, as high as 1416% were observed in full scale experiments off the coast of Troms, northern Norway, in 2009. Although greater ε in surface layers were observed in 2008 in the presence of stratification, enhancement relative to the maximum in the water column, C_C, was far greater for the unstratified conditions of 2009 than 2008. In 2009, a *Calanus* trawl located at 20 m behind the bubble plume and fishing the upper meter would have in average increased the catch by 980%.These enhancements are dramatically larger than from hull mixing, which could not elevate *Calanus* concentrations greater than elsewhere in the water column (Fig 9).

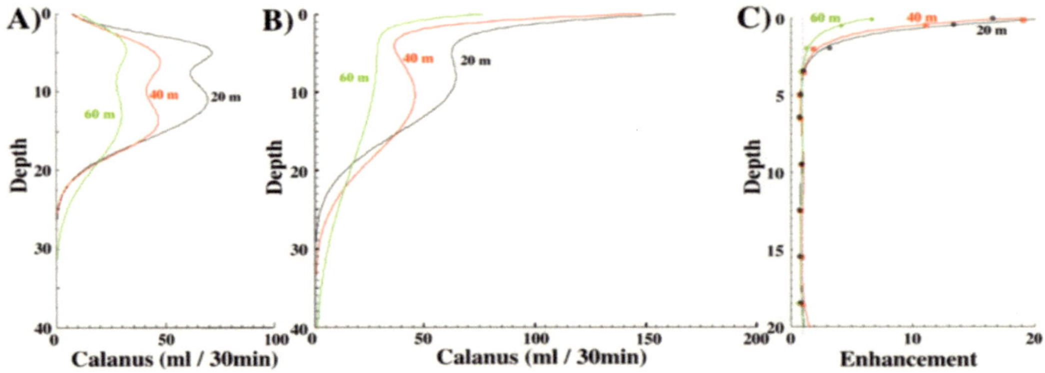

Fig. 9. Curve fits to *Calanus* sampling data for 29-30 April 2009, at 20, 40, and 60 m behind first bubble surfacing location. Data is not shown for clarity. A) Control plankton nets (C_C) B) Test plankton nets (C_T) and C) Enhancement (ε) based on curve fits (line) and data (symbols).

3.5 A continuous injection bubble pulse

The raft with tow-transverse sparging elements produced a truly novel bubble plume, continuously injecting bubbles into the same volume. Where a bubble pulse is produced from a stationary area plume and Q increases as the bubble plume rises due to decreasing hydrostatic pressure, the increase in buoyancy flux is non-linear, and except for near the sea surface, small. For example, for a 20-m plume with V_{up}=20 cm s^{-1}, Q doubles after bubbles have risen for 50 s. In comparison, the towed bubble raft's Q for the first meter doubles after 1 second during which the plume has only risen ~0.5 m, with additional increases in Q from hydrostatic expansion. On the other hand, because the bubble plume locally is a pulse, fluid acceleration is continuous. The upwelling flow generated was far less for the raft with tow-parallel sparging elements than for the raft with tow-transverse sparging elements due largely to the pulse-like nature of the bubble plume. Currently, bubble pulse behaviour, particularly for an area plume, remains completely uncharacterized, while this is the first reported continuous-injection bubble pulse.

3.6 Synergistic upwelling and flotation

Although the focus of the raft with tow-parallel sparging elements was attachment flotation and that of the raft with tow-transverse sparging elements was upwelling flotation, both processes occurred for both rafts. Moreover, the two processes are synergistic, *Calanus* with attached bubbles likely have greater difficulty escaping from the bubble plume, including when trapped in a turbulence vortex, and the added buoyancy increases their upwards velocity. Further, for *Calanus* to escape the bubble trawl, they need to jump away from the bubbles, which likely is less efficient for individuals with attached bubbles. One interesting and potentially important feature of vortex trapping is that it places zooplankton and bubbles in close proximity with numerous opportunities for bubble-zooplankton interaction and aggregation formation. Thus, significant *Calanus* could have some bubble attachment, aiding the upwelling process. Also, upwelled zooplankton likely becomes trapped in vortices, and the upwelling flow prevents downward escape. Lateral escape for zooplankton, even at the plume edge, likely is inefficient because of the inflow and the jump response is random when confronted by bubbles (own data, unpublished). Further, some fraction of *Calanus* that do escape may be re-entrained in the upwelling flow in the bubble plume.

Because the enhancement, ε, for depths shallower than ~4 m was significantly above 1, but there was no significant reduction ($\varepsilon<1$) for deeper depths, increased *Calanus* in the water column must have been primarily from lateral bubble plume entrainment of *Calanus*. Some of the lateral entrainment could have arisen from deeper water ($z>4$ m), compensating in part for the reduction in $C_T(z)$ due to vertical advection (upwelling) of a *Calanus* profile $C_C(z)$ that decreased with depth below 10-12 m. However, enhancements of 560–980% relative to the water column maximum are difficult to explain by deeper entrainment and upwelling as they would almost certainly have caused a significant deviation from $\varepsilon=1$ for $z>4$ m. Given the absence of *Calanus* in the surface layer (C_C), lateral enhancement likely included an upwards component. Also, momentum plume upwelling, which surrounds the bubble plume, could have played a role through lateral *Calanus* entrainment.

3.7 Bubble trawl bycatch reduction

The upwelling flow and the vortices appeared to be effective at trapping other copepod-sized species, although healthy fingerlings were not caught, suggesting that they were not upwelled of lifted by bubble attachment. The towed bubble plume reduced by 65% of all types of bycatch organisms, which were in the path of the bubble plume. The greatest reduction was for crabs (in different larvae stages) and fish eggs, which initially were concentrated in a shallow layer (<3 m) and likely were floated towards the sea surface by the bubble plume. At the surface, the effect of the outwelling flow apparently removed them laterally, making them unavailable for the sampling nets. Because sampling was not performed at the bubble plume edges, the fate of these bycatch organisms is unclear. The bubble plume's effect upon fish larvae and especially upon small fish seems different from that for crab larvae and fish eggs. Accordingly, while the bubble attachment processes may have enhanced the flotation of crabs and fish eggs (presumably because of hair and stickiness of fish eggs); bubble attachment to fish larvae and fingerlings seems highly improbable. Also, visual evidence of small fish swimming inside the bubble plume suggests that fish larvae and fingerlings may have actively avoided the bubble plume. The largest (~3-4 mm diameter) bubbles surfaced first and apparently were highly effective for jellyfish flotation. This not only represents a great advantage for bubble-enhanced *Calanus* harvesting, but also for conventional fine-meshed trawls for *Calanus* and other fisheries. For example, jellyfish flotation could divert jellyfish from the trawl path, avoiding associated problems with net clogging, catch damage, sorting, etc. Jellyfish flotation appears distinct from *Calanus* flotation, in that due to the morphology of the jellyfish, bubbles readily are trapped in their body, leading to more effective buoyant rise. For example, jellyfish were very common in 2008 and were observed floating at the sea surface with entrapped bubbles. Although the bubble trawls were not designed to effectively divert jellyfish through flotation; such diverters could be very useful for improving the *Calanus* fishery.

4. Conclusion

The area bubble plume-enhanced *Calanus* harvesting technology is a unique and novel design that improved copepod catch rates, reduced bycatch, and significantly decreased energy consumption during towing by allowing for a smaller collector. Results showed very strong *Calanus* enhancement relative to elsewhere in the water column in a thin surface layer during tests in the absence of stratification. Stratification was a dominant factor affecting bubble trawl performance; however, data were insufficient to characterize stratification's effect on bubble plume fluid motions beyond fluid-ambient density difference. Investigation of the bubble generation approach suggested small bubbles are problematic, particularly for deeper tow depths where dissolution becomes significant, compared to larger bubbles. Large bubble generation was effective by pressurizing a porous rubber hose - the pressure difference across the hose walls prevented hydrostatic pressure changes (swell) from causing emission variability along the sparger elements, in contrast to a drilled rubber hose. The current bubble trawl design, while appropriate for these field tests, lacked robustness for commercial application. Although highly promising, results highlighted significant areas of critical need for further study: increased sampling resolution, validation of the vortex *Calanus* trapping hypothesis, and characterization of the role of stratification in bubble plume processes related to bubble trawl performance.

5. Acknowledgment

This chapter is based on results that have been obtained through diverse research projects mainly funded by the Research Council of Norway, and the Norwegian industry. We gratefully acknowledge the inspiring discussions with many researchers involved in these projects. Particular thanks goes to Birger Enerhaug (SINTEF), Thomas McClimas (SINTEF), Ira Leifer (University of California), Snorre Angell (Calanus AS), Roger B. Larsen, Ivan Tatone and Trond Larsen (University of Tromsø), and the crews of R/V Hyas and R/V Jan Mayen.

6. References

Aksnes D.L. and Blindheim J. (1996). Circulation patterns in the North Atlantic and possible impact on population dynamics of *Calanus finmarchicus*. *Ophelia,* Vol. 44, pp. 7–28.

Aseda, T. and Imberger, J.T. (1993). Structure of bubble plumes in linearly stratified environments. *Journal of Fluid Mechanics*, Vol. 249, pp. 35-57.

Brundrett, E. (1993). Prediction of pressure drop for incompressible flow through screens. *Journal of Fluids Engineering,* Vol. 115, pp. 239–242.

Cumberbatch, E. (1982). Two-dimensional flow past a mesh. *Quarterly Journal of Mechanics and Applied Mathematics*, Vol. 35, No. 3, pp. 335-344.

Dahle, T., Kaartvedt, S. (2000). Diel patterns in stage-specific vertical migration of *Calanus finmarchicus* in habitats with midnight sun. *ICES Journal of Marine Science,* 57, 1800-1818.

Falkenhaug, T., Timonin, T., Tande, K. (1997). Abundance and distribution of numerical important species of zooplankton in summer and winter periods of 1990 and 1991 in North Norwegian fjords. *Journal of Plankton Research,* Vol. 19, No. 4, pp. 449-468.

Fanneløp, T. K., Webber, D. M. (2003). On buoyant plumes rising from area sources in a calm environment, *Journal of Fluid Mechanics,* Vol. 497, pp. 319-334.

Gibbings, J.C. (1973). The pyramid gauze diffuser. Ingenieur-Archiv Vol. 42, pp. 225–233.

Gjøsund, S.H. (2006). Flow through nets and trawls of high solidity. SINTEF Fisheries and Aquaculture report no. SFH80 A063012, ISBN 82-14-03872-3, Trondheim, January 2006.

Gjøsund, S.H., Enerhaug, B. (2010). Flow through nets and trawls of low porosity. *Ocean Engineering*, Vol. 37, No. 4, pp. 345-354.

Grimaldo, E., Leifer, I., Gjøsund, S.H., Larsen, R.B., Jeuthe, H., Basedow, S. (2010). Field demostration of a novel towed, area bubble-plume zooplankton (*Calanus sp.*) harvester. *Fisheries Rersearch,* Vol. 107, pp. 147-158.

Johnson, B. D., Cooke, R. C. (1980). Organic Particle and Aggregate Formation Resulting from the Dissolution of Bubbles in Seawater. *Limnology and Oceanography,* Vol. 5, No. 4, pp. 653-661.

Harris, R.P., Wiebe, P.H., Lenz, J., Skjoldal, H.R., and Huntley, M. (Eds.), (2000). *ICES Zooplankton Methodology Manual.* Academic Press.

Kattner, G. and Krause, M. (1987). Changes in lipids during the development of *Calanus finmarchicus* s.l. from Copepodid I to adult. *Marin Biology,* Vol. 96, pp. 511–518.

Lalli, C. M., Parsons, T. R. (1997). *Biological Oceanography.* ISBN 0750633840, The Open University.

Larsen, T. (2009). Slepemotstand, effektivitet og fangstsammensetning i fiske med trål tilpasset dyreplankton - Forsøk med trål i småskala. MSc thesis, The Norwegian College of Fishery Science, University of Tromsø, February 2009.

Leifer, I., Boles, J. (2005). Measurement of marine hydrocarbon seep flow through fractured rock and unconsolidated sediment. *Marine and Petroleum Geology,* Vol. 22, No. 4, pp. 551-568.

Leifer, I., Jeuthe, H., Gjøsund, S.H., Johansen, V. (2009). Engineered and natural marine seep, bubble driven buoyancy flows. *Journal of Physical Oceanography,* Vol. 39, No. 12, pp. 3071-3090.

Leifer, I., Clark, J.F., Chen, R.F. (2000). Modifications to the local environment by natural marine hydrocarbon seeps. *Geophysical Research Letters,* Vol. 27, pp. 3711-3714.

Leifer, I., Luyendyk, B., Boles, J., Clark, J.F. (2006). Natural marine seepage blowout: Contribution to atmospheric methane. *Global Biogeochemical Cycles,* Vol. 20, GB3008, doi: 10.1029/2005GB002668, pp. 1-9,

Leifer, I., Patro, R. K. (2002). The bubble mechanism for methane transport from the shallow sea bed to the surface: A review and sensitivity study. *Continental Shelf Research,* Vol. 22, No. 16, pp. 2409-2428.

Leifer, I., Judd, A. (2002). Oceanic methane layers: A bubble deposition mechanism from marinehydrocarbon seepage. *Terra Nova,* Vol. 16, pp. 471-485.

Lemckert, C.J., Imberger, J. (1993). Energetic bubble plumes in arbitrary stratification. *Journal of Hydraulic Engineering,* Vol. 119, pp. 680-703.

List, E.J. (1982). Turbulent jets and plumes. *Annual Review of Fluid Mechanics,* Vol. 14, pp. 189–212.

Løland, G. (1991). Current forces on and flow through fish farms. Doctoral thesis, Dept. of Marine Hydrodynamics, The Norwegian Institute of Technology.

Mao, L., Yoon, R.-H. (1997). Prediction flotation rates using a rate equation derived from first principles. *International Journal Mineral Processing,* Vol. 51, pp. 171-181.

Mari, X. (1999). Carbon content and C:N ratio of transparent exopolymeric particles (TEP) produced by bubble exudates of diatoms. *Marine Ecology Progress Series,* Vol. 183, pp. 59-71.

McDougall, T.J. (1978). Bubble plumes in stratified environments. *Journal of Fluid Mechanics,* Vol. 85, pp. 655-672.

McQueen, D.J., Yan, N.D. (1993). Metering filtration efficiency of freshwater zooplankton hausl: reminders from the past. *Journal of Plankton Research,* Vol. 15, Vol. 1, pp. 57-65.

Milgram, J.H. (1983). Mean flow in round bubble plumes. *Journal of Fluid Mechanics,* Vol. 133, pp. 345-376.

Mizukawaa, K., Takada, H., Takeuchi, I., Ikemoto, T., Omori, K., Tsuchiya, K. (2009). Bioconcentration and biomagnification of polybrominated diphenyl ethers (PBDEs) through lower-trophic-level coastal marine food web. *Marine Pollution Bulletin,* Vol. *58, No.* 8, pp. 1217-1224

Nicol, S., Endo, Y. (1999). Krill fisheries: Development, management and ecosystem implications. *Aquatic Living Resources,* Vol. 12, No. 2, pp. 105-120.

Patro, R.K., Leifer, I., Bowyer, P. (2002). Better bubble process modelling: Improved bubble hydrodynamics parameterization. In: *Gas Transfer at Water Surfaces,* M.Donelan, (Ed.). American Geophysical Union, Washington, pp. 315-320.

Persechini, M.A.M., Jota, F.G., Peres, A.E.C. (2000). Dynamic model of a flotation column. *Minerals Engineering,* Vol. 13, pp. 1465-1481.

Reynolds, A.J. (1969). Flow deflection by gauze screens. Journal of Mechanical Enginering Science, Vol. 11, pp. 290-294.

Schadlow, G. (1992). Bubble plume dynamics in a stratified medium and the implications for water quality amelioration in lakes. *Water Resources Research,* Vol. 28, No. 2, 313-321.

Schubauer, G.B., Spangenberg, W.G., and Klebanoff, P.S. (1950). Aerodynamic characteristics of damping screens. NACA Technical note no. 2001.

Singleton, V. L., Gantzer, P., Little, J.C. (2007). Linear bubble plume model for hypolimnetic oxygenation: Full-scale validation and sensitivity analysis. *Water Resources Research,* Vol. 43. Wo2405, Doi:10.1029/2005WR004836, pp. 1-12.

Skjoldal, H.R., Dalpadado, P., Dommasnes, A. (2004). Food webs and trophic interactions. *In: The Norwegian Sea ecosystem, Skjoldal HR (*Ed.*).* Tapir Academic, Trondheim, Norway, pp. 447—506

Skjoldal, H.R. (2005). *The Norwegian Sea Ecosystem,* Tapir Academic Press.

Solomon, E.A., Kastner, M., MacDonald, I.R., Leifer, I. (2009). Considerable methane fluxes to the atmosphere from hydrocarbon seeps in the Gulf of Mexico. *Nature Geoscience,* Vol. 2, pp. 561-565.

Sournia, A. (Ed.), (1978). *Phytoplankton manual.* UNESCO, Paris, 1978.

Tande, K. S., Miller, C.B. (2000). Population dynamics of *Calanus* in the North Atlantic: Results from the trans-Atlantic study of *Calanus finmarchicus. ICES Journal of Marine Science,* Vol. 57, No. 6, pp. 1527-1527.

Taylor, G.I. and Batchelor, G.K. (1949). The effect of wire gauzes on small disturbances in a uniform stream. *Quarterly Journal of Mechanics and Applied Mathematics,* Vol. 2, pp. 1-29.

Topham, D.R. (1975). Hydrodynamics of an oil well blowout. Beaufort Sea Technical Report, Institute of Ocean Science, Sidney BC, No 33.

Tranter, D.J., and Heron, A.C. (1967). Experiments on filtration in plankton nets. *Australian Journal of Marine Freshwater Research,* Vol. 16, pp. 281-291.

Unstad, K., Tande, K. (1991). Depth distribution of *Calanus finmarchicus* and *C. glacialis* in relation to environmental conditions in the Barents Sea. *Polar Research,* Vol. 10, No. 2, pp. 409-420.

Wiborg, K.F., Hansen, K. (1974). Fiske og utnyttelse av raudåte (*Calanus finmarchicus*). Fisken og Havet, Serie B, Vol. 10, pp. 1-25, Institute of Marine Research, Bergen.

Wu, J. (1995). Mechanisms of animal cell damage associated with gas bubbles and cell protection by medium additives. *Journal of Biotechnology,* Vol. 43, pp. 81-94.

12

Evidence of Island Effects in South African Enterprise Ecosystems

Daan Toerien and Maitland Seaman
Centre for Environmental Management,
University of the Free State, Bloemfontein
South Africa

1. Introduction

Living organisms and business enterprises (hereafter the term *enterprise* will be used to refer to a business enterprise) exhibit certain similarities. Beinhocker (2006) stated that: "Economic wealth and biological wealth are thermodynamically the same sort of phenomena, and not just metaphorically. Both are systems of locally low entropy, patterns of order which evolved over time under the constraint of fitness functions. Both are forms of fit order." He added that organisms and business enterprises are both subjected to Darwinian-type competition. Based on the foregoing reasoning we proposed that there should be similarities between natural ecosystems and enterprise ecosystems as described by Toerien & Seaman (2010).

Natural ecosystems have been defined as biotic communities or assemblages and their associated physical environments in specific places (Tansley, 1935). A basic structural requirement is that an ecosystem encompasses a biotic complex, an abiotic complex, interaction between them and a physical space. Natural ecosystems can be of any size as long as organisms, the physical environment and their interactions can exist within them. Therefore, natural ecosystems can be as small as a patch of soil supporting plants and microbes, or as large as the entire biosphere of the Earth (Pickett & Cadenasso, 2002).

The relationship between enterprises and towns meets the above norms for ecosystems, and towns should be considered enterprise ecosystems (Toerien & Seaman, 2010), a hypothesis tested in studies of the enterprise structures of selected South African towns by employing techniques often used in ecological studies (Toerien & Seaman, 2010, 2012a, 2012b). Towns could be clustered on the basis of their enterprise structures and the groups such identified made sense (Toerien & Seaman, 2010). Proportionalities in enterprise structures observed in more than a hundred South African towns lent further credence to the hypothesis (Toerien & Seaman, 2012a). In addition, enterprise diversity followed patterns observed for biodiversity in natural ecosystems (Toerien & Seaman, 2012b).

The *Theory of Island Biogeography* (MacArthur & Wilson, 1967) marked the coming of age of ecological science as a discipline with a theoretical/conceptual base (May, 2010). It is the dominant symbol of a transition which took place four decades ago from descriptive to analytical approaches in ecology and biogeography (Losos & Ricklefs, 2010). Islands are ecological systems and many ecological systems have island attributes (Losos & Ricklefs, 2010).

Whittaker et al. (2010) stated that: "The MacArthur-Wilson model recognizes that, for a discrete and isolated biological system, the number of species at any point in time must be a function of the number previously occurring there plus those gained through immigration and/or speciation… minus those having gone locally extinct. Their theory proposes that these three fundamental processes should vary in a predictable fashion in response to time since system initiation, and in relation to two principal controlling geographical/ environmental influences: isolation and area." These ideas could be used to test hypotheses about enterprise ecosystems.

Begon et al. (1990) stressed that there are many types of islands because the term does not only encompass islands of land in a sea of water but lakes are 'islands' in a 'sea' of land, mountain tops are high-altitude 'islands' in a low altitude 'ocean', clearings in forests are 'islands' in a 'sea' of trees, etc. Can South African rural towns (as enterprise ecosystems) also be considered islands? We think that the enterprise structures of towns should also exhibit island effects. If correct, this would extend our original hypothesis that towns are enterprise ecosystems (Toerien & Seaman, 2010) to one that states that towns are enterprise ecosystems that also exhibit island effects. If correct the finding might have socio-economic consequences for South African towns.

The specific null hypothesis posed in this chapter is that South African towns (enterprise ecosystems) do not exhibit island effects as predicted by the Species Equilibrium Model of MacArthur & Wilson (1967).

1.1 Logic of chapter

The following part of the chapter provides a conceptual argument as to why rural South African towns could be considered enterprise islands. This is followed by a brief overview of the Species Equilibrium Model of natural ecosystems (MacArthur & Wilson, 1967) and its possible use in tests of hypotheses about enterprise islands. A brief overview of the history of the Karoo, South Africa's arid heartland, and an explanation of why a part of this region is suitable for use as a case study follow. Next there is an overview of the twelve towns of the Karoo midlands of the Eastern Cape province of South Africa which have been selected as a case study for the purposes of this chapter. It includes previous research done on this region and its towns. The next section deals with six different ways in which the null hypothesis was tested. This is followed by an analysis of the relationship between town age and enterprise development in the towns of the case study. The results are then discussed and this is followed by our conclusions.

2. Island biogeography and enterprise ecosystems

2.1 South African towns as islands

Before 1652 the south-western part of South Africa had been occupied for more than 70 millennia by hunter-gatherers (Henshilwood et al., 2002; Marean et al., 2007). About two millennia ago herders arrived in south-western Africa (Giliomee & Mbenga, 2007). Neither the hunter-gatherers nor the herders established permanent villages or towns. In 1652 the

Dutch East India Company established a victualing station at what is now Cape Town and from here European farmers spread into the inland areas over time, replacing the hunter-gatherers and herders (Giliomee & Mbenga, 2007). The need of these farmers for household goods e.g. cloth, salt and sugar were largely met by travelling pedlars who visited individual farms to barter their goods for items such as livestock, skins and home-made soap (Fransen, 2006; Giliomee & Mbenga, 2007; Van Waart, 2001; Wickins, 1983). Because the basic needs of the farmers and their families were met, there were few, if any, commercial reasons for the establishment of towns in inland South Africa (Fransen, 2006).

The actual establishment of inland South African towns arose from two other needs: (i) authorities (firstly the Dutch East India Company and later the British government) needed effective control and administration of rural communities, which in turn led to the establishment of administrative towns, called drostdys (Fransen, 2006; Giliomee & Mbenga, 2007), and (ii) the need of farmers and their families for religious services over time led to the establishment of new parishes and the building of churches in the vicinity of which first villages and later towns developed (called 'church towns' by Fransen, 2006). How then did the enterprises of these towns develop?

Entrepreneurs create enterprises. Entrepreneurs had to move from elsewhere to these towns to use the 'entrepreneurial space' that had become available. The pedlars mentioned before were the initial entrepreneurs that supplied the goods needed by the farmers and their families. Many of them were Jews with experience of trading, immigrants from Lithuania, Poland, Russia or Germany. Once they made enough money, many set up shops in the towns where Jewish communities played an important commercial role, also providing valuable contact with Jewish communities elsewhere in the world (Van Waart, 2001). The immigration of knowledgeable entrepreneurs thus formed an important part of the establishment of the commercial functions of rural South African towns. This process is akin to the colonisation of island by biota from elsewhere.

Conceptually it is possible to think about the establishment of the drostdys or the building of churches as the equivalent of volcanic activity forcing a new island to rise above the sea and becoming available to be colonised, e.g. Surtsey Island (Lindroth, 1973). Once an administrative centre was established or a church built in a specific location, people came to use the services on offer and it became possible for entrepreneurs to conceive possible business opportunities and to react to them. These proto-towns were thus 'islands' in a sea of farms through which entrepreneurs bringing much-needed entrepreneurship had to pass in order to 'colonise' the towns. The question, however, is whether these 'enterprise islands' responded similarly to colonisation than natural islands?

2.2 A brief consideration of the species equilibrium model for islands

The Species Equilibrium Model was first presented as a graph of gross extinction and immigration rates against the number of species present on an island (Schoener, 2010). It was based on two assumptions: (i) the rate of immigration of new species (those not yet present on the island) decrease monotonically with an increasing number of species already present. It reaches zero when all species in the source area are on the island, and, (ii) the rate of extinction of species increases monotonically as the number of species increases. These two assumptions imply that equilibrium between immigration and extinction would eventually occur, at which time the immigration and extinction rates will have the same value, called the turnover rate at equilibrium.

In-depth studies of the re-colonisation by biota of different islands after catastrophic events such as hurricanes, however, have shown that equilibrium can be steady (a constant number of species), cyclical (a regular fluctuation in the number of species), or moving directionally (a slow unidirectional change in the number of species brought on by a systemic change in immigration and/or extinction rates, e.g. due to climate change) (Schoener, 2010).

The Species Equilibrium Model also predicts that that species lists will vary in composition even after equilibrium is reached (Schoener, 2010) because some species may become extinct or emigrate. At the same time immigration of species would still occur and consequently there would be a turnover of species. This prediction was shown to be true by calculation of the relative turnover rates of different island biotas, e.g. for birds of the Channel Islands (Jones & Diamond, 1976).

The Species Equilibrium Model also makes predictions about the effects of distance between islands and their sources of immigrants and about the effects of the area of islands. Near islands of similar size should have more species and large islands at the same distance as small islands should have more species (Schoener, 2010). These predictions have also been verified by numerous studies of island biota (Lomolino et al., 2010).

2.3 The species equilibrium model and enterprises in rural South African towns

The Equilibrium Model provides insights into ways of testing the hypothesis that South African towns are enterprise islands. Whilst ecologists study the presence/absence of species, our studies have focused on individual enterprises in towns. If towns are enterprise islands one can expect evidence that the rate of establishment of new enterprises and the rate of disappearance of previously established enterprises would even out and that enterprise equilibrium would be established in the enterprise islands.

A first test of the null hypothesis, therefore, is a comparison of the enterprise lists [determined as described by Toerien & Seaman (2010)] of selected towns and for different years in order to determine: (i) whether enterprise equilibriums exist, and, (ii) if there is significant turnover of enterprises under condition of enterprise equilibrium. Number balances based on the enterprises at time t_1, enterprises which disappear between times t_1 and t_2, enterprises which appear between times t_1 and t_2, and the enterprises at time t_2, of selected towns would indicate whether there are equilibriums and if these are linked to enterprise turnover as predicted. Answers in the affirmative would contribute to the rejection of the null hypothesis. This exercise is described more fully later.

To examine the question of whether enterprise islands also respond to the proximity of a source of immigrant entrepreneurs is a more complex task. In the case of enterprises, proximity would have to refer to the sources of entrepreneurs who bring business ideas to towns from elsewhere. This is not a property that could be easily deduced from an examination of the names of enterprises in telephone directories and this question could only be dealt with obliquely. Toerien (2012), however, noted that gentrification of Prince Albert, a Karoo town, brought new entrepreneurial skills and development to the town, which suggests that proximity to entrepreneurs might be important.

A larger island that is as far from a mainland than a smaller island should according the Species Equilibrium Model, house more species than the smaller island (Schoener, 2010). If this statement is extended to towns, the challenge is to find appropriate surrogate measures of island or town size. The simplest way to estimate this is from resident numbers. However, for a long time the dominant theory about the roles and functions of South African towns was based on Christaller's Central Place Theory (Christaller, 1966, as cited in Van der Merwe & Nel, 1975). The theory was developed in 1933 after Christaller considered economic relationships between cities and their hinterlands (Van der Merwe & Nel, 1975). Christaller asserted that settlements (towns) functioned as 'central places' providing services to surrounding areas. The general usefulness of the theory has, however, been questioned (e.g. Fujita et al., 2005). Nevertheless, the implications of Christaller's theory should be considered, i.e. if a town is an 'enterprise island', the number of enterprises in a town should be related to the area of its surrounding district and/or the population size of that district. Prudence, therefore, required that the null hypothesis be tested against a number of indicators of town (and hence enterprise island) size: (i) the resident population (or urban population) numbers of towns, (ii) the total population of town and district, (iii) the rural population numbers of the districts surrounding towns, and (iv) the areas of the districts surrounding the towns. This exercise will be described more fully later.

In the case of rural South African towns, and in contrast to many towns in Europe and Asia, the founding dates of towns are precisely known. This provided a useful adjunct to the ideas derived from the Species Equilibrium Model, namely the possibility that older towns might have first-comer benefits. This aspect was also investigated.

The selection of towns to test the null hypothesis required consideration. Ideally the towns should be located in places with fairly similar climates, and the reasons for the founding of the towns and the governmental/regulatory systems under which they resorted should have been similar. Towns in parts of the Karoo of South Africa meet these requirements, as explained later.

3. A brief overview of the history of the Karoo, South Africa

The Karoo is a semi-arid area of approximately 400,000 km^2 situated in the central and western interior of South Africa (Nel & Hill, 2008). It is a single ecosystem, sub-divided into a winter rainfall and a summer rainfall area (Cowling, 1986). Khoikhoi herders moved some 1600 to 2000 years ago from the north into south-western South Africa (Boonzaaier et al., 1996; Giliomee & Mbenga, 2007). The Khoikhoi herded sheep and cattle, also in the Karoo. Their animals were rarely slaughtered. Herds expressed personal wealth rather than commercial utility and land was communally owned (Boonzaaier et al., 1996). To find good pastures tribes of herders moved around according to climate and season.

In 1652 the Dutch East India Company established a victualing station at the Cape of Good Hope to provide fresh produce and meat to the crews of their ships. Thereby a permanent European presence in the south-western Cape was established (Giliomee & Mbenga, 2007; Guelke, 1979). Their limited capacity to raise stock soon meant that cattle and sheep had to be sourced from the Khoikhoi (Giliomee & Mbenga, 2007). The Company's thrust into the more distant domains of the Khoikhoi consisted of three distinct, though overlapping, phases (Elphick, 1979).

A 'trading frontier' to obtain livestock from the Khoikhoi expanded steadily until about 1700. However, the ability of the Khoikhoi to supply enough livestock also became limited (Elphick, 1979). Secondly, the Dutch East India Company started allocating land that had traditionally fallen under Khoikhoi control and allowed free farmers to settle there (Wickins, 1983). The third frontier was one of semi-nomadic European pastoralists (called 'trekboers') who moved inland (Elphick, 1979). The farmers adopted the agricultural technologies of the Khoikhoi, i.e. the herding of fat-tailed sheep and cattle adapted to local conditions. The 'trekboers' supplied livestock to the Dutch East India Company but were more subsistence than commercial farmers. They were not historically or otherwise inclined to be commercial entrepreneurs.

Wool production in the south-western Cape was negligible up to the end of the 18th century. By 1804 fewer than 8,000 sheep out of a total population of 1,34 million were wool-producing merinos. Burrows (1994) commented: "During the next fifty years wool farming became the staple economy of the countryside. The merino sheep provided farmers with a secure and rising income, which released them from the poverty of subsistence stock farming." In 1830 the Cape exported 15,000 kg of wool and 22 million kg by 1872. By the mid-19th century the sheep farmers were firmly linked into the Colonial and global economies and wool was the Cape Colony's major staple export. Sheep farmers in the Karoo were part of the wool production system and their need for religious services as well as the government's need for administrative control drove the establishment of new parishes and drostdys followed by the development of villages and towns (Fransen, 2006). The sheep farmers' need for household and other goods were initially satisfied by itinerant pedlars and later by traders in towns.

4. The towns of the Karoo midlands of the Eastern Cape

In South Africa there are areas within which: (i) the founding reasons and dates of towns are well documented (Fransen, 2006), (ii) where other reasons for town evolution such as the exploitation of minerals are absent, and, (iii) in which at any time a single authority has always governed the whole of the region. The Karoo midlands of the Eastern Cape Province is one such area and provides an opportunity for studying possible 'island' effects in enterprise development in South African towns. The twelve towns (Table 1, Figure 1) of the region have also been studied in reasonable detail before (Nel & Hill, 2008). This contributed to the potential of the area and its towns to serve as a suitable case study for the purposes of our study.

The twelve towns were established to serve rural communities in a pre-industrial era. The three towns of drostdy origin (Cradock, Graaff-Reinet and Somerset East) are the oldest (Table 1) with ages (in 2011) that varied from 186 to 226 years. The younger towns are all 'church towns' and their ages in 2011 varied between 135 and 157 years (Table 1). Over their history the towns and/or their hinterlands have at all times fallen under the same authorities/governments: the Dutch East India Company until 1795, the British and Batavian governments in turn until 1806, then the British government until 1910, the Cape Provincial government of South Africa until 1994, and thereafter the Eastern Cape Provincial government of South Africa. Differences in enterprise structures and dynamics should, therefore, not be ascribed to different government systems, policies and practices.

Towns	Year Founded	Total population (2004)*	Enterprises (2006/07)
Aberdeen	1858	8670	46
Cradock	1814	39296	303
Graaff-Reinet	1785	42208	356
Hofmeyr	1873	5243	18
Jansenville	1854	11452	50
Middelburg	1852	23786	166
Pearston	1859	5182	20
Somerset East	1825	30459	199
Steynsburg	1872	8437	41
Steytlerville	1876	5632	35
Ventersburg	1875	5243	18
Willowmore	1862	12358	55

* After Nel & Hill (2008)

Table 1. Selected Karoo towns of the Eastern Cape Province, South Africa.

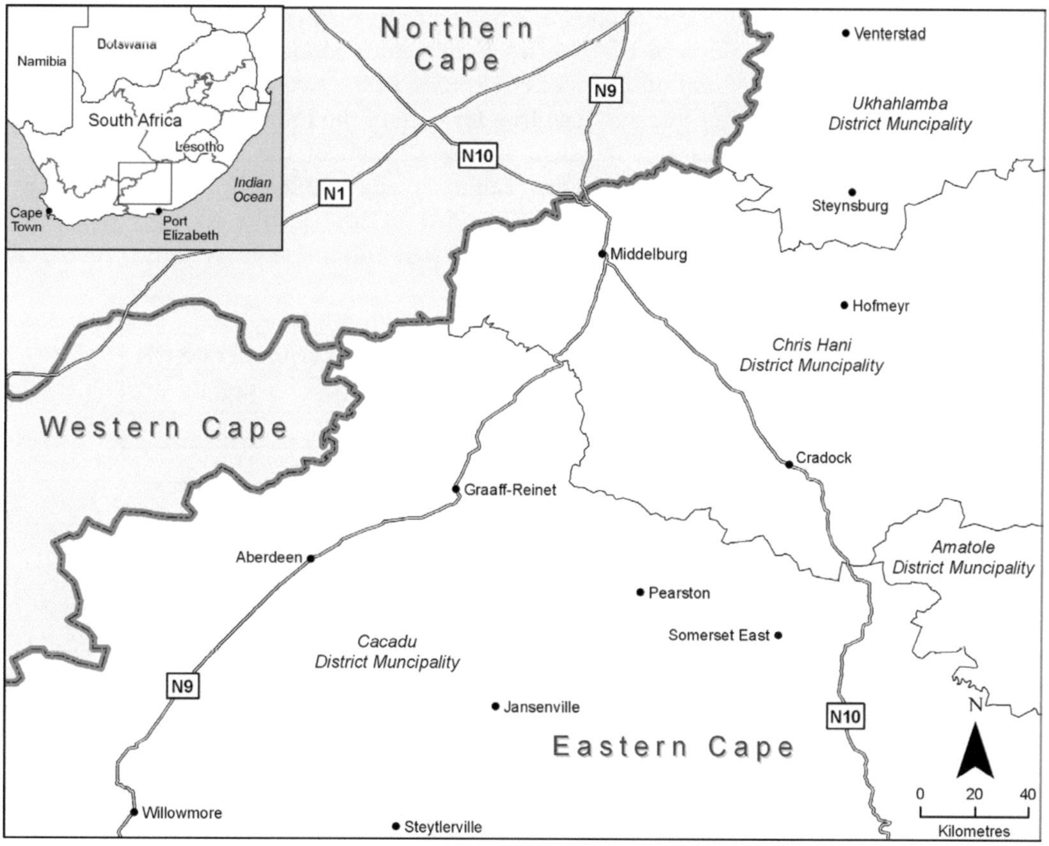

Fig. 1. A map of the selected Eastern Cape Karoo towns.

5. Testing the null hypothesis

5.1 Drawing up enterprise lists and constructing enterprise number balances

Telephone directories for Port Elizabeth and the Eastern Cape Country were used to identify the enterprises listed for the twelve towns (Table 1) for the years 2006/07 (Telkom Directory Services, 2006) and 2010/11 (Trudon, 2010). These enterprises were classified into 19 business sectors as described by Toerien & Seaman (2010). The lists were scrutinised to identify all enterprises which disappeared from or were new to the lists between 2006/07 and 2010/11. Through this process complete number balances (total enterprises as well as for each of the 19 business sectors) were constructed for each of the twelve towns. Turnover rates were calculated according to the following equation (Schoener, 2010):

$$\textit{Turnover rate} = \frac{\text{(No. of disappearences + No. of new enterprises) x 100}}{\text{(No. of enterprises in 2006/07 + No. of new enterprises in 2010/11)}} \qquad (1)$$

5.2 Enterprise equilibriums and turnover rates

The first test of the null hypothesis showed that enterprise development in the selected towns followed the predictions of the Species Equilibrium Model (see regional totals in Table 2). Not only were the total numbers of enterprises in the two different years for all of the towns almost identical, but this was also true for each of the towns. The reason was that the number of disappearances and new enterprises in each of the towns were in good balance. The turnover rates were high (Table 2) and reflected very competitive environments as predicted by Beinhocker (2006). Based on these results, the null hypothesis can clearly be rejected. South African towns are not just enterprise ecosystems (Toerien & Seaman, 2010) but also enterprise islands.

Town	2006/07	Extinct	New	2010/11	Turnover rate (%)
Aberdeen	43	5	8	46	14.6
Cradock	303	65	65	303	21.5
Graaff-Reinet	346	93	103	356	27.9
Hofmeyr	17	0	1	18	2.9
Jansenville	47	4	7	50	11.3
Middelburg	162	32	36	166	20.7
Pearston	21	5	4	20	22.0
Somerset East	198	42	43	199	21.4
Steynsburg	47	8	2	41	11.4
Steytlerville	30	8	13	35	32.3
Venterstad	20	4	2	18	15.8
Willowmore	54	9	10	55	17.4
Regional total	1288	275	294	1307	21.9

Table 2. Enterprise numbers and turnover rates for twelve Eastern Cape Karoo towns.

The same general picture emerged from number balances of different business sectors (Table 3). A few sectors showed some growth (e.g. the construction, health services, tourism and hospitality, and real estate sectors), most were very balanced and a few sectors showed some decline (e.g. legal services, processing, and the trade sectors). In general the adherence to the prediction of stable equilibriums by the Species Equilibrium Model was remarkable. These results also support the rejection of the null hypothesis.

Sector	2006/07	Extinct	New	2010/11	Turnover rate (%)
Agricultural Products & Services	106	16	15	105	14.7
Construction	43	5	15	53	20.8
Engineering & Technical Services	32	6	5	31	17.5
Factories	5	0	1	6	9.1
Financial Services	80	19	16	77	22.3
General Services	56	11	12	57	20.4
Health Services	101	21	38	118	26.9
Legal Services	21	9	5	17	36.8
Mining	0	0	0	0	0.0
News and Advertising	3	0	0	3	0.0
Personal Services	101	17	18	102	17.2
Processing	22	5	2	19	17.1
Professional Services	42	16	15	41	37.3
Real Estate Services	16	5	9	20	38.9
Telecommunications Services	14	8	5	11	52.0
Tourism and Hospitality	171	41	50	180	25.9
Trade	305	60	51	296	18.5
Transport & Earthworks	39	7	7	39	17.9
Unknown	31	13	11	29	40.0
Motor vehicles	100	16	19	103	17.2
Total enterprises	1288	275	294	1307	21.9

Table 3. Enterprise numbers and turnover rates between 2006/07 and 2010/11 of the different business sectors of twelve Eastern Cape Karoo towns.

The data presented in Table 3 allowed us to identify the business sectors that provide the 'momentum' for change in the enterprise structures of the selected towns. The 2006/07 numbers, the disappearances, new ventures and numbers for 2010/11 (Table 4) were normalised as percentages of the total enterprises given in Table 3.

Momentum for change was calculated as follows (equation 2):

$$\begin{aligned}&\textit{Sector's contribution to momentum for change} =\\&(\%\ \textit{of disappearences} + \%\ \textit{of new})(\%\ \textit{of enterprises in}\ 2006-2007 +\\&\%\ \textit{of enterprises in}\ 2010-2011)\ \textit{and expressed as}\ \%\ \textit{of total momentum}\end{aligned}$$

Business sector	2006/07	Disappear	New	2010/11	Momentum
Trade	23.7	21.8	17.3	22.6	44.7
Tourism and Hospitality	13.3	14.9	17.0	13.8	21.3
Health Services	7.8	7.6	12.9	9.0	8.5
Personal Services	7.8	6.2	6.1	7.8	4.7
Motor vehicles	7.8	5.8	6.5	7.9	4.7
Agricultural Products & Services	8.2	5.8	5.1	8.0	4.4
Financial Services	6.2	6.9	5.4	5.9	3.7
General Services	4.3	4.0	4.1	4.4	1.7
Professional Services	3.3	5.8	5.1	3.1	1.7
Construction	3.3	1.8	5.1	4.1	1.3
Unknown	2.4	4.7	3.7	2.2	1.0
Transport & Earthworks	3.0	2.5	2.4	3.0	0.7
Engineering & Technical Services	2.5	2.2	1.7	2.4	0.5
Legal Services	1.6	3.3	1.7	1.3	0.4
Real Estate Services	1.2	1.8	3.1	1.5	0.3
Telecommunications Services	1.1	2.9	1.7	0.8	0.2
Processing	1.7	1.8	0.7	1.5	0.2
Factories	0.4	0.0	0.3	0.5	0.0
Mining	0.0	0.0	0.0	0.0	0.0
News and Advertising	0.2	0.0	0.0	0.2	0.0
Total	100.0	100.0	100.0	100.0	100.0

Table 4. The contribution of the different business sectors to the momentum for change in the enterprise structures of the Karoo towns of the Eastern Cape.

More than 80% of the momentum for change in enterprise structures was contributed by only six business sectors: trading, tourism and hospitality, health services, personal services, the motor vehicle sector and the agricultural products and services sector. It seems as if an 80/20 rule (Pareto Principle) applies.

5.3 Urban population size and number of enterprises

The Species Equilibrium Model predicts that larger islands should have more species than smaller islands because they offer more space for colonisation by different species (Schoener, 2010). The first test - whether this prediction is also true for enterprise islands - was to compare total enterprise numbers and urban populations of the twelve towns, with urban populations serving as surrogate for the size of the 'islands'. In this regard we were fortunate that population and enterprise numbers had been determined for specific years over a period of about a century for the twelve towns (Nel & Hill, 2008) and we had received permission to use their data. This allowed us to extract 11 different data pairs for the same year or close together years of urban population sizes and enterprise numbers for the twelve towns. To test if enterprise numbers (dependent variable) for the twelve towns showed the same patterns of variance than urban population numbers (independent variable) Pearson correlation coefficients and regression equations were calculated with the use of Microsoft Excel software (Table 5).

Highly significant ($P < 0.01$) correlations were established for every data pair. With the exception of 1904, more than 80 and often more than 90 per cent of the variance (R^2) of the data pairs was explained. The urban populations and the number of enterprises in towns followed similar patterns. Over a period of a century larger enterprise islands housed more enterprises and vice versa; the towns as enterprise islands fitted the predictions of the Species Equilibrium Theory. Based on these results the null hypothesis could also be rejected.

Year(s)	Correlation	Slope	Intercept	% Variance explained (R^2)	Significance
1904	0.73	0.0090	32.2	54.0	P < 0.01
1911	0.97	0.0116	16.6	94.9	P < 0.01
1921	0.92	0.0091	26.8	85.5	P < 0.01
1935/36	0.90	0.0077	17.6	81.2	P < 0.01
1951	0.98	0.0072	16.6	95.9	P < 0.01
1960/61	0.98	0.0094	3.5	95.5	P < 0.01
1970	0.98	0.0076	-1.0	96.0	P < 0.01
1980	0.96	0.0094	1.0	93.1	P < 0.01
1990/91	0.98	0.0076	-7.8	96.9	P < 0.01
2000/01	0.99	0.0080	-19.4	97.3	P < 0.01
2004/06	0.99	0.0091	-29.6	97.8	P < 0.01

Table 5. A century of urban population-enterprise relationships of twelve Eastern Cape Karoo towns.

5.4 Rural population size and number of enterprises

The support in South Africa for the Central Place Theory (e.g. Van der Merwe & Nel, 1975) necessitated an investigation of the relationship between the number of enterprises in the selected towns and rural population sizes. The data of Nel and Hill allowed us to extract 11 different data pairs of rural population and enterprise numbers for the same year combinations used for the urban population-enterprise analysis. Pearson correlation coefficients and regression equations were also calculated with Microsoft Excel software (Table 6).

The rural populations and enterprise numbers of the region were with the exception of 1904 highly significantly ($P < 0.01$) correlated and from about 50 to more than 70 per cent of the variance was explained (Table 6). Larger towns were associated with larger rural populations. It is noticeable that the correlation coefficients in this exercise were generally lower than those for the urban populations (Table 5). Nevertheless the prediction of the Species Equilibrium Model about the impact of island size on species richness (Schoener, 2010) also applied to this surrogate of island size and it served to reject the null hypothesis.

Year(s)	Correlation	Slope	Intercept	% Variance explained (R^2)	Significance
1904	0.52	0.0086	1.7	26.6	Not significant
1911	0.79	0.0078	-2.2	63.0	P<0.01
1921	0.79	0.0097	-1.2	62.5	P<0.01
1935/36	0.78	0.0090	-0.8	60.2	P<0.01
1951	0.81	0.0082	3.2	66.4	P<0.01
1960/61	0.75	0.0111	-9.8	56.3	P<0.01
1970	0.81	0.0120	-16.3	65.2	P<0.01
1980	0.70	0.0161	4.8	48.7	P<0.01
1990/91	0.79	0.0202	-9.8	61.8	P<0.01
2000/01	0.81	0.0276	-28.8	65.9	P<0.01
2004/06	0.88	0.0242	-45.9	77.2	P<0.01

Table 6. A century of rural population-enterprise relationships for twelve Eastern Cape Karoo towns.

5.5 Total population size and number of enterprises

The fact that enterprise numbers in the selected towns were significantly correlated to both the rural and the urban populations (Tables 4 and 5) suggested that they should also be correlated with total population numbers for the districts in which the towns are located. This was indeed the case and, for all data pairs except for one (1904), more than 80 per cent of the variance was explained (Table 7). This analysis also supports the rejection of the null hypothesis. The results also question the modern applicability of the Central Place Theory to South African towns because enterprise numbers are not only correlated with the rural populations (Table 6), but more strongly with the total and urban populations (Tables 5 and 7).

Year(s)	Correlation	Slope	Intercept	% Variance explained (R^2)	Significance
1904	0.84	0.0070	-8.59	70.2	P<0.01
1911	0.94	0.0054	-2.20	88.4	P<0.01
1921	0.92	0.0061	2.53	83.8	P<0.01
1935/36	0.89	0.0050	2.68	78.6	P<0.01
1951	0.90	0.0042	7.30	80.5	P<0.01
1960/61	0.92	0.0056	-9.99	85.1	P<0.01
1970	0.98	0.0053	-16.74	95.6	P<0.01
1980	0.93	0.0067	-8.76	86.5	P<0.01
1990/91	0.97	0.0060	-15.18	94.4	P<0.01
2000/01	0.97	0.0065	-26.51	94.9	P<0.01
2004/06	0.97	0.0070	-36.95	94.9	P<0.01

Table 7. A century of district population-enterprise relationships for twelve Eastern Cape Karoo towns.

5.6 Island size and turnover rates

The Species Equilibrium Theory also predicts that the turnover rate in species richness is a function of island size: larger islands have larger turnover rates and vice versa. To examine this prediction, we calculated Pearson correlation coefficients between the average number of enterprises for 2006/07 and 2010/11 (surrogate for 'island size' and independent variable) and the turnover rates (dependent variable) for the different towns. This correlation was not statistically significant (P = 0.05) (Figure 2). However, an examination of the data suggested that two outlier towns (Hofmeyr and Steytlerville, see arrows in Figure 2) were primarily responsible for the negative result. When these towns were excluded in a correlation analysis, a statistically significant ($P < 0.01$) correlation coefficient of 0.76 was obtained (as suggested by the trend line in Figure 2). This suggests that the prediction of the Species Equilibrium Model might be correct, but that it requires further investigation. In this test the null hypothesis cannot be rejected.

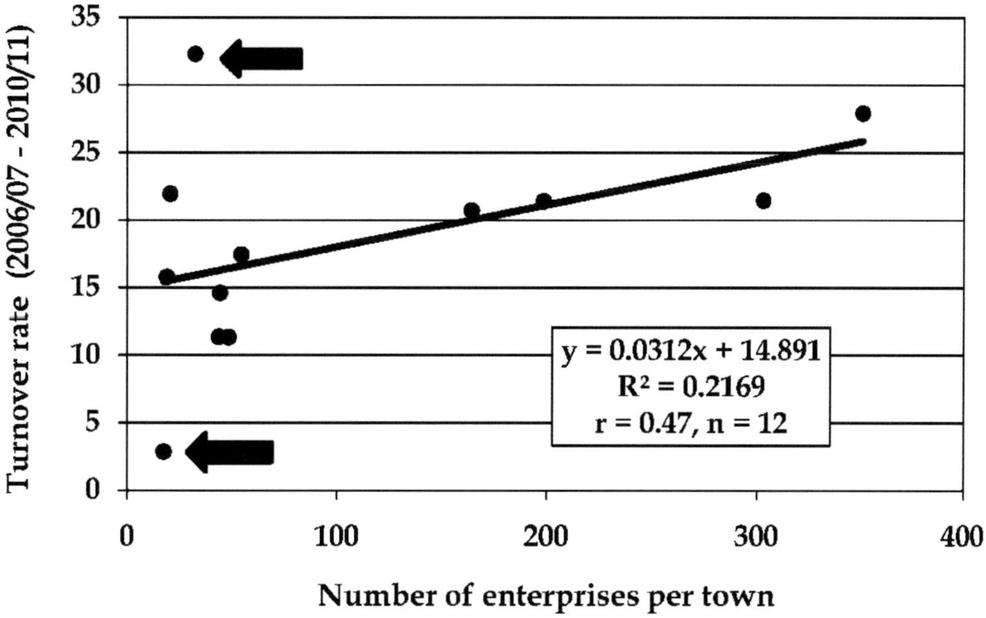

Fig. 2. Turnover rate as a function of the size of enterprise structures of towns.

5.7 District area and the number of enterprises in the selected towns

The Central Place Theory suggests that towns are very dependent on their rural hinterlands (Christaller, 1966 as quoted by Van der Merwe & Nel, 1975). District area could possibly also serve as a surrogate for 'enterprise island' size and could be used to test the null hypothesis. The data of Nel and Hill contained three different estimates (for 1938, 1981 and 1993 respectively) of the district areas surrounding the selected towns. This allowed us to examine the relationship between district area and the enterprise numbers for the closest years to the years identified above: (i) 1935 and 1938, (ii) 1980 and 1981, and, (iii) 1990 and 1991.

The district areas and enterprise numbers were significantly ($P< 0.05$) correlated and 40 to 48% of the variance was explained (Table 8). The use of this surrogate for 'island size' also supported the prediction of the Species Equilibrium Model about the impact of island size on species richness, here measured as enterprise richness, and serves to reject the null hypothesis.

Area - enterprises	Correlation	Slope	Intercept	% of variance explained (R^2)	Significance
1935-38	0.69	0.00015	-7.57	47.9	P<0.01
1980-81	0.67	0.00028	-37.20	44.3	P<0.05
1990-91	0.64	0.00026	-35.62	40.4	P<0.05

Table 8. The relationships of enterprise numbers of the twelve Eastern Cape Karoo towns and district areas at three separate times.

5.8 The impact of town age on enterprise numbers

In the case of rural South African towns, and in contrast to many towns in Europe and Asia (e.g. Braudel, 1979), the founding dates of towns are precisely known. This provided an opportunity to assess if older towns have first-comer advantages as far as enterprise development is concerned. This was indeed the case (Table 9) and for the eleven different years during the past century that were analysed there were statistically significant ($P < 0.01$) positive correlations. The first-comer advantages stayed in place for more than a century. The steady increase with time of the slopes and the decreases in the intercepts of the regression equations (Tables 7 and 9) suggest that the larger (and older) towns of the Eastern Cape Karoo have in modern times grown their enterprise numbers more rapidly than smaller towns.

Year	Correlation	Slope	Intercept	Variance explained (%)	Significance
1904	0.81	1.22	-6.4	65.8	P<0.01
1911	0.93	1.03	-9.9	86.4	P<0.01
1921	0.94	1.27	-25.7	88.6	P<0.01
1935	0.84	1.15	-39.6	69.9	P<0.01
1951	0.92	1.28	-70.8	84.9	P<0.01
1961	0.94	2.12	-163.0	88.7	P<0.01
1970	0.92	1.92	-165.5	85.0	P<0.01
1980	0.95	2.75	-266.8	90.1	P<0.01
1990	0.93	2.46	-266.9	86.6	P<0.01
2000	0.94	3.00	-369.2	87.8	P<0.01
2006	0.95	3.32	-438.9	90.3	P<0.01

Table 9. The relationships of enterprise numbers (dependent variable) and town ages (independent variable) of the twelve Eastern Cape Karoo towns.

6. Discussion

Veblen (1898, as cited in Witt, 2008) introduced the notion of evolutionary economics, an approach that would attract much support later. However, there is still little agreement among the researchers in the field when it comes to deciding what is specific about evolutionary economics (Witt, 2008). Some interpretations of evolutionary economics consider the Darwinian theory of evolution relevant for understanding economic behaviour and the transformation of economic institutions and technology. Other interpretations do not embrace, or even explicitly reject, that idea (Witt, 2008). He stressed that at the core, this is a controversy about the basic (ontological) assumption about the structure of reality. It relates to the question of whether evolutionary change in nature and in the economy represent connected spheres of reality, making them likely to mutually influence each other, an issue important to this study. In addition there is the issue of whether evolutionary theorising in economics can profit from borrowing analytical tools from evolutionary biology, e.g. models of selection processes and population dynamics (Witt, 2008).

Witt (2004) cautioned that one should be careful to accept that evolution in the economy and evolution in nature are similar or even identical. Beinhocker (2006) belongs to the school of belief of the similarities between biological and economic evolution and he discussed these in great detail in his book, *The Origin of Wealth*. His views, briefly presented earlier, will not be further examined here, except to state that Toerien & Seaman (2010) built on them by examining and accepting the hypothesis that towns are enterprise ecosystems.

The Species Equilibrium Model (MacArthur & Wilson, 1967) marked the transition of ecology from a descriptive to an analytical science (Losos & Ricklefs, 2010). On the other hand town studies in South Africa had already by the 1960s involved some quantitative approaches, e.g. applied by Davies & Cook (1968) and extensively discussed by Van der Merwe & Nel (1975). At that time the hierarchical order of towns was the primary focus (e.g. Davies & Cook, 1968). There was little or no appreciation of the analogy between the systems in which organisms compete, survive or die and the systems wherein enterprises compete, survive or die, as discussed by Beinhocker (2006).

We believe that the analogy provides a new way of thinking about and analysing South African towns. To do this we first conceptually developed the hypothesis that many South African towns (i.e. enterprise ecosystems) could also be viewed as enterprise islands in seas of farms. If the contention is correct, the Species Equilibrium Model (perhaps in modified form) could be applied in analyses of town enterprise structures. We posed a null hypothesis that the towns are not enterprise islands. Overwhelming evidence was obtained that the null hypothesis could be rejected. As suggested by the model, quantification of enterprise numbers for 2006/07 and 2010/11 for selected case study towns indicated stable equilibriums in regional as well as individual town enterprises despite high turnover rates (Table 2). The same general picture emerged for enterprise numbers of different business sectors (Table 3). In addition some evidence, not statistically significant, were obtained that the magnitude of turnover rates is higher in larger enterprise islands (Figure 2).

The Species Equilibrium Model also predicts that larger islands will have more species than smaller islands when equidistant from a mainland (source of immigrants). Different surrogates of island size (or town size) were used in statistical examinations of this prediction for enterprise numbers in towns and all of the comparisons supported the rejection of the null hypothesis (Tables 5 to 8). In fact, it was demonstrated that some of these relationships or proportionalities have existed in the case study towns for a period of more than a century. In addition, there is a positive link between the number of enterprises and the ages of towns (Table 9), indicating fist-comer advantages.

Based on the evidence presented as well as on the Species Equilibrium Model it is possible to think of an Enterprise Equilibrium Model for towns that could explain the enterprise proportionalities observed in the enterprise structures of 140 South African towns (Toerien & Seaman, 2012a) or the century-long proportionalities between different population group and enterprise numbers in the Eastern Cape (Tables 5 to 7).

Acceptance of such a model would be dependent on an understanding of what the units of evolutionary selection in economic evolution are or, in other words, what the economic equivalent of a gene is. Beinhocker (2006) expounded on this issue. He stated that businesses are the interactors that struggle in survival-of-the-fittest competition in economic evolution but they are not the units of selection (similar to organisms in biology). Modules (developed by entrepreneurs or managers) are components of business plans (explicit or implicit) that provide a basis for the differential selection between businesses in competitive environments. They are the units of selection in economic evolution as genes are in biological evolution. Successful genes increase in frequency in biological populations and successful business plan modules extend their influence over resources over time (Beinhocker, 2006). In other words, these successful modules are repeated more often in business plans as more and more entrepreneurs/managers apply them.

The Species Equilibrium Model recognises that three fundamental processes, i.e. species gained through immigration and/or speciation minus those having become extinct locally, determine the number of species at any point in time and result in species equilibriums. In addition, two principal controlling geographical/environmental influences, isolation and area, contribute to the magnitude of the equilibriums achieved (Whittaker et al., 2010). this study showed that enterprise equilibriums are also determined by significant turnover rates resulting from the appearance of new ventures and the disappearance of existing enterprises. The resulting equilibriums are proportional to the size of the respective towns (enterprise islands).

Two of these natural processes deserve further comment: isolation and speciation, especially in terms of what their equivalents may be in economic evolution. In the case of towns new ventures depend, if only in part, on their proximity to (or isolation from) entrepreneurs, whilst the ability of managers to adopt their business plans to change business operations might be the equivalent of speciation on natural islands. Isolation and speciation are, albeit in modified form, probably also important in economic evolution.

An important difference in this regard needs to be noted. Whereas with natural islands it would be impossible to physically change isolation, the same is not true as far as isolated enterprise islands are concerned. Local authorities and development agencies can use (and have used) different tactics to bring entrepreneurs in contact with existing or potential business opportunities in cities/towns. The familiar view of job creation is that business location is largely a function of traditional economic values such as tax structure and cost of doing business. However, Johnson & Rasker (1995) found that positive economic growth also resulted from the maintenance of a high environmental quality. This suggests some level of entrepreneurial activity may be encouraged through efforts which maintain local environmental and quality of life amenities. There are probably many different creative ways by which towns can expose their entrepreneurial opportunities to entrepreneurs from elsewhere. In addition existing enterprises use tactics such as training and consulting to improve the skills of their staff and their abilities to strategize and compete more effectively.

Finally, one further issue raised by Beinhocker (2006) deserves attention. He noted that as complex adaptive open systems the natural world as well as the economic world is subject to entropy. In open systems there is a never-ending battle between energy-powered order creation and entropy-driven order destruction. Enterprise ecosystems, therefore, are driven by energy, which costs money. Toerien & Marais (2012) argued that the money that flows to or circulates in a town determines: (i) the number of businesses that can be supported, and, (ii) the number of people that can exist in the town. Therefore population numbers serve well as surrogate measures (Tables 5 to 7) of such money flows. The entrepreneurial spaces of enterprise islands ultimately probably depend on money.

7. Conclusions

The rejection of the null hypothesis that towns are not enterprise islands because of the good fit of rural South African towns to predictions derived from the Species Equilibrium Model supports contentions: (i) that natural and economic systems are very similar, and, (ii) the ecology and the economy can learn from each other (Beinhocker, 2010). Rural South African towns are not just enterprise ecosystems (Toerien & Seaman, 2010) but also enterprise islands.

We believe that the similarities between biological evolutionary systems and evolutionary economic systems, as pointed out by many including Beinhocker (2006), result in the possibility of applying methods and approaches used in the one successfully in the other. For instance, this study has shown that it is possible to develop new insights about entrepreneurship in rural towns by applying the Species Equilibrium Model developed for island biogeography. As a consequence it is possible to think about the development of an Enterprise Equilibrium Model for towns and its application in decisions about local economic development.

8. Acknowledgements

We thank Ettienne Nel, Trevor Hill and Doreen Atkinson for permission to use their population, enterprise and area data of the case study towns. Frank Sokolic was responsible for the map of the study area and helped with table outlays and Figure 2. Annamarie du Preez provided library support, Marie Toerien analytical support and Max Döckel valuable comments about the manuscript. Marilie van der Walt of Indigo Language Solutions provided language editing. Financial support was received from the Centre for Environmental Management, University of the Free State, Bloemfontein, South Africa.

9. References

Beinhocker, E.D. (2006) The Origin of Wealth: Evolution, Complexity, and the Radical Remaking of Economics. Harvard Business School Press, ISBN 1-57851-777-X, Boston.

Begon, M., Harper, J.L., & Townsend, C.R. (1990) Ecology: Individuals, Populations and Communities, 2nd Edition, Blackwell Science, ISBN 0-86542-111-0, Cambridge.

Boonzaaier, E., Malherbe, C., Smith, A. & Berens, P. (1996) The Cape Herders: A History of the Khoikhoi in Southern Africa, David Philip, ISBN 0 86486 311 X, Cape Town.

Braudel, F. (1979) Civilization and Capitalism. 15th–18th Century. Volume 1. The structures of Everyday Life: The Limits of the Possible. Translation from the French revised by S. Reynolds, University of California Press, ISBN 0-520-08114-5, Berkeley.

Burrows, E.H. (1994) Overberg Outspan: A Chronicle of People and Places in the South Western Districts of the Cape, Facsimile Edition, Swellendam Trust, ISBN 0 620 12930 1, Swellendam.

Cowling, R.M. (1986) A Description of the Karoo Biome Project. South African National Scientific Programmes Report No. 122, Foundation for Research Development, ISBN 0 7988 3803 5, Pretoria.

Davies, R.J. & Cook, G.P. (1968) Reappraisal of the South African urban hierarchy, South African Geographical Journal, Vol. 50, No. 2, 116-132.

Elphick, R. (1979) The Khoisan to c. 1770. In: The Shaping of South African Society, 1652–1820. Ed. by R. Elphick & H. Giliomee, pp. 3–33, Maskew Miller, ISBN 0 636 00516 X, Cape Town.

Fransen, H. (2006) Old Towns and Villages of the Cape: A Survey of the Origin and Development of Towns, Villages and Hamlets at the Cape of Good Hope, Jonathan Ball Publishers, ISBN 1 86842 227 5, Johannesburg.

Fujita, M., Krugman, P. & Venables, A.J. (2005) The Spatial Economy: Cities, Regions, and International Trade, MIT Press, ISBN 978-0-262-06204-6, Cambridge.

Giliomee, H. & Mbenga, B. (2007) New History of South Africa. Tafelberg, ISBN-13: 978 0 624 04358 4, Cape Town.

Guelke, L. (1979) The white settlers, 1652–1780, In: The Shaping of South African Society, 1652–1820. Ed. by R. Elphick & H. Giliomee, pp. 41–74, Maskew Miller, ISBN 0 636 00516 X, Cape Town.

Henshilwood, C.S., d'Errico, F., Yates, R., Jacobs, Z., Tribolo, C., Duller, G.A.T., Mercier, N., Sealy, J.C., Valladas, H., Watts, I, & Wintle, A.G. (2002) Emergence of modern human behavior: Middle Stone Age engravings from South Africa, *Science,* Vol. 295, No. 5558, 15 February 2002, 1278–1280.

Johnson, J.D. & Rasker, R. (1995) The role of economic and quality of life values in rural business location. Journal of Rural Studies, Vol. 11, No. 4, 405-416.

Jones, H.L. & Diamond, J.M. (1976) Short-time-base studies of turnover in breeding bird populations on the Californian Channel Islands, The Condor Vol. 78, No. 4, 526–549.

Lindroth, C.H. (1973) Surtsey, Iceland: The Development of a New Fauna, 1963-1970. Terrestrial Invertebrates. Munksgaard, Copenhagen.

Lomolino, M.V., Riddle, B.R., Whittaker, R.J. & Brown, J.H. (2010) Biogeography, 4th ed. Sinauer Associates, ISBN-10: 0878934944, Sunderland.

Losos, J.B. & Ricklefs, R.E. (eds.) (2010) The Theory of Island Biogeography Revisited, pp. xi–xiv, Princeton University Press, ISBN 978-0-691-13653-0, Princeton.

MacArthur, R.R.H. & Wilson, E.O. (1967) The Theory of Island Biography, Princeton University Press, ISBN 0-691-08836-5, Princeton.

Marean, C.W., Bar-Matthews, M., Bernatchez, J., Fisher, E., Goldberg, P., Herries, A.I.R., Jacobs, Z., Jerardino, A., Karkanas, P., Minichillo, T., Nilssen, P.J., Thompson, E., Watts, I. & Williams, H.M. (2007) Early human use of marine resources and pigment in South Africa during the Middle Pleistocene. Nature, 449, 18 October 2007, 905-908.

May, R.M. (2010) Foreword, In: The Theory of Island Biogeography Revisited, Ed. by Losos, J.B. & Ricklefs, R.E., pp. vii-ix, Princeton University Press, ISBN 978-0-691-13653-0, Princeton.

Nel, E.L. & Hill T.R. (2008) Marginalisation and demographic change in the semi-arid Karoo, South Africa. Journal of Arid Environments, Vol. 72, No. 12, 2264-2274.

Pickett, S.T.A. & Cadenasso, M.L. (2002) The ecosystem as a multidimensional concept: meaning, model, and metaphor. Ecosystems Vol. 5, No. 1, 1–10.

Schoener, T.W. (2010) The MacArthur-Wilson equilibrium model: A chronicle of what it said and how it was tested. In: The Theory of Island Biogeography Revisited, Ed. Losos, J.B. & Ricklefs, R.E., pp. 52-87, Princeton University Press, ISBN 978-0-691-13653-0, Princeton.

Tansley, A.G. (1935) The use and abuse of vegetational concepts and terms. Ecology Vol. 16, No. 3, 284-307.

Telkom Directory Services (2006) Port Elizabeth & Eastern Cape Country 2006/2007, Telkom Directory Services, Johannesburg.

Toerien, D.F. (2012) Prince Albert: A fourth economic bubble or sustainable development? In: Small Town Geographies in Africa in press, Ed. Donaldson, R. & Marais, L., Nova Science Publishers, Hauppauge, N.Y.

Toerien, D.F. & Marais, L., (2012) Classification of South African towns revisited. In: Small Town Geographies in Africa, in press, Ed. Donaldson, R. & Marais, L., Nova Science Publishers, Hauppauge, N.Y.

Toerien, D.F. & Seaman M.T. (2010) The enterprise ecology of towns in the Karoo, South Africa. South African Journal of Science, Vol 106, No. 1, 24–33.

Toerien, D.F. & Seaman M.T. (2012a) Proportionality in enterprise development of South African towns. South African Journal of Science (in press).

Toerien, D.F. & Seaman M.T. (2012b) Enterprise diversity in selected South African towns. South African Journal of Science (submitted).

Trudon, (2010) Port Elizabeth & Eastern Cape 2010/2011. Trudon, Bruma.

Van der Merwe, I.J. & Nel, A. (1975) Die stad en sy omgewing: 'n studie in nedersettingsgeografie (in Afrikaans), Universiteitsuitgewers, Stellenbosch.

Van Waart, S. (2001) Swartberg en Sy Mense & Paleise van Pluime (in Afrikaans), Lapa, Pretoria.

Whittaker, R.J., Triantis, K.A. & Ladle, R.J. (2010) A general dynamic theory of oceanic island biogeography, In: The Theory of Island Biogeography Revisited, Eds. Losos, J.B. & Ricklefs, R.E., pp. 88-115, Princeton University Press, ISBN 978-0-691-13653-0, Princeton.

Wickins, P.L. (1983) Agriculture. In: Economic History of South Africa, Ed. Coleman, F.L., pp. 37-85, HAUM, Pretoria.

Witt, U. (2004) On the proper interpretation of 'evolution' in economics and its implications for production theory, Journal of Economic Methodology Vol. 11, No. 2, 125–146.

Witt, U. (2008) What is specific about evolutionary economics? Journal of Evolutionary Economics Vol. 18, No. 5, 547–575.

13

Toxicity Analysis of Effluent Released During Recovery of Metals from Polymetallic Sea Nodules Using Fish Haematological Parameters

Huma Vaseem and T.K. Banerjee
Department of Zoology, Banaras Hindu University, Varanasi
India

1. Introduction

To meet its domestic demands India has to import many of the economically important metals like Manganese, Iron, Cobalt, Zinc, Nickel and others. As the land based resources of these metals are depleting very fast, considerable efforts have been made to extract metals from pollymetallic sea nodules (PMN) during the past four decades all over the world. To attain self sufficiency scientists at National Metallurgical Laboratory, Council of Scientific and Industrial Research, Jamshedpur, India has developed an indigenous process to recover some of these metals from PMN. The PMN are rock concretions on the sea bottom formed of concentric layer of iron and manganese hydroxides around their core. They are small, slightly flattened, dark-brown coloured balls measuring 5 to 10 centimeters in diameters. The chemical composition of the nodules varies considerably according to the kind of minerals and the size and characteristics of the core. Those of the greatest economic interest contain Mn (27-30%), Ni (1.25-1.5%), Cu (1-1.4%), and Cobalt (0.2-0.25%). Other constituents include Fe (6%), Si (5%), Al (3%) with lesser amounts of Ca, Na, Mg, K, Ti, and Ba along with hydrogen and oxygen. For metal extraction purpose nodules are dried at 110^0 C, grounded and treated with reducing agents, followed by ammonia leaching. Separation of metals is done by the process of solvent extraction and electrowining (Jana *et al.*, 1990; Kumar *et al.*, 1990; Agarwal & Goodrich, 2008; Biswas *et al.*, 2009). During the process of metal recovery highly contaminated effluent is generated that still retains substantial amount of metals (Vaseem & Banerjee, 2011a). These metals are highly toxic and are one of the main causes of environmental pollution. Two most important factors that contribute to the deleterious effects of heavy metals as pollutants are their non-degradation in the nature (unlike organic pollutants) and their tendency to bioconcentrate and settle at the bottom of water bodies. Hence our main aim has been to monitor the toxicity rendered to the aquatic ecosystem by this highly contaminated effluent (Table 1) using fish as an experimental model. *Labeo rohita* (commonly known as Rohu), a major Indian carp of great nutritional importance has been selected for the toxicity analyses of the effluent because fishes have widely been used as effective bioindicator. This graceful Indo-Gangetic riverine species is one of the three important major carps of the Indian subcontinent belonging to the family cyprinidae. It is the natural inhabitant of the wetlands of northern and central India, and the rivers of Pakistan, Bangladesh and Myanmar. It is a diurnal, herbivore and generally solitary species. It attains sexual maturity within two years. In nature it spawns in the marginal areas of flooded rivers. Due to its wider feeding niche, rohu is usually stocked at relatively greater quantity than the other two carps *Catla catla* and *Cirrhina mrigala*. Higher consumer preference and market demand for rohu during recent years have also led to the increased practices of culture of this fish species. There has always been a chance of metal toxicity of this fish through contamination of the water bodies by various polluting agents generated through anthropogenic activities.

Heavy metals	**Conc. (mg/l) In raw effluent**	**Standard (EPA) 2003 (mg/ml)**
Mn	4.957 ± 0.130	0.2
Cu	1.432 ± 0.013	0.5
Zn	0.816 ± 0.013	2.0
Fe	0.762 ± 0.02	2.0
Pb	0.655 ± 0.017	0.05
Cr	0.07 ± 0.001	0.05
Cd	0.018 ±0.0004	0.01
BOD	182 ± 3	40
pH	5.2 ± 0.5	5-9
Sodium	130.26 ± 1.96	200
Potassium	3.42 ± 0.07	n.a
Sulphate	2300 ± 4.515	750
DO (mg/l)	18.8 ±0.821	n.a
CO_2	28.60 ± 0.418	n.a
Carbonate	296 ±2.2	n.a

Table 1. Physicochemical properties of sea nodule effluent.
Source: Vaseem and Banerjee (2011 a).
Abbreviations. n.a: data not available; Values are given in mean ± SD

Amongst the various tissue components employed for toxicity estimation, the blood parameters of the fish have been selected for the study because ambient contaminants often produce rapid changes in the blood characteristics (Carvalho & Fernandes, 2006). Several haematological indices haematocrit (Ht), haemoglobin (Hb), total erythrocyte count (TEC), total leukocyte count (TLC) have successfully been applied in the past to assess the functional status and oxygen carrying capacity of blood stream of variously exposed fishes (Shah & Altindag, 2004). Kori *et al* (1991) observed decreased Hb, Ht, RBC counts in copper exposed *Clarias isheriensis*. Das and Mukhrjee (2000) observed decreased Hb, TEC and serum protein content and increased TLC and blood glucose concentration of the quinolphos exposed carp, *L. rohita*. Decreased Hb, TEC, packed cell volume (PCV) and mean corpuscular haemoglobin concentration (MCHC) and increased white blood cell count and MCV have been noticed in *L. rohita* collected from the polluted lakes of Bangalore, Karnataka India (Zhushi *et al.*,2009). Kavitha *et al* (2010) observed decrease in various haematological parameters (Hb, Ht, RBC, WBC, plasma glucose, plasma protein) in arsenate treated Indian major carp, *Catla catla*. They however observed increased corpuscular indeces like MCV, MCH and MCHC in the same fish. Hence in the present study analyses of these blood parameters of *L. rohita* have been applied to evaluate the toxicity of the sea nodule effluent.

2. Materials and methods

2.1 Experimental design

Healthy specimens of the freshwater teleost *Labeo rohita,* belonging to the family Cyprinidae, were collected from the hatchery at Banaras Hindu University, Varanasi, India. The fish were acclimatised to the laboratory conditions (temperature 26.0^0 C ± 2.0) for one month in tap water (pH: 7.2 ± 0.3; salinity: 0.3 ± 0.08 pg/l, dissolved oxygen: 6.6 ± 0.06 mg/l, total alkalinity: 21.0 ± 8.0 mg/l, and total hardness: 16.0 ± 0.04 mg/l) in 40 litre plastic tubs. The fish were fed *ad libitum* with rice bran and groundnut oil cake (2:1) twice daily. Water in the tubs was aerated and changed after every 24 hrs. Batches of 15 fish (weight of 28-30 g and 11-12 cm in length) were exposed to the 20 litre of the raw effluent. Parallel batches of control fish (15 in each batch) were exposed to plain tap water under identical laboratory conditions. Ten randomly selected fish each from experimental as well as control aquaria were sacrificed after 10 and 20 days of exposure for haematological as well as biochemical analyses.

2.2 Collection of blood

Blood from cold anaesthetised fish was collected from the cardiac region by puncturing the heart using a plastic disposable syringe fitted with a 26-gauge needle, moisturised with heparin and was immediately transferred to separate heparinised chilled plastic vials and immediately returned to the ice box.

2.3 Haematological analysis

Blood samples from the control as well as experimental groups of fish were subjected to determination of Hb, Ht, RBC and WBC count. Haemoglobin was estimated by using the Sahli's hemoglobinometer. Oxygen carrying capacity of the fish blood was calculated by multiplying the haemoglobin content by 1.25 oxygen combining power of Hb/g (Johansen, 1970). Erythrocyte and leukocyte counts were studied by Neubauer's improved hemocytometer using Hayem's and Tuerk's solution as a diluting fluid, respectively (Samuel, 1986). Hematocrit values were measured by Wintrobe's methods. Mean corpuscular haemoglobin, mean corpuscular haemoglobin concentration, and mean corpuscular volume were calculated by the standard formulae suggested by Dacie and Lewis (1991). The remainder of the blood was centrifuged for 15 min for biochemical analysis of the serum.

$$\text{MCV} = \frac{\text{PCV/1,000 ml blood}}{\text{RBC in millions/mm3}} = \text{fl}$$

$$\text{MCH} = \frac{\text{Hb in g/1,000 ml blood}}{\text{RBC in millions/mm3}} = \text{pg}$$

$$\text{MCHC} = \frac{\text{Hb in g/100 ml blood}}{\text{PCV/100 ml}} \times 100 = \text{g/dl}$$

2.4 Biochemical analysis

The concentrations of serum glucose, cholesterol and protein were estimated by the methods of Seifter *et al.* (1950), Zlatkis *et al.* (1953), and Lowry *et al.* (1951) respectively.

2.5 Statistical analysis

For statistical analyses one-way analysis of variance (ANOVA) was performed to determine significance of differences ($p< 0.05$) between the pairs of means. Duncan's multiple range test (DMRT) was also applied ($p< 0.05$) to find out which means are significantly different from others. Since there were no significant variations between the values obtained from various control groups, the average value of all the control groups was taken into account. In table 2, 3 & 4 alphabets denote results of DMRT. Different alphabets show significant changes in various parameters in control and exposed fish ($p< 0.05$).

3. Results

The Hb content (g/dl) of the unexposed control fish (*L. rohita*) was 8.367 ±0 .694. Following exposure it started decreasing and was 6.2± 0.216 after 10 days and 4.43 ± 0.464 after 20 days. Both the decreases were statistically significant ($p<0.05$) (Table 2).

The haematcrit value (%) of control fish was 30.66±0.339. In exposed fish it decreased significantly ($p<0.05$) becoming 0.26± 0.163 after 10 day and 19.5± 0.637 after 20 days (table 2).

Following exposure the total erythrocyte count (10^6 mm^{-1}) decreased significantly ($p<0.05$) and became 2.367 ± 0.205 from 3.1± 0.163 (control) after 10 days. It continued to decrease further becoming 1.14 ± 0.043 after 20 days (Table 2).

The total leukocyte count increased significantly ($p<0.05$) following exposure (Table 2). After 10 days of exposure TLC of the fish became 25.32 ± 0.891 (from 19.23 ± 0.418 in control). It increased further and became 28.73 ±0.309 after 20 days.

The MCV, MCH, MCHC and O_2 carrying capacity of blood deduced from the above mentioned haematological data has been detailed in table 3.

Pattern of changes in the serum glucose, cholesterol and protein levels have been shown in table 4. The serum glucose content (mg/dl) increased significantly ($p<0.05$) from 73.19 ± 1.617 in control to 77.733 ± 0.329 after 10 days and 93.006 ± 2.163 after 20 days of exposure.

The serum cholesterol level (mg/dl) also increased from 49.04 ± 1.043 in control to 56.52 ± 1.79 and 73.446 ± 2.011 after 10 and 20 days of exposure respectively.

Total serum protein level (mg/dl) in control fish was 3.266± 0.169. While it decreased to 2.9 ± 0.081 on 10 days of exposure. It became 1.166 ± 0.124 after 20 days. The decrease in protein concentration was statistically significant ($p< 0.05$) for both the values: after 10 and 20 days of exposure.

Parameters	Control	10 days exposed	20 days exposed
Total erythrocyte count (×106 mm^3)	3.1 ± 0.163^a	2.367 ± 0.205^b	1.14 ± 0.043^c
Total leukocyte count (×103 mm^3)	19.233 ± 0.418^a	25.32 ± 0.081^b	28.73 ± 0.309^c
Haemoglobin (g/dl)	8.367 ± 0.694^a	6.2 ± 0.216^b	4.43 ± 0.464^c
Haematocrit (%)	30.66 ± 0.339^a	26 ± 0.163^b	19.5 ± 0.637^c

Values are given in mean ± SD

Table 2. Alteration in haematological parameters in sea nodule exposed *L. rohita*.

Parameters	Control	10 days exposed	20 days exposed
Glucose (mg/dl)	73.187 ± 1.617[a]	77.73 ± 0.329[b]	110.08 ± 3.16[c]
Cholesterol (mg/dl)	49.04 ± 1.043[a]	56.52 ± 1.795[b]	73.446 ± 2.011[c]
Protein (mg/dl)	31.23 ± 1.827[a]	29 ± 2.32[a]	4.16 ± 0.812[b]

Table 3. Alteration in Biochemical parameters of blood in sea nodule exposed *L. rohita.* Values are given in mean ± SD

Parameters	Control	10 days exposed	20 days exposed
MCV (fl)	99.276 ± 6.809[a]	110.73 ± 10.09[b]	171.356 ± 8.77[c]
MCH (pg)	27.047 ± 1.703[a]	26.313 ± 2.16[a]	38.873 ± 3.779[b]
MCHC (g/dl)	27.28 ± 1.60[a]	23.853 ± 1.89[a]	22.71 ± 2.145[a]
O2 carrying capacity Of blood (ml $O_2/g^{-1}/Hb$)	10.458 ± 0.690[a]	7.75 ± 0.568[b]	5.54 ± 0.580[c]

Table 4. Alteration in calculated haematological indices in sea nodule exposed *L. rohita.* Values are given in mean ± SD

4. Discussion

Blood parameters are considered as a patho-physiological indicator of the entire body and therefore are important in diagnosing the structural and functional status of fishes exposed to toxicants (Adhikari & Sarkar, 2004; Maheswaran et al., 2008).

Estimation of haemoglobin was employed because this blood component is a part of the sophisticated oxygen delivery system that provides the desired amount of oxygen to the tissues under a wide variety of circumstances (Voet & Voet, 1990). The oxygen transport function of blood is the product of a complex integration of the effects of various physicochemical factors such as temperature and the concentrations of allosteric co-factors, dissolved gases, protons and other ions on the oxygen binding properties of haemoglobin (Weber & Lykkeboe, 1978; Weber, 1982). According to Blaxhall and Daisley (1973) the determination of haemoglobin concentration can be a good indicator of anaemic conditions in fish. A review of table 2 suggests that the PMN effluent causes loss of Hb leading to anaemic condition to the fish after 20 day of exposure. Cyriac et al. (1989) considered decreases in haemoglobin concentration as a contribution to haemodilution. Haemodilution is a mechanism that reduces the concentration of the pollutants in the circulatory system (Smit et al. 1979). Similar haemodilution has also been observed in fish contaminated with aluminium, copper, manganese and zinc (Torres et al., 1986; Wepener, 1990; Nussey, 1994; Coetzee, 1996; Barnhoorn, 1996). The decrease in haemoglobin concentration signifies that the fish's ability to provide sufficient oxygen to the tissues is restricted considerably and results in decreased physical activity (Grobler, 1988; Wepener, 1990; Nussey, 1994). According to Reddy and Bashanihideen (1989) this significant decrease in the haemoglobin concentrations of fishes under toxic stress might be due to either an increase in the rate at which the haemoglobin is destroyed or due to decreased rate of haemoglobin synthesis. Other reason for the progressive reduction in the haemoglobin content might be attributed to depression/exhaustion of haemopoietic potential of the fish (Sawhney & Johal, 2000).

Suppression of haemopoietic activity of the kidney in addition to the increased removal of dysfunctional red blood cells might be the third reason for the decreased Hb content (Stormer et al., 1996). Devi and Banerjee (2007 a, b) also noticed anaemic condition of ammonia and lead exposed *Channa striata* due to deceased levels of Hb, TEC, TLC, Ht and cellular degeneration of RBCs.

Haematocrit (measurement of packed erythrocytes) is an important instrument for determining the amount of plasma and corpuscles in the blood and used to determine the oxygen carrying capacity of the blood (Larsson et al., 1985). It is also defined as the volume occupied by erythrocytes in a given volume of blood. In fish the haematocrit reading is valuable in determining the effect of stressors on their health. (Munkittrick & Leatherland, 1983). Significant decreases in the haematocrit values (Table 2) following exposure to the sea nodule effluent also suggests anaemia and haemodilution possibly due to gill damage or/and impaired osmoregulation (Larsson et al., 1985).

Erythrocytes are produced in the haematopoietic tissue, which is situated in the spleen and head kidney (Smith, 1982; Grey & Meyer, 1988; Kita & Itazawa, 1989; Heath, 1995). It is well known that a reduced quantity and quality of erythrocytes and a decreased haemoglobin level as also noticed in the present study could lead to deleterious oxygen transport. Extensive reduction in haemoglobin content due to any blood dyscrasia and degeneration of the red blood cells could be ascribed as pathological conditions in fishes exposed to toxicants leading to deteriorated oxygen supply (Buckley et al., 1976). Decrease in TEC in the present study (Table 2) might be due to inhibition of RBC production or Hb synthesis. Exposure to other toxicants also causes decrease in erythrocyte counts in fishes (Van der Merwe, 1992, Omoregie *et al.*, 1990; Das & Mukherjee, 2000).

The calculated haematological indices, MCHC, MCH, and MCV are other important indicators in the diagnosis of anaemia in most animals (Coles, 1986). Alterations in these haematological parameters (increase MCV and MCH, decrease of MCHC) might be due to a defence against the toxic effect of the effluent through the stimulation of erythropioesis or due to the decrease in RBCs, Hb and Hct values following disturbances in both metabolic and haemopoietic activities of the fish exposed to different concentrations of pollutants (Abd-Alla *et al.*, 1991; Mousa, 1994). The increase in MCV value might be due to increased number of immature RBC (Carvalho & Farnandes, 2006). Increase in MCV and MCH along with slightly diminished MCHC values (Table 3) suggest the macrocytic nesmochromic type of anaemia. Similar toxicological response is also recorded in common carp caused by acute effect of phenitrothion, imidan, and dichlorvos (Svobodova., 1971) and Svobodova and diazinon (Svoboda et al., 2001). Changes in haematological parameters of *C. gariepinus* due to stress caused by environmental pollutants, diseases or attack by pathogens have also been reported by a number of workers (Ezeri, 2001, Gabriel et al., 2001).

White blood cells (WBC), or leukocytes, are important component of the immune system involved in defending the body against both infectious diseases and foreign materials. Five different and diverse types of leukocytes exist, but they are all produced and derived from a multipotent cell in the bone marrow known as a hematopoietic stem cell. Leukocytes are found throughout the body, including the blood and lymphatic system. The increased total leucocytes count of the exposed fish in the present study indicates increased defensive reaction against the stressors. Increased WBC count have also been reported in fishes exposed to other certain xenobiotics like endosulfan (Abidi & Srivastava,1988), aflatoxin B (Lovell & Jantrarotai,1991), crude oil (Khadre & Shabana,1991) and to industrial effluents (Wahbi, 1992).

While analysing different haematological parameters in the freshwater fish, *Heteroclarias* sp. (Osteichthyes: Clariidae) exposed to sublethal concentration of zinc, Kori-siakpere *et al* (2008) noticed decreased values of Hb, Ht, RBC, WBC, protein and glucose and calculated hametogical parametres. Gabrial et al (2007) observed haematological changes in the catfish *Clarias gariepinus* following 14-days of exposure to refined petroleum oil, kerosene. The results include decreased values of Hb, Ht, WBC and MCV and increased levels of MCHC, MCH, neutrophils, monocytes and thrombocytes.

The presence of toxicants in aquatic ecosystem also exerts its effect at cellular or molecular levels which results in significant changes in biochemical compositions of the organisms. Due to metal complex formation, normal functioning of cells are disturbed that in turn results in disturbed physiological and biochemical equilibrium of animals (Gagnon et al, 2006; Vaseem & Banerjee, 2011b). The influence of the stressors on carbohydrate metabolism of fish includes alterations in glucose, glycogen and lactic acids contents. Among these parameters the analyses of blood glucose level (Table 4) have been used as an effective indicator to monitor the stress condition of the animal including fish. The elevated glucose level in the blood stream in this study might be due to gluconeogenesis to supplement additional energy needed to meet the increased metabolic demands (Zutshi *et al.*, 2009; Kavitha *et al.*, 2010). While estimating the carbohydrate, lipid and protein concentrations in six vital organ system of *Labeo rohita* exposed to sea nodule effluent, Vaseem and Banerjee (2011b) noticed depletion of these macromolecules in all the six tissues after 20 days of exposure. They postulated that the decrease might be due to mobilization of these macromolecules to maintain steady supply of energy in the serum to negotiate the stress during the entire period of exposure. This could also be the reason for increased level of glucose in the serum of sea nodule effluent exposed *Labeo rohita* (Table 4).

The reduction in serum protein levels in sea nodule effluent (table 4) exposed *L. rohita* might be due to breakdown of proteins and other macromolecules for several known (e.g. to meet the higher energy demand during the prevailing stress (Zutshi *et al.*, 2009) or might be due to liver cirrhosis or nephrosis or due to alteration in enzymatic activity involved in protein biosynthesis as suggested by Nandi *et al.*, 2005, Yousef *et al.*, 2008, and Palaniappan & Vijjayasundarum. 2009) or unknown reasons.

The blood cholesterol level in the present study increased significantly after both (10 as well as 20) days of exposure (Table 4). Increased cholesterol level might have occurred due to dysfunction of liver causing release of additional quantities of cholesterol into the blood. Increased level of cholesterol has also been reported in the serum of *Channa punctatus* (Kaur & Kaur 2006) and *Cirrhina mrigala* (Kumar *et al.*, 2005) due to exposure to other toxicants.

5. Conclusion

The results of the present investigation indicate that exposure to sea nodule effluent induces significant changes in the haematological and biochemical profile of the Indian major carp *L. rohita*. Prolonged exposure of sea nodule effluent also affected the survival of fish. Our data illustrated the toxicological impact of effluents having a variety of contaminants especially the toxic metals in different concentrations.

6. Acknowledgements

This work was financially supported by the Department of Ocean development, Government of India. The senior author thanks the University Grant Comission for providing a Junior Research Fellowship.

7. References

Abd-Alla, G.A., Ibrahim, M.S., Bahnasawy, M.H. & Abdel-Baky, T.E. (1991). Toxic effects of some water pollutants (gallant and mercury) on blood parameters of catfish *Clarias lazera* .J. Egypt Ger.Soc.Zool. 6:201-209.

Abidi, R. & Srivastava, U.S. (1988). Effect of endosulfan on certain aspects of hematology of the fish, *Channa punctatus* (Bloch). *Proc. Natl. Acad. Sci. India* 58:55–65.

Adhikari, S. & Sarka, B. (2004). Effects of cypermehrin and carbofuran on certain haematological parameters and prediction of their recovery in fresh water teleost, *Labeo rohita* (Ham). Ecotoxicol. Environ. Safety .58: 220-226

Agarwal, H. P. & Goodrich, J.D. (2008). Extraction of Copper, Nickel, and Cobalt from Indian ocean polymetallic nodules. The Canadian J. Of Chemical engineering, 81:303-306.

Barnhoorn, I.E.J. (1996). Effects of manganese on the haematology of *Oreochromis mossambicus* and the bioaccumulation of metals in *Labeo umbratus*. M.Sc. Thesis, Rand Afrikaans University, Johannesburg, South Africa.

Biswas, A., Chakraborti, N. & Sen, P. K. (2009). Multiobjective optimization of manganese recovery from sea nodules using genetic algorrithms. Materials and manafacturing process. 24:22-30.

Blaxhall, P. C. & Daisley, K. W. (1973). Routine haematological methods for use with fish blood. J. Fish Biol. 5:771 -781.

Bond, C.E. (1979). Biology of Fishes. Saunders College Publishing, Philadelphia, USA p. 514

Buckley, J.A., Whitmore, C.M. & Matsuda, R.I. (1976). Changes in blood chemistry and blood cell morphology in coho salmon, *Oncorhynchus kisutch* following exposure to sublethal levels of total residual chlorine in municipal wastewater. J. Fisheries Res. Board, Canada. 33: 776 -782.

Carvalho, C.S. & Fernandes, M.N. (2006). Effect of temperature on copper toxicity and hematological responses in the neotropical fish *Prochilodus scrofa* at low and high pH. *Aquaculture*, 251:109–17.

Coetzee, L. (1996). Bioaccumulation of metals in selected fish species and the effect of pH on Aluminium toxicity in a cichlid *Oreochromis mossambicus*. M.Sc. Thesis, Rand Afrikaans University, Johannes burg, South Africa.

Coles, E.H. (1986). Veterinary Clinical Pathology (4th Ed); W. B. Saunders Co. Philadelphia.

Cyriac, PJ., Anthony, A. & Nambison, P.N.K. (1989). Haemoglobin and Haematocrit values in the fish *Oreochromis mossambicus* (Peters) after short term exposure to copper and mercury. Bulletin of Environmentalm Contamination and Toxicology. 43:315 - 320.

Dacie, J. A. & Lewis, S. M. (1991). *Practical haematology A* (7th ed.). London: Churchill Livingstone.

Das, B. K. & Mukherjee, S. C. (2000). Sublethal effect of quinalphos on selected blood parameters of *Labeo rohita* (Ham.) fingerlings. Asian Fisheries Science, 13: 225-233.

Devi, R. & Banerjee, T.K. (2007 a). Haematological disturbances due to ammonia toxicity in the air-breathing fish *Channa stiata. Biochem. Cell. Arch.* 7:217-223.

Devi, R. & Banerjee, T.K. (2007 b). Estimation of the sublethal toxicity of lead nitrate in the air- breathing fish *Channa striata* employing certain haematological parameters. *Biochem. Cell. Arch.* 7:185-191.

Ezeri, G.N.O. (2001). Haematological response of *Clarias gariepinus* to bacteria infection and prophylactic treatment with antibiotics. J. Aquat. Sci. 16: 22-24.

Gabriel, U.U., Alagoa, J.K. & Allison, M.E. (2001). Effects of dispersed crude oil water dispersion on the haemoglobin and haematocrit of *Clarias gariepinus*. J. Appl. Sci. Environ. Manage. 5:9 -11.

Gabriel, U.U., Ezeri, G.N.O. & Amakiri, E.U. (2007). Haematology and gill pathology of *Clarias gariepinus* exposed to refined oil, kerosene under laboratory conditions *J. Anim. Vet. Advances.* 6: 461-465.

Gagnon, A., Jumarie, C. & Hontela, A. (2006). Effects of Cu on plasma cortisol and cortisol secretion by adrenocortical cells of rainbow trout, *Oncorhynchus mykiss*. Aquat. Toxicol. 78:59–65.

Grey, B.J. & Meyer, S.V. (1988). Rooibloedselle: telling, morfologie, hemoglobien, metabolisme en lewensduur. In: Die Fisiologiese Basis van Geneeskunde 4de uitgawe. Ed. Meyer B. J. Haum Uitgewry, Pretoria South Africa pp. 24.1 - 24.13.

Grobler, E. (1988). Die Effek van Atrasien, Sink in Yster op die Hematologie en Suurstofverbruik van *Tilapia sparrmanii* (Cichlidae). M.Sc. Thesis, Rand Afrikaans University, Johannesburg, South Africa.

Heath, A.G. (1995). Water Pollution and Fish Physiology. Second Edition. CRC Press Inc., Florida p. 359.

Hoffbrand, A.V & Pettit, J.E. (1980). Essential Haematology, Blackwell Scientific Publications, Oxford p. 227.

Jana, R. K, Pandey, B. D. & Premchand. (1999). Ammonial leaching of roast reduced deep-sea Manganese nodules. Hydrometallurgy, 53:45-56.

Johansen, K.(1970). Air-breathing fishes. In:Hoar, W. S., Randall, D.T. (Eds), Fish physiology, Academic press, New york, London.4: 361-411.

Kaur, A. & Kaur, K. (2006). Impact of nickel-chrome electroplating effluent on the protein and cholesterol contents of blood plasma of *Channa punctatus* (Bl.) during different phases of the reproductive cycle. Journal of Environmental Biology, 27(2), 241-245.

Kavitha, C., Malarvizhi, A., Senthil Kumaran, S.& Ramesh, M. (2010). Toxicological effects of arsenate exposure on haematological, biochemical and liver transaminases activity in an Indian major carp, *Catla catla*. 48:2848-2854.

Khadre, S. E. M. & Shabana ,M. B. (1991). Histopathological and physiological disorders of acute crude oil toxicity in *Mugil cephalus*. Inter.Conf.Marine fish Manag. and Develop.Alexandria Egypt.

Kita, J. & Itazawa, Y. (1989). Release of erythrocytes from the spleen during exercise and splenic constriction by adrenaline in fusion in rainbow trout. Japn. J. Ichthyol. 36: 48-52.

Kori-Siakpere, O. & Ubogu, E. O. (2008). Sublethal haematological effects of zinc on the freshwater fish, *Heteroclarias* sp. (Osteichthyes: Clariidae) African Journal of Biotechnology. 7: 2068–2073.

Kori-Siakpere, O. (1991). Chronic sublethal haematological effect of copper in a freshwater teleost: *Clarias isheriensis*. Nig. Annals Nat. Sci. 1:37-43.

Kumar, P. Y.G., Gupta, V. T., Shakti, T. & Singh, A. (2005). Haematological and biochemical abnormalities in *Cirrhinus mrigala* (Ham.) induced by lead. J. Ecophysiol. Occupl. Hlth. 5:213-216. 2005.

Kumar, V., Pandey, B. D. & Akerkar, D.D. (1990). Electrowinning of nickel in the processing of polymetallic sea nodules. Hydrometallurgy. 24:189-201.

Larsson, A., Haux., C & Sjobeck, M. (1985). Fish physiology and metal pollution. Results and experiences from laboratory and field studies. Ecotoxicol. Environ. Saf. 9:25 -281.

Lovell, R.T. & Jantrarotai, W.(1991). Subchronic toxicity of dietary aflatoxin B1 to channel catfish. Journal of Aquatic Animal Health. 2:248-254.

Lowry, O.H., Rosenbrough, N. J., Ferry, A. L. & Randall, R. J. (1951). Protein measurement with folin-phenol reagent. *Journal of Biological Chemistry*. 193:265–275.

Maheswaran. R., Devapaul, A., Velmurugan, B. & Ignacimuthu, S. (2008). Haematological studies of freshwater fish, *Clarias batrachus* (L.) exposed to mercuric chloride. Inter. J. Integr. Biol. 2: 49 - 54.

Moussa, F.I., Abou Shabana, M.B. & El-Toweissy, M.Y. (1994). Short and long term effects of acid stress on Survival, behavioral and some cellular blood constituent in the catfish *Clarias lazera*. Bull. Nat. Inst. Oceanogr. and Fish.,A.R.E. 20:229-248.

Munkittrick, K.R.& Leatherland, J.F. (1983). Haematocrit values in feral goldfish, *Carasius auratus* L., as indicators of the health of the population. J. Fish Biol. 23: 153-161.

Musa, S .& Omoregie, E. (1999). Haematological changes in mudfish, *Clarias gariepinus* (Burchell) exposed to malachite green. J. Aquat. Sci. 14:32 - 37.

Nandi, D., Patra, R.C. & Swarup, D. (2005). Effect of cysteine, methionine, ascorbic acid and thiamine on arsenic-induced oxidative stress and biochemical alterations in rats. Toxicology 211:26–35.

Nussey, G. (1994). The effect of copper on blood coagulation and general haematology of *Oreochromis mossambicus* (Cichlidae). M.Sc. Thesis, Rand Afrikaans University, Johannesburg, South Africa.

Omoregie, E., Ufodike, E.B.C. & Keke, R.I.(1990). Tissue chemistry of O. *niloticus* exposed to sublethal concentrations of Gammalin-20 and Acetellice 25EC. J. Aquat. Sci. 5:33 - 36.

Palaniappan, P.L.R.M. & Vijayasundaram, V. (2009). The effect of arsenic exposure and the efficacy of DMSA on the proteins and lipids of the gill tissues of *Labeo rohita*. Food Chem. Toxicol. 47:1752–1759.

Reddy, P.M. & Bashamohideen, M. (1989). Fenvalerate and cypermethrin induced changes in the haematological parameters of *Cyrinus carpio.* Acta Hydrochim. Hydrobiologia 17:101-107.

Samuel, R. M. (1986). Haematology. In: *Notes on clinical lab techniques* (4th ed.). Madras: Tailor.

Sawhney, A.K. & Johal, M.S. (2000). Erythrocyte alterations induced by malathion in *Channa punctatus* (Bloch). Bull. Environ. Contam. Toxicol. 64: 398 - 405.

Seifter, S., Novie, B., & Muntwyer, E. (1950). The estimation of glycogen with the Anthrone reagent. *Archives of Biochemistry, 25*:191–200.

Shah SL. & Altindag A (2004). Haematological parameters of tench (*Tinca tinca* L.) after acute and chronic exposure to lethal and sublethal mercury treatments. Bull. Environ. Contam. Toxicol. 73: 911 - 918.

Smit, G.L., Hatting, J. & Burger, A.P. (1979). Haematological assessment of the effects of the anaesthetic MS 222 in natural and neutralized form in three fresh water fish species: Interspecies differences. J. Fish Biol. 15: 633 - 643.

Smith, L.S. (1982). *Introduction to Fish Physiology.* TFH Publications, Inc., Neptune, USA. p. 346.

Stormer, .J, Jensen, E.B, & Rankin, J.C. (1996). Uptake of nitrite, nitrate and bromide in rainbow trout, *Onchorhynchus mykiss:* effects on ionic balance. Can. J. Fisheries Aquat. Sci. 53: 1943 - 1950.

Svoboda, M., Luskova, V., Drastichova, J. & Iabek, V. (2001). The effect of diazinon on hematological indices of common carp (*Cyprinus carpio* L.). *Acta Vet. (Brno)* 70,. 457–465.

Svobodova, Z. (1971). Some hematological and metabolic changes in fish occurring after pesticide intoxication. Bull. VUR Vodany 7. 29–36.

Torres, P., Tort, L., Planas, J. & Flos, R. (1986). Effects of confinement stress and additional zinc treatment on some blood parameters in the dogfish *Scyliorhinus canicula.* Comp. Biochem. Physiol. 83: 89 -92.

Van Der Merve M. (1992). Aspects of heavy metal concentration in the Olifants River, Kruger National Park and the effect of copper on the haematology of *Clarias gariepinus* (Clariidae). M.Sc. Thesis, Rand Afrikaans University, South Africa.

Vaseem, H. & Banerjee, T.K. (2011 a). Phytoremediation of the toxic effluent generated during recovery of Precious metals from polymetallic sea nodules. Inter. J. Phytoremediation. 14:457-466, 2012

Vaseem, H. & Banerjee, T.K. (2011 b). Toxicity analyses of effluent released during recovery of precious metals from polymetallic sea nodules using biochemical parameters of Indian major carp, *Labeo rohita.* Communicated for publication.

Voet, F. & Voet, J.G. (1990). Biochemistry. John Wiley and Sons, New York, USA. pp. 425 - 457.

Wahbi, O. M. (1992). Effect of Abu Qir Fertilizer Industrial effluent on the physiological characteristics of different stages of *Sparus auratus* . M.Sc. Thesis, Faculty of Science, Alexandria University

Weber, R.E. & Lykkeboe, G. (1978). Respiratory adaptations in carp blood. Influences of hypoxia, red blood cell phosphates, divalent cations on haematology-oxygen affinity. J. Comp. Physiol. 128: 127-137.

Weber, R.E. (1982). Interspecific adaptation of haemoglobin function in fish to oxygen availability. In: *Exogenous and Endogenous Influences on Metabolic and Neural Control.* Eds. A. D. F. Addink and N. Spronk, Pergamon Press, Oxford, UK. : 87 - 101.

Wepener, W.(1990). The effects of heavy metals at different pH on the blood physiology and metabolic enzymes in *Tilapia sparmanii* (Cichlidae). M.Sc. Thesis, Rand Afrikaans University, Johannesburg, South Africa.

Wepener, W., Van Vuren, J.H.J. & Dupreez, H.H. (1992). Effect of manganese and iron at neutral pH values on the haematology of the banded tilapia (*Tilapia sparrmanii*). Bull. Environ. Contam. Toxicol. 49: 613 - 619.

Yousef, M.I., El-Demerdash, F.M. & Radwan, F.M.E. (2008). Sodium arsenite induced biochemical perturbations in rats: ameliorating effect of curcumin. Food Chem. Toxicol. 46: 3506–3511

Zlatkis, A., Zak, B. & Boyle, A. J. (1953). A new method for direct determination of serum cholesterol. *Journal of Laboratory and Clinical Medicine,* 41: 481–492.

Zutshi, B., Raghu, S.G. & Nagaraja, R. (2009). Alteration in the haematology of *Labeo rohita* under stress of pollution from lakes of Bangalore, Karnataka, India. Environ Monit Assess, 168, 11-19.

14

Environmental Assessment of Brick Kilns in Chihuahua State, México, Using Digital Cartography

Alba Yadira Corral Avitia and Antonio De la Mora Covarrubias
Universidad Autónoma de Ciudad Juárez
México

1. Introduction

The world's population is concentrated in urban areas, which are characterized by being highly dynamic and complex. There are several enterprises responsible for high levels of pollution, one of them is the brick kiln. One standard traditional brick kiln may produce 86 kg of particulate matter on a 12 hours burn required to produce 10,000 bricks. There are many brick kilns of this size distributed around urban areas of the state of Chihuahua. These brick kilns represent an environmental problem due to their negative health effects if the amount of particulate matter is extrapolated to the total number of burns carried out. There is a need for a proper environmental management of natural resources and urban planning. This chapter presents two objectives using digital cartography as a tool and considering the complete and complex state of Chihuahua. One of the objectives of this study was to identify the location of the brick kiln in all the municipalities of the state and the other objective was to determine the best places, around the state of Chihuahua, to relocate the brick kiln industry in order to minimize environmental risk. Multicriteria Analysis (MCA) based on digital cartography and a Geographical Information System (GIS) were used. The factors selected for analysis included biophysical, social and economic aspects. In order to estimate the effects of brick kiln operation, a nine-level estimated risk scale was defined using hierarchical MCA and the modified Delphi technique. The weighted risk factors were mapped in the GIS using the Spatial Analysis® extension. Polygons showing the degree of population and ecosystem vulnerability were obtained by superposing the maps. Currently, 78.8 % of kilns are located in areas of medium to high vulnerability. Using the data obtained in this study, brick kilns in the state of Chihuahua can be relocated on the basis of reliable technical criteria. Risk maps are presented for those regions that contain brick kilns inside or close by the city.

2. Metropolis and pollution

From the perspective of landscape ecology, cities are mosaics of high heterogeneity due to the different materials, whether natural (rocks, soil, vegetation, water) or made by humans (concrete, asphalt, metal, plastic, glass), which are arranged according to a specific use: housing projects, transport system, commercial and recreational areas. Urban settlements occupy a small fraction of the surface of the earth in terms of area, but rapid urbanization and industrialization are key factors that depend on social and economic development of regions as well as environmental changes (Ramadan et al., 2004). On the other hand, a high percentage of the world's population is concentrated in urban centers, which are characterized by being highly dynamic and complex. This may cause tension in the use and exploitation of natural resources, but also creates pressure on management systems and urban planning in the use of ground, transportation and environmental quality (Van der Sande et al., 2003).

Urban air pollution poses a major threat to human health and the environment in most countries of the world (De Vives, 2007). Rapid urbanization has resulted in increased air emissions due to transportation, energy production and industrial activity concentrated in densely populated areas. This topic is of increasing importance, since most of the world's population lives in urban centers and demands a healthier environment (Campos et al., 2008). Among urban air pollutants most common and important for their health risks are total suspended particulates, heavy metals, nitrogen oxides, carbon monoxide, sulfur dioxide, hydrocarbons and ozone. Ozone is produced by the reaction to the sun´s light and nitrogen oxides and volatile organic compounds (Dalmasso et al., 1997). In humans the result of environmental degradation has been reflected by premature mortality due to exposure to pollutants and chronic effects, including reduction in fitness and permanent lung damage (Escobedo & Chacalo, 2008). Specifically, the World Health Organization (2002) considered air pollution as one of the most important global health priorities. It has been estimated that particulate pollution is responsible for 1.4% of all deaths worldwide (Cohen et al., 2003). Studies conducted in industrialized countries have shown that short-term changes in pollution levels are associated with changes in daily death rate. Damage was also expressed in economic terms as absenteeism, increased medical costs and, consequently, productivity losses (Escobedo & Chacalo, 2008). Although there are several enterprises responsible of high levels of pollutions, this chapter will focus on one of them: brick kilns.

3. The brick industry ... between right and wrong

Brick kiln are small businesses distributed along the poorest sides of the cities. Bricks are fabricated using clay, water, silt, and fuel. On one side these business provide food, education and housing to thousands of families. On the other side its operation generates pollutants to the environment. In México brick production, contrary to other Latin-American countries, still is a handmade craft activity, 90% of the producer does not have state of the art technology to fabricate the bricks. The different activities involved in the brick production include mixing, molding, drying, burning, and selling the final product as shown in Figure 1. Most of the work is hard and it is made under harsh conditions everywhere in México (González, 2008). Therefore, it is important to contribute with some ideas to improve the quality of life of the workers and the families that lives close by.

There is evidence that brick kilns emit a considerable amount of pollutants and they are included in national inventories (SEMARNAT, 2011). These pollutants are created due to the incomplete combustion process and the different types of fuel used in the kiln. The typical brick kiln is built open to the atmosphere; therefore consume large amounts of

Fig. 1. Brick fabrication process at Nuevo Casas Grandes, Chihuahua.

fuels. Fuel used varies according to what is available for brick makers and includes wood, recycled motor oil, coal, fuel oil, diesel, tires, trash and plastics among others. The inefficient combustion of these furnaces favors the emission of solid particles and greenhouse gases, among which are the oxides of nitrogen, sulfur, and carbon (Table 1). González et al. (1998), in Spain, reported the presence of fluoride, chlorine and bromine in addition to the mentioned emissions from the open brick kilns. And Bruce et al. (2007) reported the presence of carcinogenic polychlorinated organic compounds and heavy metals by burning some of the fuels mentioned above. Recent studies have revealed that a traditional oven emits about 863 pounds of pollutants for each production burns covering approximately 10,000 bricks (TCEQ, 2002). The problem is increased by the process. One clear example is the combustion of heavy metals. Heavy metals can remain in its original form during incineration or may react to form new compounds such as metal oxides, chlorides or fluorides (Brigden et al., 2000). It has been shown that in the incineration process mobility and bioavailability of toxic metals is greatly increased in comparison with the original waste (Brigden et al., 2000; Martirena & Martinicurena, 2001). The most common form of transport of the metals released in these and other processes is by air. The metals can thus travel thousands of miles to be deposited in the soil (Martirena & Martinicurena, 2001) affecting people's health.

Brick fabrication process	**Activities that may generate pollutants**	**Pollutant**
Clay extraction	Manual tools used for extraction	Suspended Particles
Mixing	Sieving and selection Mixing of clay, water and silt	Suspended Particles
Molding		None
Drying	Water vapor	None
Kiln fill in		None
Cooking	Fuel: wood and saw dust	Suspended Particles Nitrogen Dioxide Carbon Monoxide Volatile Organic Compounds Polychlorinated organic compounds
Classification	Discarding broken, un-cooked and deformed products	Solid waste
Transport	Discarding broken products	Solid waste

Table 1. Pollutants associated to the brick kiln industry.

The pollution generated by the brick industry is a recurring problem worldwide. However, this micro-industry provides income to thousands of families and cannot be removed easily. Each country, city or town must look for better alternatives to address this problem. Corral Avitia et al., 2009 reported the environmental impact of brick industry from the construction site to the abandonment. The results showed that a clean technology should be adapted to minimize the critic effect of air pollution. However, prior to the use of clean technologies; studies must be performed to determine, geographically, the best sites for the relocation or construction of the kilns. However, there are different factors that must be addressed in each region. First of all, the location of the total number and type of brick kilns should be documented. The best places for relocation should be known to determine if the kilns require to be relocated. Also, techniques that will improve the brickmakers quality of life (appropriate technology) must be located and implemented. And finally, environmental policy strategies to change attitudes and activities with the support of various economic, social and governance entities must be designed.

4. Risk mapping and multi-criteria analysis

In recent decades, the mapping of risk has gained importance as an evaluation tool to natural or technological disasters. Therefore, research has focused on studying physical factors that influence the onset of disasters and on identifying temporal recurrence, in order to define the potentially affected area and assess possible damage. In the case of technological risks, studies have focused mainly on territorial vulnerability and spatial distribution of potentially hazardous facilities or activities (Bosque Sendra et al., 2004). Cartographic expression is a useful tool in assessing and managing risks. It establishes the zoning from the combination of levels of risk mapping, vulnerability and exposure. The execution risk mapping land proposed to avoid damage to vulnerable communities. The risk map is a resultant guidance tool for land use planning (Aguirre-Murua, 2005, Bosque-Sendra et al., 2005). From the theoretical-scientific, environmental maps are invaluable in projecting the field of practical applications for solving problems which commonly involve the interaction of different variables. The multicriteria analisys (MCA) enables the integration and interrelationships search for making good decisions (García-Abad, 2002).

During the last decade has increased the use of geomatics, which integrates geographic information systems (GIS), remote sensing and geo-positioning systems (GPS) in order to perform land management. Besides, incorporating MCA has proven to be efficient in certain geographical areas and it is ideal to respond to environmental impact studies and risk assessment for industry and agricultural activities (Ciminari et al., 2003; Diaz-Muñoz & Diaz-Castillo, 2002; Frau Mena et al., 2006; Malczewsky, 1999). It is also useful to relocate undesirable facilities with two objectives: to avoid negative impacts and to protect the industries operational costs. For the first objective, it is required to build the industry as far as possible of urban areas to avoid negative impacts over the population's health. But the second requires that the facilities are located as close as possible to their suppliers to achieve operational efficiencies (Medina & Cerda, 2008). Therefore, the results must present alternative solutions to the multiple actors involved in decision-making (Ciminari et al., 2003; Diaz-Muñoz & Diaz-Castillo, 2002; Malczewsky, 1999). However, it is difficult to determine optimal sites for industrial development, because it involves a complex array of critical factors in several disciplines such as economic, technical, social and environmental (Jun, 2000).

There are some research published in scientific literature addressing the application of information technology techniques, multi-objective and MCA to relocate industries, landfills, power plants and solid waste disposals. Lowry et al., (1995) developed a GIS mapping analysis to estimate the degree of potential risk associated with an industrial area between México and the U.S. border. He used vulnerable areas and exposure zones to create a risk scenario assessing the most sensitive areas in the event of an accident. Tudela-Serrano et al. (2000) conducted a search for suitable areas for building thematic developments by local analysis tools, reclassification and overlaying maps, based on physical and economic criteria. Vatalls and Manoliadis (2002) and Frau Mena et al. (2006), reported the best area to build a landfill using environmental factors, socioeconomic and technical-operational areas. Zambon et al. (2005) applied the technique of MCA associated with a GIS to assess the territory for the purpose of locating power plants by reducing the environmental impact to a minimum. Using the same technology tools, Gallardo et al. (2005) and Silva et al. (2006) described a methodology to determine the most appropriate location of sites for solid waste disposal. Corral Avitia et al. in 2010, publish the first study using these integrated techniques to find the best places to relocate the brick industry in Ciudad Juárez. This study was used for the State and Municipal Government on 2010 to manage the purchase of land for the relocation of brick kilns. The relocation of the owners of the kilns is still in process due to the actual unsafe conditions of the city.

Based on all the antecedents, the contribution of the authors pretends to be useful for government and brikmakers. This chapter presents two objectives considering the complete and complex state of Chihuahua. The first objective of this study was to identify the location of the brick kiln in all the municipalities of the state and it is presented in section 5. The second objective, presented on section 6, was to determine the best places in the state of Chihuahua to relocate the brick kiln industry in order to minimize environmental risk.

5. Brick kilns in Chihuahua State

Research indicates that the brick industry is distributed in several countries but there is a lack of real information. México has a total of 32 states, however there is only a few of them who have reported inventories of kilns that are normally operating (SEMARNAT, 2011). These small businesses are hardly registered on economical census. One clear example is Chihuahua state, there are 277 brick kilns registered in the 2009 Economic Census of the National Institute of Statistics and Geography (INEGI, 2009) but in Juárez alone there are 312 Brick kilns (Corral Avitia, 2005). Therefore, there is a need to know the real problem. During 2010 and 2011 a census was carried out in all the municipalities in the state of Chihuahua in order to determine the total number of kilns, brick-burning capacity, type of furnace, fuel used and exact geographical location (González et al., 2011).

Chihuahua is the largest state located at 31° 54' - 25° 29' N latitude and longitude of 103° 16' - 109° 17' O North México. It has a total population of 3,406,465 inhabitants. Its total extension is 247,460 km^2. It is divided on 64 municipalities being the most populated in decreasing order Juárez, Chihuahua, Cuauhtémoc, Delicias, Hidalgo del Parral, Nuevo Casas Grandes, Camargo, Jiménez, Guerrero and Saucillo with population higher than 30,000 inhabitants. The municipalities of the state in which the brick industry has developed in an intensive manner are presented in Table 2.

The census (González et al., 2011) indicates that there are 959 brick kilns distributed in 16 municipalities and these can be observed in Figure 2. Most of these are located within these municipalities: Juárez (312), Chihuahua (296), Cuauhtémoc (88) and Hidalgo del Parral (85).

Figure 3 presents the location of the brick kilns with respect to the urban area on these four municipalities. Figure 4 presents the urban areas of Delicias, Camargo, and Nuevo Casas Grandes, with 49, 48 and 33 brick kilns, respectively. On a recent effort, the Secretary of Urban Development and Ecology successfully relocate the brick kilns in Cuauhtémoc and Camargo as can be observed on Figure 3c and 4a, respectively. However, there are still municipalities that have most of the kilns inside the cities; such as Juárez, Chihuahua and Nuevo Casas Grandes. On the other hand, two others, Delicias and Hidalgo del Parral, are under the risk to be reached by the urban area. Therefore, the risk cartography will be presented for those five municipalities in the next section.

Municipality	Total population	Brick kilns
Ascensión	23,975	2
Buenaventura	22,378	10
Camargo	48,748	48
Casas Grandes	10,587	11
Cuauhtémoc	154,639	88
Chihuahua	819,543	296
Delicias	137,935	49
Galeana*	5,892	6
Guerrero	39,626	2
Hidalgo del Parral	107,061	85
Janos	10,953	1
Jiménez	41,265	9
Juárez	1,332,131	312
Nuevo Casas Grandes	59,337	33
Rosales	16,785	3
Saucillo	32,325	4

*State Housing Program. The brick kilns were built to provide the bricks required for the program. They were temporary.

Table 2. Chihuahua's municipalities with brick industry. Relation between population and number of kilns.

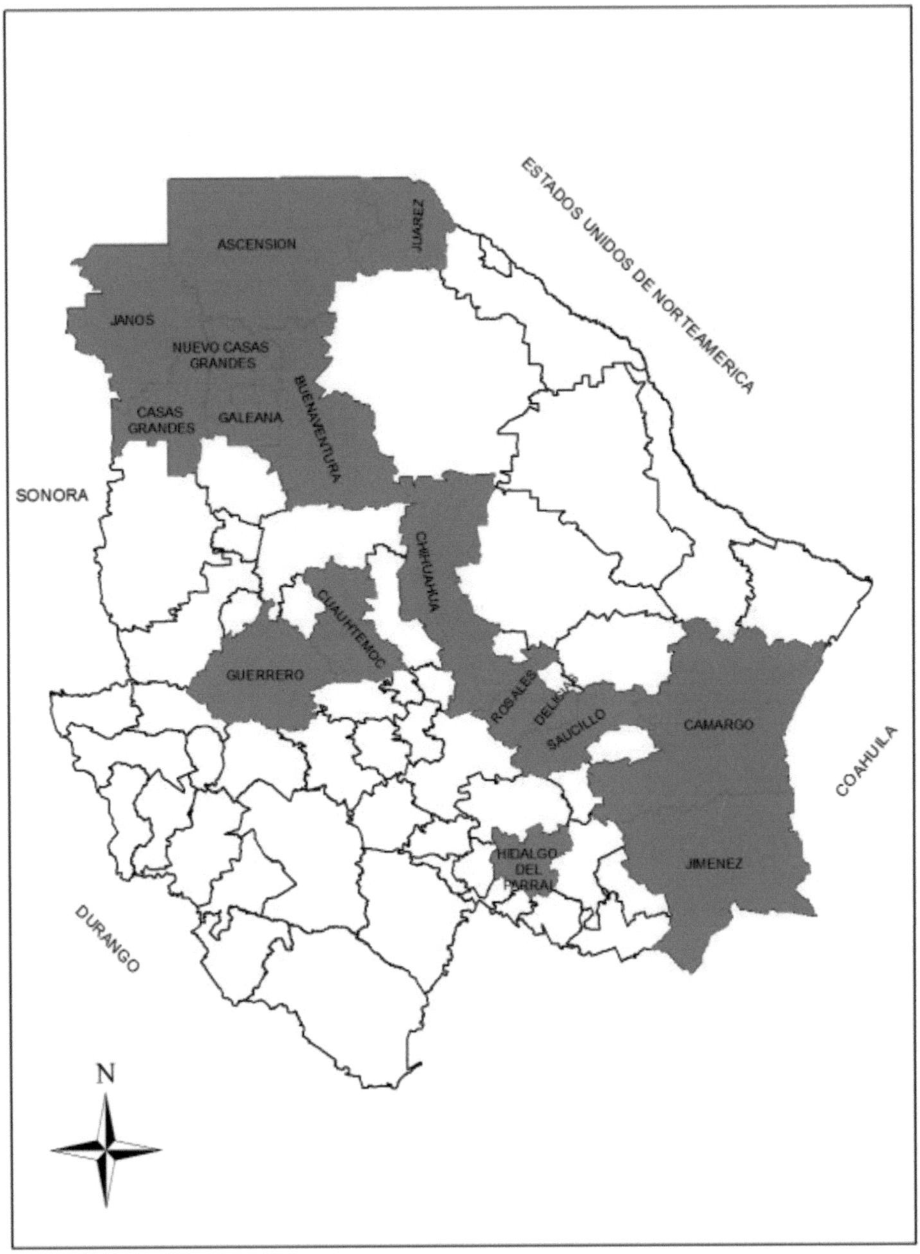

Fig. 2. State of Chihuahua´s municipalities with brick industry.

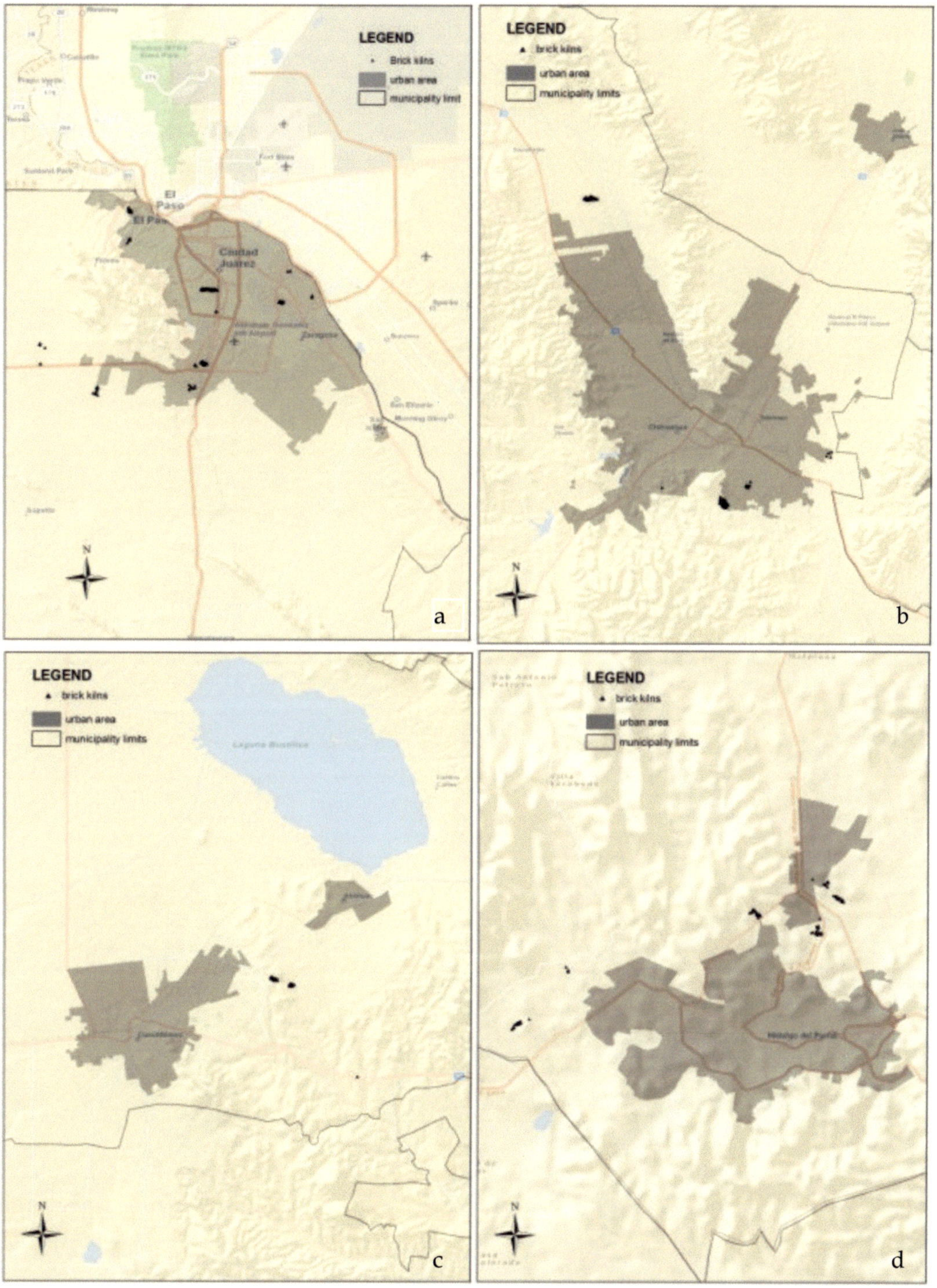

Fig. 3. Municipalities with more than 80 kilns. a) Juárez, b) Chihuahua, c) Cuauhtémoc and d) Hidalgo del Parral.

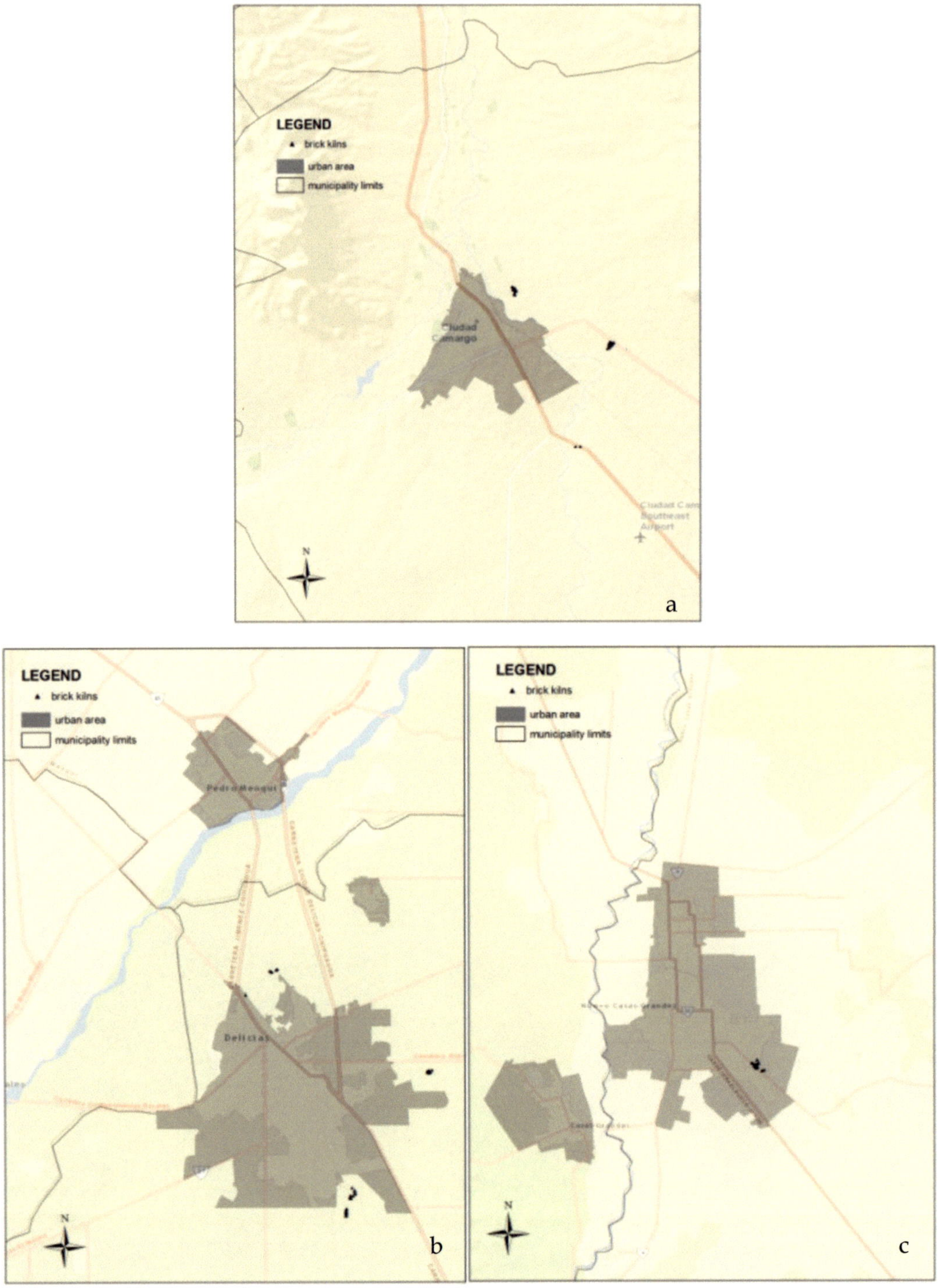

Fig. 4. Municipalities with more than 30 kilns. a) Camargo, b) Delicias and c) Nuevo Casas Grandes.

6. Risk mapping for brick kilns in Chihuahua State

This session provides the methodology to determine the best places in the state of Chihuahua to relocate the brick kiln industry. Being the ultimate objective to minimize environmental risk for the population living close to this micro-industry. The mapping was developed for five municipalities Juárez, Chihuahua, Hidalgo del Parral, Delicias and Nuevo Casas Grandes.

Since none of the five municipalities had legislation on the installation and operation of brick kilns, the factors considered in environmental impact analysis were determined by literature review and consultation with specialists in the environmental area. It was used as a reference the Ecological Law of the State of Guanajuato in the technical standard and factors proposed by NTE-IEG-001/98. The working group members are specialists in various environmental areas such as biota, water, soil, air and energy. The selection of a risk factor was based on the fact of being able to be represented spatially, so the databases used were those of the National Institute of Statistics and Geography as is the National Census of Population and Housing 2005, Digital Cartography Unit of the Biology Program at the Biomedical Science Institute of the Autonomous University of Juárez City and the National Biodiversity Commission. Maps used were soil science, geology, hydrology, vegetation and other landscape units (scale 1:250,000). The risk factors considered to evaluate each municipality are described next.

a. Vulnerable species. The mappings used were those corridors and polygons with the presence of species with ecological or economic interest.
b. Ecological zones. The information used was about natural and protected areas, such as national parks, biosphere reserves and buffer zones of flora and fauna, considered as priority areas for conservation.
c. Groundwater hydrology. It was considered the average of the mirror as groundwater wells in the state. It was assumed that shallow groundwater levels are more vulnerable to contamination than deeper.
d. Surface hydrology. Buffer zones were established. Special care was placed on surface water flows to ensure a minimal impact on hand contamination and to prevent the destruction of the kilns for potential flooding.
e. Urban areas. Urban areas are considered as exclusion areas for the brick industry; but should be considered on a relative proximity to allow the movement of employees into the workplace as well as its proximity to points of sale.
f. Elevation. The minimum and maximum desirable elevation was established in relation to the relative height above sea level for the installation of the brick industry. This was performed independently for each municipality.
g. Pedology. Soil, structure and type, are important factors that allow an efficient construction and a long-term performance of the kilns. Besides it provides resistance to infiltration of contaminants.
h. Pending. The slope of the land has a strong influence on the selection of sites for construction of kilns. It helps to reduce costs associated with it and avoid losses by landslides or mudslides.
i. Geology. Gives stability to the kilns and prevents pollutants infiltration.
j. Roads. The proximity to roads reduces the costs associated with freight transportation as well as raw material to finished product.

k. Land use. Some municipalities have land use plans that consider defined areas for the implementation of dangerous or hazardous industry. Therefore, they must be taken into account as a priority for the development of the brick industry.
l. Archaeological areas. Archaeological areas with buffer areas were considered exclusion areas for the final maps.
m. Raw materials depots. The proximity to the banks of material decrease production costs of the brick, however the environmental impact caused by removing the soil must also be considered.
n. Airport. In those municipalities that have an airport, the polygon area was considered for exclusion.
o. International and municipal limits. To avoid conflicts between countries or municipal authorities it was decided to respect the boundary edges using buffer zones.
p. Transmission lines. For safety reasons, the construction of brick kilns is prohibited in areas where transmission lines, gas, pipelines or aqueducts are present.

Multi hierarchical criteria (Malczewsky, 1999) and a modified Delphi technique (ESRI, 1991) were applied in order to establish the importance criteria and weighting of the analysis. The risk factors (a to p) cited above were classified within the corresponding category and subcategories as shown on Table 3. The participation or not of the risk factors was depended on the municipality natural and anthropogenic characteristics. Once weighted (Wc y Wf), digital map were generated with the extension Spatial Analysis (ArcGIS 9.3®) with normalized weighted data (N) according to risk factors. Using algebra and overlap maps, the polygons with different degrees of vulnerability were presented on each map. This allows to have a better view about the zones were the brick industry can be relocated without harming the environment and the brickmakers budgets.

CATEGORY	Wc	SUBCATEGORY	FACTOR	Wf	N
Environmental	Wcenv	Biological	B1	WfB1	WcenvWfB1
			B2	WfB2	WcenvWfB2
			Bn	WfBn	WcenvWfBn
		Physical	Ph1	WfPh1	WcenvWfPh1
			Ph2	WfPh2	WcenvWfPh2
			Phn	WfPhn	WcenvWfPhn
Economical	Wceco	Construction	C1	WfC1	WcecoWfC1
			C2	WfC2	WcecoWfC2
			Cn	WfCn	WcecoWfCN
		Access	A1	WfA1	WcecoWfA1
			A2	WfA2	WcecoWfA2
			An	WfAn	WcecoWfAn
Social	Wcsoc	Population	P1	WfP1	WcsocWfP1
			P2	WfP2	WcsocWfP2
			Pn	WfPn	WcsocWfPn
		Sensitive features	SF1	WfSF1	WcsocWfSF1
			SF2	WfSF2	WcsocWfSF2
			SFn	WfSFn	WcsocWfSFn

Table 3. Evaluation matrix by weight values by category (Wc) and factor (Wf) and the resulting normalized value (N) according to experts consulted. Factors vary depending on each municipality conditions.

6.1 Juárez

Juárez has 312 brick kilns distributed in 13 different sectors all around the city (Corral Avitia, 2005; Romo et al., 2004) For the municipality of Juárez shown in Figure 4, it was concluded that 90.3% of kilns are located in areas of medium to high vulnerability (Corral Avitia et al., 2010). The state acquired a land on 2010 for the relocation of brick kilns based on this map, but the project still in process and the relocation has not been achieved to this date.

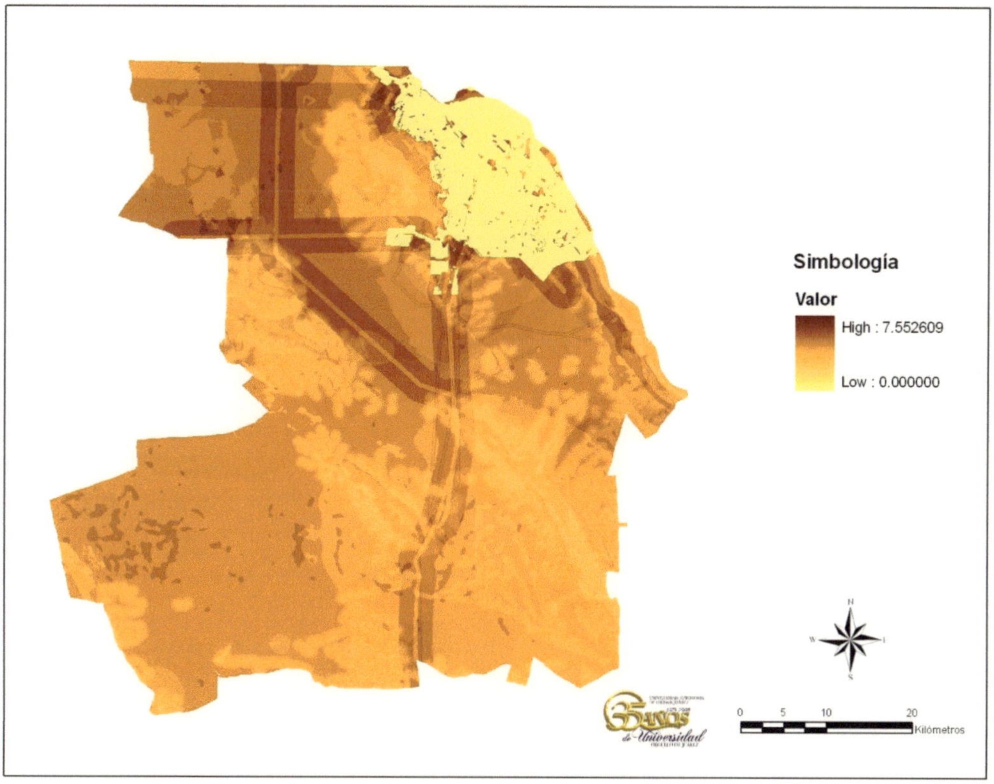

Fig. 5. Suitability map for the establishment of the brick industry in Ciudad Juárez, Chihuahua, México.

Most of the kilns in Juárez are built in the traditional way. However, there are 24 MK kilns in three of the sectors built by El Paso Electric on 2003 (TCEQ, 2002). These MK kiln have a dome that minimizes 80% of the emissions to the atmosphere. The burns are made more efficiently, because it keeps the heat inside minimizing the amount of fuel required and allowing 100% of finished product. Although these kilns offer economic and environmental advantages, the brickmakers are resistant to build more because they are waiting for the relocation.

6.2 Chihuahua

Chihuahua has 296 brick kiln distributed in five sectors. These kilns are called "camperos" because they are built with the bricks that are going to be fired. This kind of kilns represents a problem for an inventory because they may be or not in the site from one day to the other. They burn saw dust and wood in most of the cases. Relocation of kilns was made on the 90's but the urban growth is reaching them again as can be seeing on Figure 3b. There are more kilns on the Southwest side of the city close to the hills. The kilns in this area have a capacity of 20 to 30,000 bricks per kiln. This can also be translated as higher pollution levels due to the amount of fuel required to obtain that amount of product. The resulting risk map on Figure 6 can be used if a new relocation wants to be made. The best polygons are towards the Northeast side.

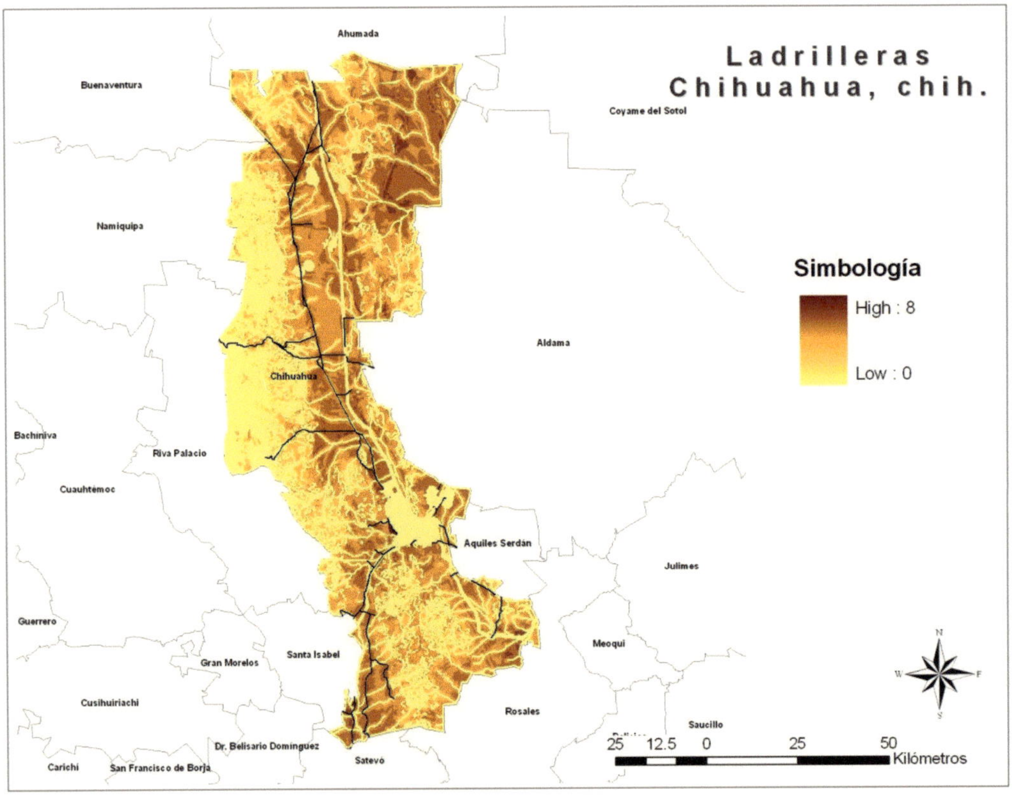

Fig. 6. Suitability map for the establishment of the brick industry in Chihuahua, Chihuahua, México.

6.3 Nuevo Casas Grandes

In Nuevo Casas Grandes, brick industry consists of several individual producers engaged in the manufacture and sale of bricks of different sizes with the objective of providing the capital city and conurbation areas. Currently, there are 33 producers operating for the last 25 years individually. All brick kilns are immersed in the urban area which is a health problem for the inhabitants of the area. As shown in the suitability map in Figure 7 the best areas would be located north and east of the municipality. This map was presented to the municipal authorities who are interested in the community health. They propose to donate some land for the relocation of brick kilns. The proposed site is marked (black square) in the map with a medium high category suitable for this use.

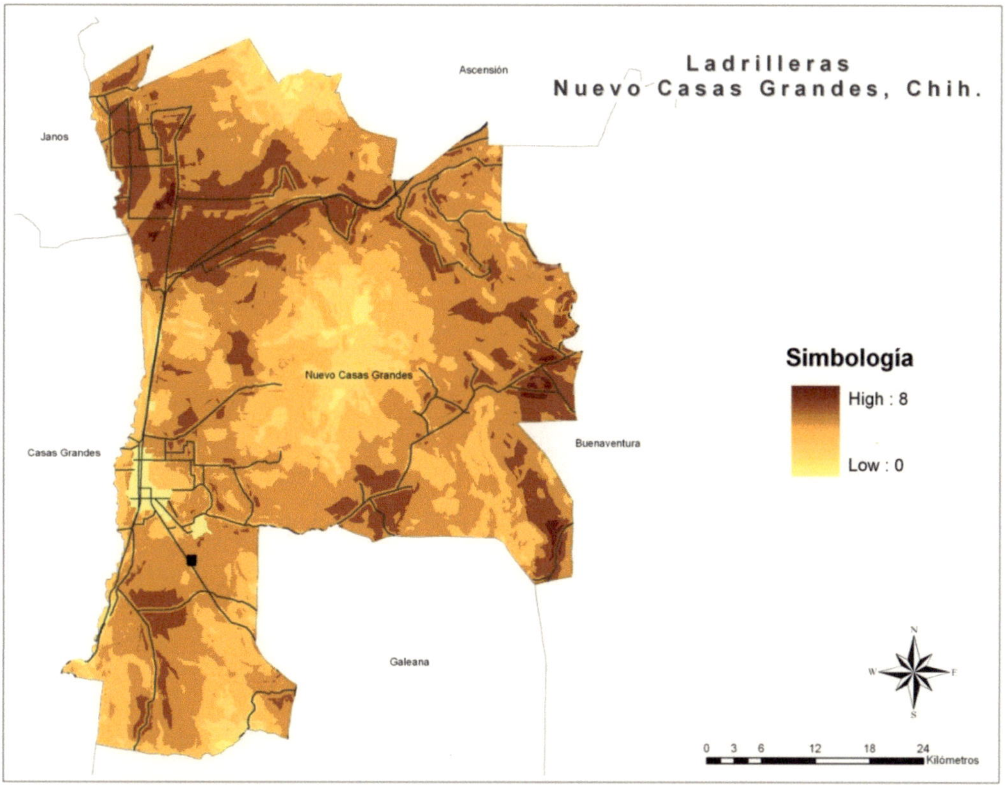

Fig. 7. Suitability map for the establishment of the brick industry in Nuevo Casas Grandes, Chihuahua, México.

6.4 Hidalgo del Parral

There are 85 kilns distributed in 7 sectors at Hidalgo del Parral, but only two of them are far from the urban area. The other 5 sectors are close and soon will be reached by human developments. Therefore, the map can be useful to make the correct decision on where they can be relocated.

The kilns are static and the production is adequate to use the MK kiln. The brickmakers have the advantage of selling the product to people from communities that are close to Hidalgo del Parral. That justifies the lack of brick kilns in its vicinity.

6.5 Delicias

Delicias has 49 brick kilns in 5 different sectors. The kilns in this areas are static, they may be counted for the inventory and signs will be left they move to different places. Most of the kilns were located close to streams of water that are used to irrigate the crop fields. They are located outside the urban area but they may be reached if urban development is not controlled. This map may be used to orientate the location of the kilns in the future.

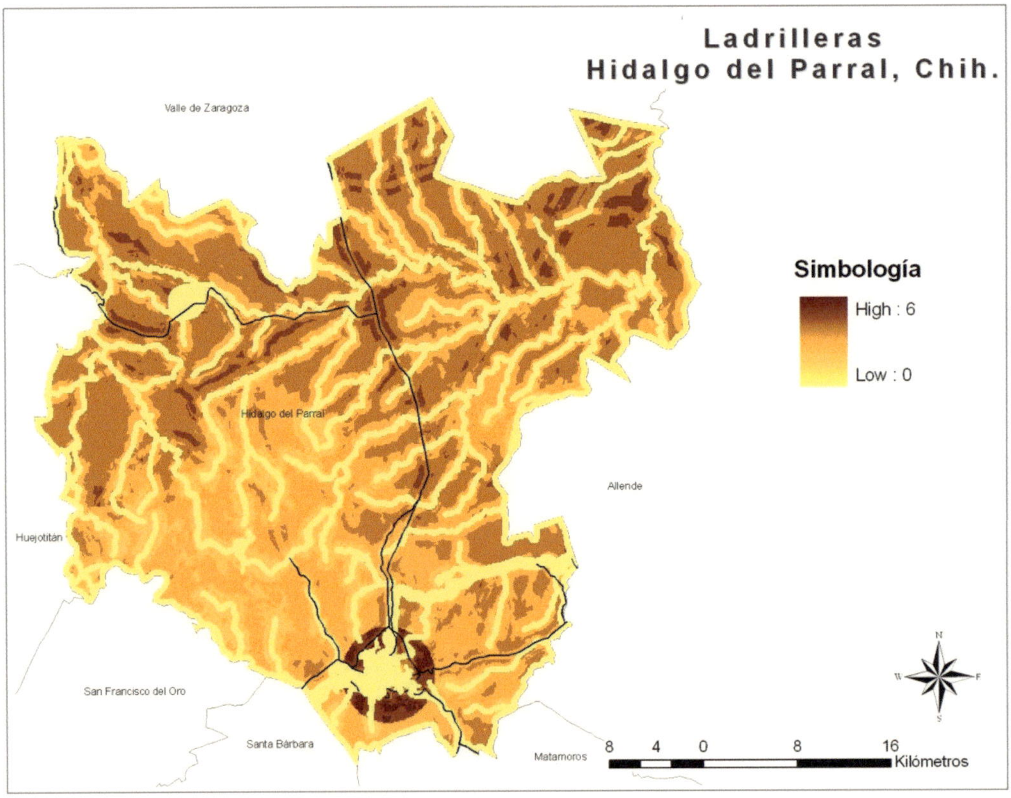

Fig. 8. Suitability map for the establishment of the brick industry in Hidalgo del Parral, Chihuahua, México.

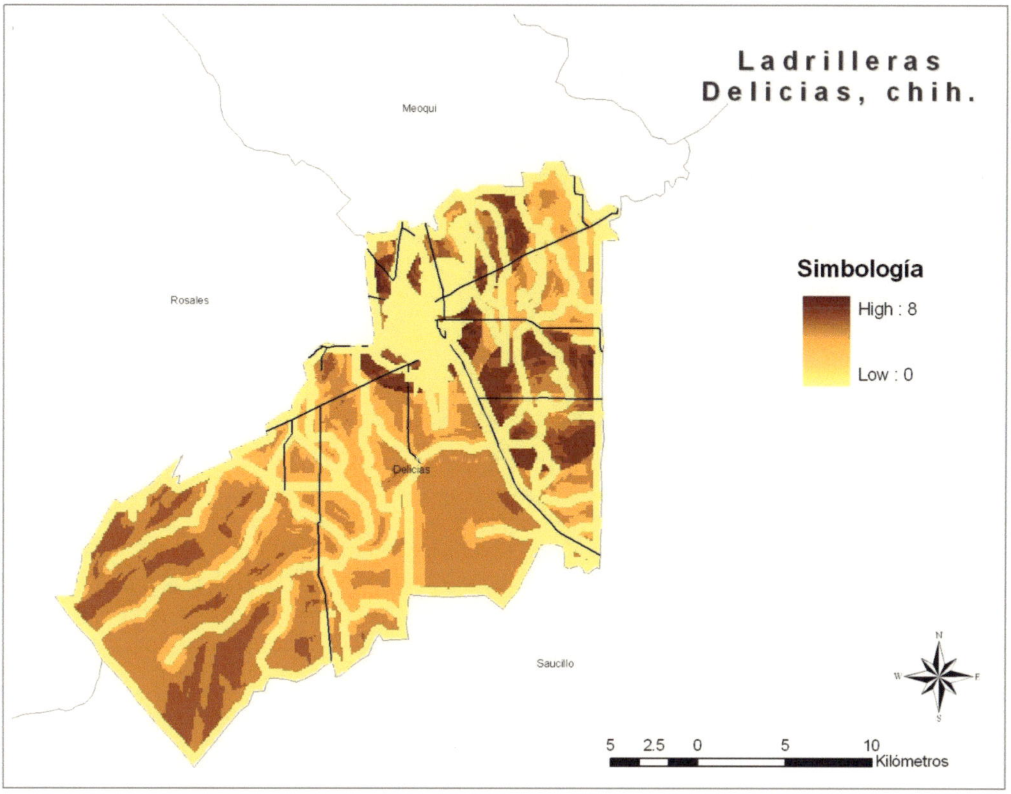

Fig. 9. Suitability map for the establishment of the brick industry in Delicias, Chihuahua, México.

7. Forecast in Chihuahua State

Finally Figure 10 shows the ability to forecast statewide. In this case, exclusion parameters were applied in municipalities with over 2,500 inhabitants, protected natural areas, and archaeological sites. The resolution of the resulting map of the general cartographic product is compromised due to the size of the state. The pixel size was too thick so it would be preferable to use the municipal level models instead. There is a natural correlation between the risk maps with that obtained after the census. Those zones where no kilns were found, correspond to excluded areas.

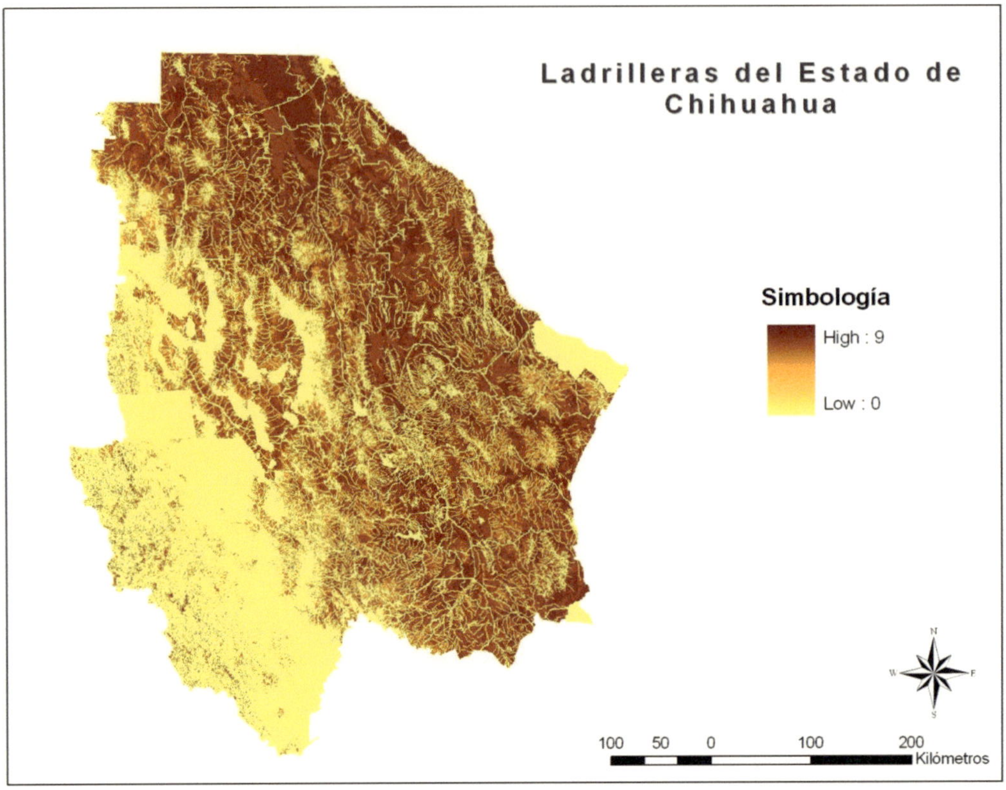

Fig. 10. Chihuahua State apptitude map.

8. Conclusion

The environment has been changing lately. Although, there are multiple factors, one of them is the anthropogenic activity performed recently. As the population grows, there is an increase on needs of housing. There are different kinds of materials used to construct houses; however bricks are traditional in most of the places and are always on demand. The problem is that some human needs, leads to air pollution being the case for brick production. However, the real dimension of the problem is still unknown. This chapter is an example of an initiative to know the root of the problem and to propose a solution. There were two objectives presented at the beginning of the chapter: to know the amount and location of the kilns in Chihuahua State and to elaborate risk maps for those sites with the highest amount of kilns. Both objectives were achieved. First, it was found that the number of kilns officially reported is far from reality. This difference leads to underestimate the problem of air pollution derived from this micro-industry. Second, environmental risk maps resulting from the combination of GIS and the AMC are valuable tools for solving the problem of the brick industry location. Their advantage lies in the exploration time savings through digital geographical reading of the variables included in the study to evaluate at one time the entire territory.

While this study presents a methodology with sound technical criteria for selecting areas of low vulnerability for the relocation of brick kilns, it is necessary to regulate the operation of this industry. The traditional type of kilns must be moved towards more efficient models as the one presented for Juárez (MK kilns), the type of fuel must be regulated, dates and hours of operation and maximum production allowed while it is established. All should be done by competent authority. Besides it is recommended to perform frequent inspection of the emission's monitoring and periodicity of medical checks of workers, among others, to achieve the overall goal, improve the quality of life of people involved in the process.

9. Acknowledgment

The authors acknowledge the support of Mixed Fund National Council of Science and Technology (CONACYT) and Chihuahua State Government under contract CHIH-2009-C01-117781 and Autonomous University of Juárez City. Special thanks to Sac-Nicte Guadalupe González Posada, and her classmates of the program in Master of Science in Biological Chemistry for the field work developed for the present work.

10. References

Aguirre-Murua G. (2005). La valorización de los riesgos en la ordenación del territorio: metodología práctica. Boletín A.G.E. 40, 393-405.

Bosque Sendra, J.; Díaz Castillo, C.; Díaz Muñoz, M. A.; Gómez Delgado, M.; González Ferreiro, D.; Rodríguez Espinosa, V. M. & Salado García, M. J. (2004): "Propuesta metodológica para caracterizar las áreas expuestas a riesgos tecnológicos mediante SIG. Aplicación en la Comunidad de Madrid", *GeoFocus* (4) 44-78.

Bosque-Sendra J.; Ortega-Sisquez, A. & Rodríguez-Espinosa V.M. (2005). Cartografía de riesgos naturales en América Central con datos obtenidos desde Internet. *Doc. Anal. Geogr.* 45, 41-70.

Brigden, K.; Labunska, I.; Stringer, R.; Johnston, P.; Santillo, D. & Ashton, J. (2000). Identificación de contaminantes orgánicos y metales pesados en muestras recolectadas del incinerador de desechos peligrosos y hospitalarios Trieco, Provincia de Buenos Aires, Argentina. Laboratorios de Investigación de Greenpeace, Departamento de Ciencias Biológicas, Universidad e Exeter, Exeter, Reino Unido. 34p

Bruce, C.W., Corral, A.Y. & Lara, A.S. (2007). Development of cleaner burning brick kilns in Ciudad Juárez, Chihuahua, México. *J. Air & Waste Manage. Assoc.* 57, 444-456.

Campos, A.; Gómez, R.; Licon, L.; Carrillo, J.; Ramírez, E. & Herrera, E. F. (2008). Monitoreo de contaminantes atmosféricos en la Ciudad de Chihuahua (Norte de México) como una herramienta para la gestión de la calidad del aire. *Revista Latinoamericana de Recursos Naturales*. 4(3): 357-366.

Codes, M. I.; Robledo, S. B.; Alessandro, M. & Maffei A. (2002). Impacto ambiental de las ladrilleras en el Algarrobal. In: *Departamento de las Heras, Mendoza, Argentina*. August 10 2011. Available from: http://ffyl.uncu.edu.ar/IMG/pdf/Impacto_de_las_ladrilleras_El_Algarrobal.pdf

Ciminari, M.; Torrens, C. & Jurio E. (2003). Los sistemas de información geográfica: una herramienta eficaz para el análisis ambiental. Memorias. Primer Congreso de la Ciencia Cartográfica y VIII Semana Nacional de Cartografía. Buenos Aires, Argentina. 25 al 27 de junio, 2003. pp 1-11.

Cohen, A.J.; Anderson, H.R.; Ostro, B.; Pandey, K.D.; Kryzanowsky, M.; Kuenzly, N.2003. Mortality impacts of Urban Air Pollution. En: Ezzati M, Lopez AD, Rodgers A, Murray CJL (editores). Comparative Quantification of Health Risks: Global and Regional Burden of Disease Attributable to Selected Major Risk Factors. Ginebra: Organización Mundial de la Salud.

Corral Avitia A. Y. (2005). "Padrón de ladrilleras 2005 en Ciudad Juárez, Chihuahua". Instituto de Ciencias Biomédicas. Departamento de Ciencias Básicas. Programa de Química. Reporte de Difusión.

Corral-Avitia A.Y., De la Mora-Covarrubias A. y Carrasco-Urrutia K.A. (2009). Impacto ambiental de las ladrilleras en Ciudad Juárez. En *Lazos Verdes: primer encuentro académico en pro del medio ambiente*. Minguez H. Coordinador. 248 pp. 1-248. Editorial de la Universidad Autónoma de Ciudad Juárez. ISBN: 978-607-7623-45-8. Ciudad Juárez, Chihuahua, México.

Corral Avitia, A. Y.; De la Mora Covarrubias, A.; Cota Espericueta, A. D.; Corral Díaz, R.; Carrasco Urrutia K. A.; Santana Contreras, L. E. (2010). "La cartografía de riesgo como instrumento técnico para la reubicación de la industria ladrillera."Revista Internacional de Contaminación Ambiental. No. 2. Vol. 6.

Dalmasso, A.; Candia, R. & Llera, J. (1997). La Vegetación como Indicadora de la Contaminación por Polvo Atmosférico. *MULTEQUINA* 6: 91-97.

De Vives, A. L. (2007). Species Arboreal as a Bioindicator of the Environmental Pollution: Analysis by SR-TXRF. *Nuclear Instruments and Methods in Physics Research A* 579: 494–498.

Díaz-Muñoz, M.A. & Díaz-Castillo, C. (2002). El análisis de la vulnerabilidad en la cartografía de riesgos tecnológicos. Algunas cuestiones conceptuales y metodológicas. Serie Geográfica 10, 27-41.

Escobedo, F. & Chacalo, A. 2008. Estimación preliminar de la descontaminación atmosférica por el arbolado urbano de la ciudad de México. *INTERCIENCIA* 33(1): 20-33.

Frau Mena, C. J.; Gajardo, V. & Ormazabal, Y. (2006). Modelación espacial mediante geomática y evaluación multicriterio para la ordenación territorial. *Rev. Fac. Ing.* 14(1):81-89

Gallardo, A.; Bovea, M.D. & Bordás, R. (2005). LIGRE: una herramienta para la generación de mapas de orientación a la ubicación de instalaciones de gestión de residuos. Aplicación al emplazamiento de vertederos en la provincia de Castellón. *Residuos* 83. Marzo-Abril.

García-Abad J.J. (2002). Cartografía ambiental. Desarrollo y propuestas de sistematización. Observatorio Medioambiental 5, 47–78.

González, I.; Galán, E. & Fabbri, Y.B. (1998). Problemática de las emisiones de flúor, cloro y azufre durante la cocción de materiales de la industria ladrillera. *Bol. Soc. Esp. Ceram. Vidr.* 37, 307-313.

González, M. (2008). "Proceso de fabricación de ladrillos" Ingeniería Electrónica y Telecomunicaciones. Universidad Politécnica de Puebla.

González P. S. G., Delgado R. M., González G. N. A., Rea L. M. A., & Corral A. A.Y. (2011). Situación actual de la industria ladrillera en el Estado de Chihuahua. Proceedings of X Congreso Internacional y XV Congreso Nacional de Ciencias Ambientales. Santiago de Querétaro, Querétaro, México. August 2011.

Instituto Nacional de Estadística y Geografía. (2009). Censo Económico. México.

Jun, C. (2000). Design of an Intelligent Geographic Information System for Multi-criteria Site Analysis. *URISA Journal*. Vol. 12, No. 3.

Lowry, J.; Harve, J.; Miller, J. & Hepner, F. (1995) "A GIS-Based Sensitivy Analysis of Community Vulnerability to Hazardous Contaminants on th México/U.S. Border" *Photogrammetric Engineering & Remote Sensing*, Vol. 61, No. 11, pp. 1347- 1359.

Malczewsky, J. (1999). GIS and multicriteria decisión analysis. John Wiley & Sons, Inc. United States of America.

Martirena, J. & Martinicurena, A. (2001). Contenido de metales pesados en suelos forestales de Navarra. *Gorosti*. 18-24.

Medina, M. & Cerda, J., (2008) Modelo de localización óptima de actividades no deseadas aplicado a los residuos sólidos en la región metropolitana, Santiago de Chile, Ingeniare.

NORMA TÉCNICA ECOLÓGICA NTE-IEG-001/98. (1998). Instituto de Ecología. Guanajuato, Gto.

Ramadan, E., Feng X., Cheng Z. 2004. Satellite remote sensing for urban growth assessment in Shaoxing City, Zhejiang Province *J Zhejiang Univ SCI* 5(9):1095-1101.

Romo, L.; Cordova, G. & Cervera, L. (2004) Estudio Urbano-Ambiental de las Ladrilleras en el Municipio de Juárez. Estudios Fronterizos, Vol. 5, Núm. 9. Pág. 10-33.

SEMARNAT, 2011. Emissions Inventories.

Silva, J.T.; Estrada, F.; Ochoa, S. & Cruz, G. (2006). Propuesta metodológica para la ubicación de áreas de disposición de residuos sólidos urbanos. *Rev. Int. Contam. Ambien.* 22(4) 147-156.

TCEQ, Texas Commission on Environmental Quality 2002. "A Study of Brick-Making Processes along the Texas Portion of the U.S.-México Border: Senate Bill 749". Border Affairs Division. SFR-081/02. December 2002.

Tudela-Serrano, M.L.; Molina-Ruiz, J. & Albaladejo, M.V. (2000). Metodología práctica de aplicación de un proyecto SIG en el término municipal de Cehegín (Murcia). *Papeles de Geografía*. 3, 155-167.

Valenzuela E. (2005). Fundamentos de estadística, Introducción a la estadística. Universidad Técnica Federico Santa María.

Van der Sande, C.J.; de Jong, S.M. & de Roo, A.P.J. (2003). A segmentation and classification approach of IKONOS-2 imagery for land cover mapping to assist flood risk and flood damage assessment International *Journal of Applied Earth Observation and Geoinformation*. 4:217–229

Vatalls, K. & Manoliadis, O. (2002). A two-level multicriteria DSS for Landfill Site Selection Using GIS: Case Study in Western Macedonia, Greece. *Journal of Geographic Information and Decision Analysis*. Vol. 6, No. 1, pp. 49-56

Zambon, K. L.; Carneiro, A. & Negri, J.C. (2005). Análise de decisão multicritério na localização de usinas termoelétricas utilizando SIG. *Pesquisa Operacional*. 25(2):183-199.

15

Challenges to the Expansion of Ethanol Production in Brazil

Ester Galli
KTH – Royal Institute of Technology
Sweden

1. Introduction

In the early 1990s, climate change came onto the agenda and was identified as being one of the most serious environmental problems of our time (UN 1992, article 2). The United Nations Framework Convention on Climate Change (UNFCCC) is an international environmental treaty produced at the United Nations Conference for Environment and Development (UNCED, also known as the Rio Earth Summit) held in Rio de Janeiro in 1992. The objective of the treaty is to stabilise greenhouse gas concentrations in the atmosphere at a level that would prevent dangerous anthropogenic interference with the climate system. The main outcome of the treaty is the Kyoto Protocol, which sets mandatory emission limits on the individual signatory countries (Kyoto, 1997).

Since the introduction of the Kyoto Protocol in 1997, world-wide concern about climate change and its impact on global warming has motivated unprecedented discussions on energy sustainability (Cox et al., 2000; Hansen et al., 2005). It is generally agreed that current energy resources, largely based on fossil fuels, are not sustainable for the long term (FAO, 2008). To meet these new challenges, research on the possible contribution of biomass to future global energy supply was initiated at international level (e.g. Hoogwijk et al., 2003; Berndes et al., 2003) and new policies to promote renewable energy in general, and its use in the transport sector in particular, were launched.

Policies to promote renewable energy existed in a few countries in the 1980s and early 1990s, but emerged in many more countries, states and provinces between 2005 and 2010. By 2009, 85 countries had some type of policy target, compared with 45 in 2005 (REN 21, 2010). Mandates for blending biofuels into vehicle fuels have been enacted in at least 41 states/provinces and 24 countries at the national level. Most mandates require blending 10-15% ethanol with petrol. There were 53 states/provinces/countries with biofuel mandates in 2007, 55 in 2008 and 65 in 2009 (REN 21, 2010).

Following this tendency, in 2007 the EU set the ambitious goal of obtaining 10% of its transport sector's energy consumption from biofuels by 2020, as a method of reducing emissions of greenhouse gases. Based on estimates of future petrol consumption, these 10% represent 10-14 billion litres of ethanol. The US Energy Mandate set a target for the production of 68 billion litres (15 billion gallons) from maize using current technologies by 2015 and an additional 95 billion litres (21 billion gallons) from *advanced biofuels* by 2022.

Many other countries have already adopted blends of 2% (E2), 5% (E5) or 10% (E10) ethanol with petrol, among them China, Canada, Australia, and Vietnam. Others, such as Colombia, Argentina, the Philippines and South Africa, will adopt E2 or E10 by 2010-2012 (Licht, 2008a, 2008b; Goldemberg, 2008).

It was in this scenario that Brazil's long-standing biofuel programme attracted world-wide attention. The programme has reduced the country's dependence on petrol (by 2009 the nation's petrol consumption had been cut by 50%) and significantly cut emissions of CO_2 (Macedo et al., 2008). A sign of the world interest in the Brazilian biofuel programme was the visit in 2007 of the US president George W. Bush to Brazil to establish cooperation for developing a global market for ethanol. The US pursuit of alternatives to oil led the country to search for technical and commercial cooperation with the 'world leader in biofuel' (The Washington Post, 2007). The visit resulted in the *Memorandum of Understanding between the United States and Brazil to Advance Cooperation on Biofuels* being signed by the two countries, with the aim of hosting initiatives to promote biofuels around the world.

However, it was only for a short while as by 2007, a range of divergent views had made media headlines: policy-makers, energy producers, academics and environmentalists in many countries expressed their concerns about the sustainability of Brazilian sugarcane ethanol as a biofuel. In response to the divergent views about ethanol production, the study described in this chapter was set up in order to examine the state-of-the-art of the agricultural stage of ethanol production and its impacts on the biodiversity of the two main Brazilian biomes: Amazon and Cerrado.

2. Theoretical framework

Three theoretical concepts guided this study. The concept of 'risk perception' is central to the understanding of what people mean when they say that something is (or is not) risky. 'Frame' is the core concept. It is used to understand how people make sense of the world around them, how they make sense of issues in a conflicting situation. The concept of sustainable development is used to clarify what people mean when they say that a biofuel is (or is not) sustainable.

2.1 Perception of risk

Risk perception is the term commonly used to refer to judgements made when people are asked to evaluate hazardous activities and technologies (Slovic, 1992). The importance of the knowledge of how people and experts perceive risk lies in the contribution that such knowledge may give to decision-makers who seek to avoid both the creation and the progress of conflicts.

The study of risk perception can be traced back to the late 1960s. It started as a response to the public opposition to nuclear technologies. The seminal work of Starr (1969) about the public attitude towards technologies showed that risk acceptance was related to subjective dimensions of risk and not only to technical estimates of risk. Since then, different approaches have been developed in order to discover what people mean when they say that something is or is not risky and to determine what factors underlie these perceptions.

The debate about the concept of risk perception is well known and has been well documented by Slovic (2000). It was started by scholars of psychology (Fischhoff et al., 1978; Slovic, 1992) and was broadened to include scholars of sociology (Jasanoff, 1997; Otway, 1992) as well as those of anthropology (Douglas and Wildavsky, 1982; Thompson et al., 1990).

In the course of the debates, social constructionists asserted that it is not possible to evaluate the acceptability of a risk within a society solely on the basis of the risk perceptions of the individual members (Cvetkovich and Earle, 1992), a concept that has guided much of the psychological approaches. On the contrary, the social constructionists give great importance to social aspects of risk. It is in this respect that the view of other scholars (Lewicki et al., 2003; Schön and Rein, 1994) and our own view came into consideration. In our understanding the comprehension of a conflict depends on a combination of the social and political context of the conflict, understanding of the characteristics of the individuals involved in the conflict, and how these characteristics may influence their reactions. Considering these assumptions, frame analysis was selected as the main tool in this chapter.

2.2 Frames and framing

Framing has its roots in psychology and sociology. The psychological origins of framing lie in the experimental work by Kahneman and Tversky (1979, 1984), for which Kahneman received the 2002 Nobel Prize in economics (Kahneman, 2002). They examined how different presentations of essentially identical decision-making scenarios influence people's choices and their evaluation of the various options presented to them. In sociology, the concept of frame was inspired by Goffman's work (1984) and employed in making sense of people's behaviour, especially in situations where decisions have to be made. Goffman (1984:24) argues that individuals apply interpretative schemes or 'primary frameworks' to classify information and interpret it meaningfully.

Another way of explaining framing is that a person constructs frames when trying to organise complex phenomena into coherent and understandable categories. When a phenomenon is labelled, meaning is imparted to some aspects of what is observed, while other aspects are discounted because they appear irrelevant or counter-intuitive (Schön and Rein, 1994; Kaufman et al., 2003). As interpretive lenses, frames help people make sense of complex situations in ways that are internally consistent with their world views; they give meaning to events in the context of life experience and understanding.

From the works cited above, it can be deduced that people use frames to help them to make sense of the world around them, the complex reality, abounded by complex and often divergent information. Individuals construct frames in order to comprehend new, complex and sometimes complicated events; to identify and interpret specific aspects that seem to be key to understanding the situation, and to communicate that interpretation to others (Schön and Rein, 1994). Through this process, individuals select the issues they give attention to, 'imparting meaning and significance to elements within the frame and setting them apart from what is outside the frame' (Buechler, 2000:41). The processing of framing may or may not create a conflict situation, which arises when the parties involved observe different aspects of the same situation, or observe the same aspects but interpret them differently. In such cases the parties frame the situation in conflicting ways, i.e. they hold conflicting frames (Schön and Rein, 1994).

Various types of frames underlie the origins of conflicts. As for instance, divergent frames of risk perception directly affect preferences for specific decisions in a conflict, as well as how the conflict should be managed (Elliot, 2003). The importance of how people perceive risks influencing the decision making process is given in this chapter by the set of public policies related to the conservation of the biome Cerrado.

2.3 Sustainable development as a political concept

Sustainable development can be seen as a 'political concept' shaped by political controversy and power struggle. In recent years, issues of environment and human development have been gathered together under the integrative framework of sustainability (sustainable development). Major policy processes have been constructed and are producing new and supposedly far-reaching agreements, policies and strategies.

Since it first appeared in the Brundland report (1987), 'sustainable development' has been defined a number of times (Lélé, 1991; Carter, 2001). There are those who believe that one should not try to define sustainable development too rigorously and claim that, to some extent, the value of the phrase lies in its broad vagueness. It allows people with hitherto irreconcilable positions in the environment development debate to search for common ground without appearing to compromise their positions (Lélé, 1991).

According to others (Dryzek, 2005), the proliferation of meaning of sustainable development is not just an exercise in academic or practical clarification: 'It is also an issue of different interests with different substantive concerns trying to stake their claim in the sustainable development territory' (p. 146). As sustainable development has become more important, key interests have tried to define sustainable development to suit their own purposes. 'Thus an African government might emphasise the need for global redistribution of wealth from North to South in order to eliminate poverty, while a transnational corporation might insist that sustainability is impossible without vibrant economic growth to conquer poverty, stabilise population levels, provide for human welfare and, of course, maintain profit levels (Carter, n/d). To Carter (n/d), sustainable development, like beauty, is in the eye of the beholder; it therefore promises something for everyone. As Lélé (1991) has put it:

'Sustainable development is a 'metafix' that will unite everybody from the profit-minded industrialist and risk-minimizing subsistence farmer to the equity-seeking social worker, the pollution-concerned or wildlife-loving First Worlder, the growth-maximizing policy maker, the goal-oriented bureaucrat and, therefore, the vote-counting politician' (p. 613).

These chameleon characteristics attract a wide array of supporters, but they also make sustainable development a highly contestable concept that not only deals with interdependencies between economy and ecology, but also combines the ecological question with the social question on a global scale. Some aims of sustainable development are radical: the elimination of poverty, the pursuit of global equity, wider use of appropriate technologies, and a shift away from consumerist lifestyles. Other themes, such as the preference for the capitalist economic system and the need for continued economic growth, seem to call for the acceptance of the status quo. Old questions need to be answered: What are basic needs? Should they reflect the needs of citizens in the USA and/or Bangladesh? How far will the living standards of rich industrialised nations have to be adjusted to achieve sustainable consumption patterns? Different answers to these questions produce conflicting interpretations of sustainable development (Carter, n/d, 2001).

For example, those defending the rights of less developed countries may have cause for complaint, particularly about a widespread attitude among Europeans, who tend to see sustainable development as an exercise in the conservation of nature and in environmental management, while forgetting about equitable distribution and economic growth in less developed countries (Huber, 2000, 2005). What comes to mind at this point is that any discussion of sustainable development, and of sustainability, must first answer the questions:

'What is to be sustained? For whom? How long?' The value of the concept of sustainability, like that of sustainable development, however, lies in its ability to generate an operational consensus between groups with fundamentally different answers to these questions, i.e., those concerned either about the survival of future human generations, or about the survival of wildlife, or human health, or the satisfaction of immediate subsistence needs (food, fuel, fodder) with a low degree of risk. It is therefore vital to identify those aspects of sustainability that do actually cater to such diverse interests, and those that involve tradeoffs' (Lélé, 1991:615).

In this chapter we show how differently people 'frame' issues of sustainability, for example, how they perceive the risks concerning the expansion of ethanol production to the Amazon and to the Cerrado biome.

3. Methodology

This chapter describes part of a larger study on the developments, divergent views and conflicts concerning the agricultural phase of the production and expansion of ethanol in Brazil. The approach adopted here combined a literature review with field work carried out in the north-east of São Paulo state, where 30 ethanol and sugar industries/plants are located. Data were obtained through interviews with representatives of the parties and through listening to their talks at conferences, seminars, workshops and on TV, reading their communications and watching their behaviour, and studying items in the media. The field work was complemented by consulting scientific literature, official statistics, non-governmental organisations and official reports. The research was carried out from October 2007 to June 2011.

Exploratory interviews were carried out with around 20 key actors such as sugarcane and ethanol producers, engineers and other experts on the production of ethanol and with academics knowledgeable about natural ecosystems. In these interviews, as is typical of exploratory interviews, just a few questions were used to structure the interview. The questions differed according to the expertise of the interviewee.

Data were also obtained at several academic seminars promoted by the School of Economics, Business and Accounting of the University of São Paulo (FEA/USP) and Institute of Advanced Studies of the USP (IEA/USP), and during the Ethanol Summit 2009 and 2011, organised by the Sugarcane Industry Association (UNICA). The latter events can be considered *ambulant bibliotheca,* since so many national and international scholars, government representatives, NGOs, industry representatives, students and representatives of civil society were present. They represented Brazil, European Union, Sweden, Belgium,

Netherlands, United Kingdom, USA and Japan. The presentations revealed their position in relation to global ethanol production and expansion in general, and Brazilian ethanol production in particular. Attendance at such events, as observer and interviewer, was an essential part of the data gathering process and the frame construction. The actors were asked about how they included the dynamics of land use in Brazil in models to evaluate land competition and how they related the expansion of sugarcane to the Brazilian main biomes.

The methods of analysis used were concerned with the issue of how actors' constructions of meaning (frames) were constructed and represented on the basis of the available material. A combination of analytical tools was used for analysis of the texts, as proposed by Kvale (2008). This approach gave us the possibility to freely change between different approaches, an eclectic form of generating meaning that is a common mode of interview analysis, in contrast to systematic analytical modes. Bricologe, as the mode was called by Kvale (2008), involves interplay of different techniques during the analysis of interviews or other text. As proposed by Schmidt (2005) and Kvale (2008), all the texts were read through once to gain a first overall impression. Repeated readings of the texts were then carried out to allow interesting passages that had gone unnoticed in the previous reading to be identified. As advised by Böhn (2005), in the first stages of the analysis as many different people, situations and documents as possible were selected to obtain data covering the complete spectrum of the research question. Subsequently, data were sought that confirmed or modified the first categories of analysis that had been developed.

4. Background

Frame analysis requires familiarity with the object of study: the divergent views over the expansion of sugarcane plantation to increase ethanol production in Brazil. In turn, analysis of the divergent views and the conflicts they have created requires a broad understanding of the context in which the conflicts take place. To fulfil such requirements, a detailed background description is given here. This consists of: description of land use in Brazil and of the instruments concerning its control, the evolution of Amazonian deforestation, the peculiarities and importance of the Cerrado biome and a description of the proposal for the expansion of ethanol production in the country. This description consists of what I, the researcher, after an intense analysis of the data, considered to be the background to the divergent views and the divergence in the perception of risks associated with the expansion of sugarcane for ethanol production.

4.1 Land use

At present, some 11% (1.5 billion ha) of the global land surface (13.4 billion ha) is used in crop production (arable land and land under permanent crop) and 26% (3.5 billion ha) for pasture land (United Nations Food and Agriculture Organisation [FAO], 2002; Hoogwijk et al., 2003). According to FAO, a lack of land is not forecast for the period until 2030. Suitable land for rain-fed crop production is almost three times the capacity currently used (FAO, 2002).

Concerning Brazil, the total area of the country's territory is 851 million hectares. In 2007, the total area of agricultural land occupied 355 million ha (42% of territory), with pasture land occupying 172 million ha and arable land 77 million ha, subdivided as shown in Table 1.

	Area 10^6 ha	% of Brazil's territory	% of agricultural land
Brazilian territory	851	100	--
Total agricultural land (1+2+3)	355.0	42.0	100.0
1. Total crop land	77.0	9.0	21.0
Soybean	21.0	2.4	5.8
Maize	14.0	1.6	3.9
Sugarcane	7.8	0.9	2.2
sugarcane for ethanol	3.4	0.4	1.0
Oranges	0.9	0.1	0.3
2. Pasture	172.0	20	49
3. Available land	106.0	12	30

Source: MAPA 2007

Table 1. Land uses in Brazil (10^6 hectares; percentage of total territory) in 2007.

In 2007, for the global production of 50 billion litres of ethanol that replaced 3% of the world consumption of petrol, 15 million ha were used, which represented 1% of agricultural land world-wide (Goldemberg, 2008, 2009). As regards Brazilian sugarcane-based ethanol, sugarcane plantations for the production of 22.5 billion litres of ethanol used 3.4 million hectares, which represented 0.4% of the total area of the country's territory and 5% of the 77 million ha of arable land. This volume would be enough to substitute about 1% of global demand for petrol and replace 50% of the petrol consumption in Brazil (Goldemberg, 2009).

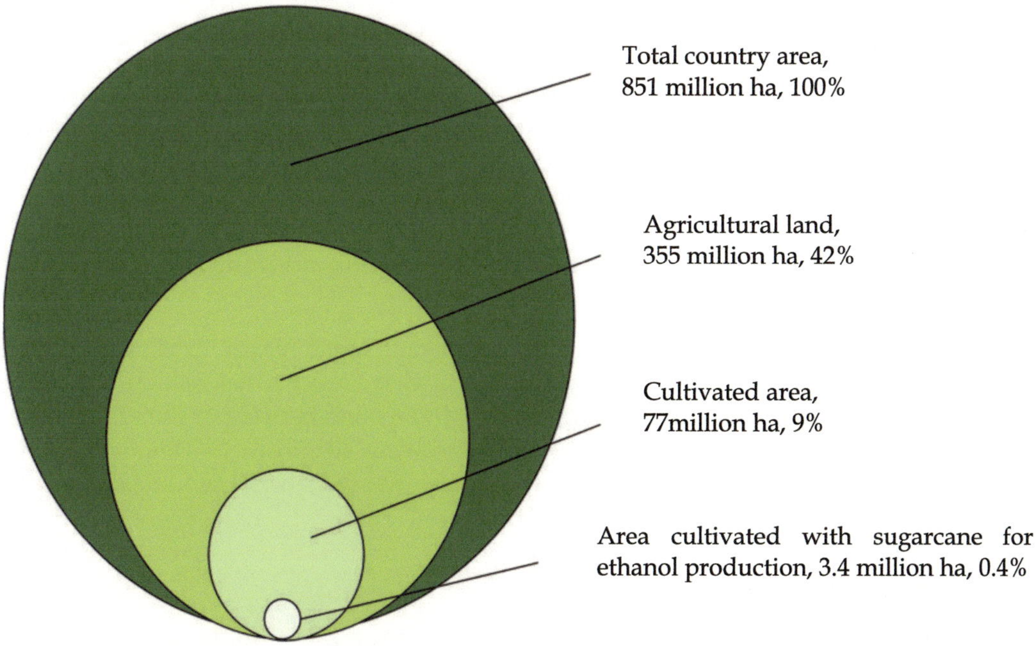

Source: Adapted from The Brazilian Ministry of Agriculture Livestock and Supply [MAPA] 2007 and from the Brazilian National Economic and Social Bank [BNDES]/Center for Strategic Studies and Management [CGEE], 2008.

Fig. 1. Illustration of the percentage of land use in Brazil.

4.2 Instruments for the control of land use

There are various sections of legislation and public policies that influence the land use in the country. Within this set of restriction and obligation, 25% of the Cerrado receives some form of protection, as defined in the Forestry Code (Smeets et al., 2008). However, only seven per cent of the biome is designated as Protected Areas for Conservation (Conservation International, n/d).

4.2.1 National Forestry Code

Brazilian environmental legislation is based on the National Forestry Code (Federal Law 4771/65), and the Environmental Crimes Law (Federal Law 9605/98). According to the Forestry Code, a legal reserve (LR) of 80% is required for rural properties in the Amazon region, 35% in the Amazonian Cerrado (savannah) and 20% in the rest of the country, including São Paulo State. The LR regulation means that all properties must guarantee at least the required percentage of forestry cover with native trees (or reforested with native trees).

The Forestry Code also designated areas close to rivers and water streams, as for example riparian forest, as Areas of Permanent Protection (APPs), which means that these areas must be maintained in, or restores to their natural state. APPs, as well as LRs, are protection areas within a property. LRs are estimated to cover 32% of the territory and APPs 17% (Miranda, 2009).

4.2.2 ZAE Cana

Another government instrument was launched in 2010 to avoid the destruction of valuable forest and in response to international criticism (Galli, 2011). The countrywide agro-ecological land use zoning for sugarcane will restrict sugarcane growth in or near environmentally sensitive areas (MAPA, 2009a). The no-go areas for expansion are Amazonian and Pantanal biomes, and areas of high biodiversity. Although this zoning has no status in law, the expansion of sugarcane to *non-suitable* land is restricted through state credits, i.e. state credits are unavailable for such expansion.

The ZAE Cana includes a set of mandatory environmental, economic, social, climate and soil restrictions, limiting future expansion of sugarcane to around 8% (65 million ha) of the nation's territory, most within the Cerrado biome. Today, one per cent of Brazil's territory landmass, equivalent to 8 million ha, is used for growing sugarcane (with great regional variation); half of this area is used for growing sugarcane for ethanol production and the other half for sugar production.

4.2.3 National Action Plans

Among the national policies are the Action Plan to Prevent and Control Deforestation in the Legal Amazon, starting in 2004. At first, the main concern was with loss of biodiversity, but by 2007, carbon emissions from deforestation in the Amazon had become the main cause for concern. The Action Plan is a set of cross-government policies and measures to improve monitoring, strengthen enforcement, define conservation areas and foster sustainable activities in the region, and it uses the images from INPE's near real-time deforestation detection system.

Again at national level, the Action Plan to Prevent and Control Deforestation and Fire in the Cerrado (PPCerrado) was launched in 2010. Its main goals are to decrease deforestation in this biome by 40% by 2020 in order to fulfil the requirements of the National Climate Policy and to increase the protected area for biodiversity. It should be noted that while the Action Plan in the Legal Amazon used images from the INPE deforestation detection system, which has shown great efficiency in the control of fire and deforestation in the area, for PPCerrado a detection system is not yet available.

4.2.4 The Brazilian protected areas programme

The Brazilian strategy to select new protected areas includes two key considerations: a focus on areas of high biological importance and prioritisation of those areas under strong anthropogenic pressure (Silva, 2005). In 2000, the Sistema Nacional de Unidades de Conservação da Natureza (SNUC) was made law. It focuses on the creation and management of federal protected areas. The SNUC divides protected areas into two main categories: (1) strictly protected, with biodiversity conservation as the principal objective; and (2) sustainable use, allowing for varying forms and degrees of exploitation, with biodiversity protection as a secondary objective. Analogous instruments for creating and managing protected areas exist at the state and municipal levels (Silva, 2005).

The Federal Conservation Units (FCUs) defined and protected by legislation cover 8.6% of the country's territory. Indigenous land (IL) also defined and protected by the legislation covers 12.8% of the territory. The State Conservation Units, defined by state governments (SCUs), cover 7%of the territory. In total, these protected areas cover 28% of Brazilian territory (Miranda, 2009), unevenly distributed in the biomes, with most of them located in the Amazon Biome, followed by the Cerrado. All legally protected areas, including the LRs and APPS, account for 70% of the country's territory, resulting in land legally available for agricultural use occupying an estimated 30% of Brazilian territory (Miranda, 2009).

4.3 Brazil's second biggest biome: Cerrado

The Cerrado is a complex ecosystem of grass and shrub land that extends over ten Brazilian states. It is the second largest vegetation formation in Brazil, after the Brazilian Amazon. Originally the Cerrado occupied an area of around 200 million ha, 24% of Brazil's territory (Figure 2), but by 2009 it occupied around 12%. It is known as the world richest savannah in terms of biodiversity. It houses more than 10,000 known species of plants, of which 4,400 are endemic; 837 species of birds, 160 species of mammals, 1,200 species of fish, and many others kinds of animals (EMBRAPA, n.d). The Cerrado has a dry landscape with small and twisted trees, and it supports several species of wild cats, such as puma and jaguar, as well as maned wolves, giant armadillos and giant anteaters. In addition, the Cerrado region shelters the headwaters and the important part of South American watersheds; it borders and holds areas of transition with other major Brazilian ecosystems, e.g. the Amazon, and is essential to their ecological equilibrium (EMBRAPA, n.d).

Fig. 2. Original areas of biomes in Brazil, expressed in terms of percentage of Brazil's territory.

The Cerrado has supported human populations for a very long time. In the 16th century, Portuguese settlers mainly colonised the Brazilian coastline, and the interior was spared any major development. However, in the mid-20th century the Brazilian government started to build a new capital city in the state of Goiás, in the centre of the Cerrado. The construction of Brasília was accompanied by large-scale infrastructure development, which encouraged movement into the interior of the country, in the remote areas of the North and parts of the Central West regions. It was also when Legal Amazonia, a socio-geographical and political division in the country, was created by the government in 1965. This covers 60% of Brazilian territory and had an estimated 21 million inhabitants in 2000 (IBGE, 2001). It contains the Amazon forest, parts of the Cerrado biome and of other biomes and an area of intense agricultural activities.

Because the soil was considered poor, it was not until the 1980s that agricultural use of the Cerrado biome started to become more intensive. This was the result of a government policy to develop remote areas of the country, in combination with extensive scientific research in soil science which received the 2006 World Food Prize (laureates Lobato and Paolinelli of Brazil and McClung of the USA). To make space for the expansion of arable and pasture land, the rates of deforestation in the Cerrado have steadily increased since the 1980s. The Cerrado became Brazil's new agriculture frontier. This was accompanied by an increase in population and trade development based on agricultural production. In 2000, the Cerrado region was responsible for 35% of all crops produced in Brazil. Currently, 58% of all soy bean production comes from the Cerrado.

4.4 Proposed expansion of sugarcane for ethanol production

Sugarcane plantations require flat land (allowing for mechanical harvesting), good soil quality (lower quantity of fertiliser needed) and a water supply (no or minimum irrigation), among other things. These factors were considered in a formal analysis of Brazilian ethanol production conducted by researchers at the Interdisciplinary Center for Energy Planning (NIPE) at UNICAMP and summarised by Cerqueira Leite et al. (2009). The country's potential to replace 5% (102 billion litres) of the world demand for petrol with sugarcane ethanol by the year 2025 was evaluated. This scenario is the basis for the federal government's expansion policy for ethanol. The analysis estimated that it would take 17 million hectares (ha) of sugarcane to provide the projected demand of 102 billion litres of ethanol, plus 20% required by law for legal reserves, making in total 21 million ha. It also considered that sugarcane plantations will replace degraded areas or areas currently used as low productivity pasture, through the conversion of pasture and the intensification of cattle grazing, following a trend in the country in recent years (Nassar et al., 2008). This model of expansion will avoid the main environmental threat posed by expanding the amount of land under cultivation for energy or any other use through irreversible conversion of virgin ecosystem, guaranteeing a green fuel (Goldemberg et al., 2008). The analysis estimated that 30% of Brazilian pasture land is degraded (more than 50 million ha), most within the Cerrado biome.

Projections from the Brazilian Ministry of Agriculture indicate that if Brazilian production doubled by 2017, no more than 1.7 per cent of the land would be used (MAPA, 2009a). Projections developed for the study carried out by Nassar et al. (2008) indicated that harvested sugarcane area in Brazil will reach 11.7 million ha in 2018, departing from 7.8 million ha in 2008. Area allocated for crops (soybean, maize, cotton, rice and dry beans) is expected to grow from 37.8 million ha to 43.8 million ha. Pasture land will move to the opposite direction, being reduced from 165 to 162 million ha.

The proposed expansion of sugarcane plantations is to occur only on degraded pasture land (although not well defined in the literature, degraded pasture land has been considered areas with natural vegetation that have been used for cattle breeding and abandoned after 4 or 5 years of use due to low productivity). Such land can readily be found in the Cerrado biome (Cerqueira Leite et al., 2009), confirming a trend in recent years (Nassar et al., 2008). This, combined with higher productivity of cattle farming, will make more land available for the expansion of sugarcane for ethanol production, avoiding indirect land use change.

5. The study

The study described below is divided in two main parts: frame analysis of the Brazilian ethanol programme and perceptions of risks concerning the expansion of ethanol production in Brazil. The frame analysis elicited conflicting and non-conflicting frames related to the ethanol programme. In order to understand the conflicting frames, mainly the frames *ethanol as a brown fuel* and *ethanol as green fuel,* the perceptions of risks were also analysed.

5.1 Frame analysis of the Brazilian ethanol programme

The results showed that in the past four decades there has been a continuous process of framing and reframing regarding the issue of ethanol production in Brazil. Although the Brazilian ethanol programme was established more than 30 years ago, it was not until the first decade of the 21st century that the production of ethanol became an international issue. The prospect of future large-scale world production of ethanol raised conflicts in terms of social, economic and environmental factors, the latter especially related to the loss of biodiversity in Brazil's two largest biomes: Amazon and Cerrado, the latter being the focus of this chapter.

5.1.1 The Brazilian independence frame, 1975-2000

For centuries, sugar was Brazil's main product from sugarcane. However, there is information that ethanol (then termed 'alcohol') was already being produced from sugarcane in the early 1900s, when Brazil experienced a rapid increase in the use of by-products from oil. However, it was in the 1970s that ethanol started being more intensively produced as a fuel to make the country independent in energy. Frame analysis showed that an *energy independence frame* was adopted by the Brazilian government. The nationalistic values of the military government ruling Brazil at that time, and their belief in a country self-sufficient in energy and economically strong, were decisive for the creation of the National Alcohol Programme (Proálcool) in 1975 (Decree 76.593 signed by President Enerst Geisel). The programme was a response to the sharp increase in the world price of oil in the 1970s, from $2.90 to $12.45 a barrel. At that time, over 80% of Brazilian petroleum came from imported oil. In 1973, the cost of oil represented 10% of the total import costs of the country and this had increased to 32% by 1974. Brazil's payments for imported petroleum increased from US$600 million in 1973 to US$2.5 billion in 1974, severely affecting the balance of trade. These results came to weigh heavily on Brazilian foreign debt and inflation over the course of the following years. The development of a substitute for petrol was one of the initiatives taken in response to high oil prices (BNDES/CGEE, 2008).

The Proálcool programme addressed two problems simultaneously: it replaced oil consumption and helped to stabilise the international sugar market, where prices had fallen from US$639 a ton in 1974 to US$ 176 in 1979, an example of the large oscillations which were not uncommon in the sugar market at that time (Cerqueira Leite et al., 2009). Given low international sugar prices, the Brazilian government sought to utilise domestic sugar supplies to produce ethanol and, consequently, to reduce the country's dependence on foreign oil.

The second global oil crisis in 1979 exposed the continuing vulnerability of Brazil to international oil shocks and reinforced the political will to continue with the Proálcool programme. The price of oil rose from US$25 in 1979 to US$40 in 1981. From 1979 to 1984, the Brazilian government directed more funds to the construction of new distilleries, capable of producing hydrated alcohol for fuel engines, and the vehicle industries were stimulated to construct 100% pure ethanol-fuelled vehicles (Natale Neto, 2007).

The 1980s was a decade of important economic and political changes in Brazil. The economy had serious problems, with very high inflation rates and foreign debt. Politically, the country began a process of redemocratisation. This coincided with low international oil prices and consequently oil supply became less of a problem. In 1985, under the new administration (the first civilian government administration after a period of two decades of military government), public investment in the Proálcool programme was gradually cut back, for example for the building of new distilleries. In 1988, 63% of all vehicles (around 100% of cars) produced in the country were powered by ethanol. That number declined to 47% in 1989, 10% in 1990, 0.4% in 1996 and less than 0.1 in 1997 (Natale Neto, 2007).

The high price of imported oil on the international market after the oil crisis stimulated the development of Petrobras, the Brazilian Petroleum Company founded in 1953 by President Getúlio Vargas. New incentives were given to the company to drill petroleum off the Brazilian coast. The good results achieved by the company in the 1990s brought Brazil to self-sufficiency in oil in the first decade of the 21st century (for more detail refer to Galli, 2008).

In 1999 the cost of ethanol production was still high compared with the price of petrol manufactured from imported oil. Despite the decrease in government incentives, new technologies for ethanol production were being developed, mainly by the private sector, and the cost-effectiveness of the programme improved significantly (La Rovere, 2004). However, it was not until 2003 with the introduction of the flex fuel engine, which can use ethanol and petrol in any proportions, that the use of this biofuel started increasing again. By 2007, flex fuel cars accounted for 90% of all new cars registered in Brazil (ANFAVEA, 2010). All these changes in combination significantly improved what was re-named the Brazilian Ethanol Programme in the late 2000s.

5.1.2 Framing ethanol as a tool for economic development

The international visibility of the Brazilian ethanol programme increased with the need of different countries to meet their emissions targets relating to climate change agreements. In order to fulfil these targets, global ethanol production has increased sharply, from less than 20 billion litres in 2000 to 53 billion litres in 2007, 69 billion litres in 2008 and 76 billion litres in 2009. The US and Brazil accounted for 90% of this production, with the US producing more than 50% and Brazil about 35% (REN21, 2010). Brazil exported 3.5 billion litres of ethanol in 2007, compared with 227 million litres in 2000 (Secex/UNICA).

Brazil saw its competitive advantage as an opportunity to bring its ethanol to the international market. The *energy independence frame* that had orientated the Brazilian government's actions during the 1970s and 1980s was replaced by an *economic frame* strongly orientated towards the international market. The belief of national and international actors in ethanol as a possible alternative to fossil fuel acted as a stimulus to the Brazilian government to plan ethanol expansion for the coming years, aiming at doubling national production by 2017, to 63 billion litres annually (Inter-American Development Bank [IDB], 2010). Bioethanol's share of the Brazilian National Economic and Social Development Bank (BNDES) total loans for capital investment went from 1.5% in 2004 to 7% in 2008 (BNDES in Ethanol Summit 2009), providing a strong indicator of future development. This was reflected in the number of new distillery projects, with 136 new distilleries to be ready by

2014 (Nastari, 2006). By 2010 there were 423 distilleries in operation, compared with 350 in 2006 (MAPA 2008, 2010).

Other responses were an expansion of the area planted with sugarcane, as illustrated in Figure 3, and an increase in sugarcane production (volume of sugarcane harvested) and productivity (volume of sugarcane per ha), as shown in Figure 4. The increase in productivity was the result of plant breeding programmes, the introduction of biological pest control, improved management and greater selectivity regarding production sites. As a result of these improvements, in 2008 the area planted with sugarcane for the production of ethanol was 38% less than it would have been had the land use and production procedures of the 1970s remained in place (BNDES/CGEE, 2008).

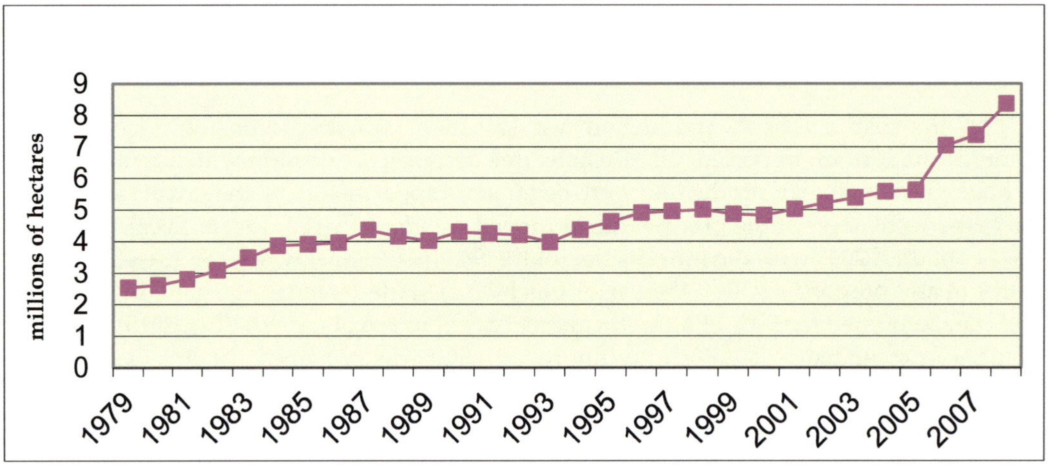

Source: MAPA (2009b).

Fig. 3. Annual expansion of the area (10^6 hectares) planted with sugarcane in Brazil, 1979-2008.

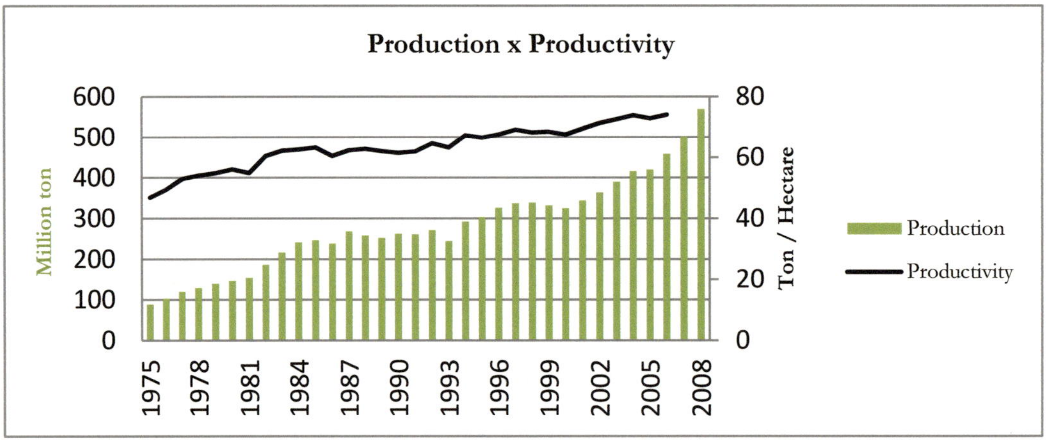

Source: MAPA (2009b).

Fig. 4. Annual changes in the production and productivity of Brazilian sugarcane, 1975-2008.

5.1.3 Framing and reframing the impacts of ethanol on natural vegetation

By 2007, a wave of criticism was being directed at the production and consumption of biofuels. A reframing process was going on: from being a sustainable and effective alternative to fossil fuel, ethanol turned into a monster that destroys the Amazon forest, competes with food production and increases CO_2 emissions; the 'presidential diplomacy that celebrated biofuels in March 2007 ignited a firestorm of opposition to ethanol and its production in both the US and Brazil' (Langevin, 2008).

In 2008, a second wave of criticism of biofuels in general, and ethanol in particular, was published, mostly in Europe, but also in the US and Brazil. One possible explanation is that two articles published in Science Magazine in early 2008 (Fargione et al., 2008; Searchinger et al., 2008) received a great deal of public attention because they claimed that widespread adoption of biofuels, including ethanol, will produce more greenhouse gas (GHG) emissions, especially CO_2 emissions, than conventional petroleum-based petrol due to indirect land use changes. For Brazil, the first assumption in those articles was that new agricultural land would come from forest. Deforestation of Amazon forest and CO_2 emissions from indirect land use changes have since been the focus of attention in scientific and commercial debates about Brazilian ethanol. Concerns that sugarcane will invade the Amazon region or push soybean plantations and cattle into virgin forests, causing more deforestation, also featured in major media headlines (e.g. Financial Times, 2008; The Guardian, 2008).

The fear of increasing CO_2 emissions and deforestation of Amazonia strongly affected the international negotiations for Brazilian ethanol. The European Union's representative at the Ethanol Summit (2009) argued that EU consumers 'need to have a guarantee that Brazilian ethanol is sustainable', reflecting the EU Directive specifying that biofuel feedstock must not come from areas of high carbon stock and must conserve biodiversity and that the resulting biofuel must reduce CO_2 emissions by at least 35%. In the USA the Californian Air Resources Board (CARB) and Environmental Protection Agency (EPA) required that CO_2 emissions due to direct and indirect land use changes be included in the life cycle analysis of ethanol (for more about the debate over CO_2 emissions due to direct and indirect land use changes, see Galli, 2011; Galli, 2013).

Once again, the way their views (that sugarcane plantations will cause deforestation of Amazonia and increase CO_2 emissions) were expressed in the media captured the attention of the public and policy-makers, and led them to view ethanol as *a* fuel destructive for nature, *framing ethanol as a brown fuel.*

However, many of the criticisms were denied by important Brazilian actors (e.g. BNDES/CGEE, 2008; Cerqueira Leite et al., 2009; Goldemberg et al., 2008) who argued that the expansion of sugarcane is not a threat to Brazilian natural vegetation. These actors were holders of what we named an *ethanol as green fuel frame* (for a more detailed frame analysis see Galli, 2011). One explanation for such divergent views on the expansion of sugarcane is the way the actors perceive the risks concerning this expansion, as described below.

5.2 Perceptions of risks

We were first alerted to the threat posed by the expansion of sugarcane to the Cerrado biome during exploratory interviews with academics knowledgeable about natural ecosystems when this study had just started in 2007. Their perceptions of the risks to the Cerrado biodiversity due to this expansion stimulated us to carry out a broad literature review on the subject.

5.2.1 Risks deriving from expansion of sugarcane to the Amazon forest

A study conducted by Miranda (2007a) showed that Brazil has mostly preserved its primary forest (70%), while Europe (excluding Russia), the worst case, has preserved 0.3%. It is estimated that 75% of the world original forests have disappeared. However, the international frame concerning Brazil's forest has not been one of preservation, but rather one of destruction. Paradoxically, this is in spite of the fact that Brazil is one of the countries that has destroyed least forest (Miranda, 2007b).

On the other hand, in the last three decades, Legal Amazonia has suffered high deforestation rates (Figure 5), with 372 thousand km^2 of rainforest - an area almost the size of Japan[1] - being converted either to pastoral or arable land in this period. One hypothesis is that these high deforestation rates have attracted the world's attention and have given some international actors cause to fear that the expansion of sugarcane would induce displacement of natural vegetation. However, deforestation rates in the Amazon have declined markedly since 2004, by 75% between 2004 and 2009 (National Institute of Space Research [INPE], 2010). It is reported that the slowing of deforestation levels is primarily a result of the Action Plan for Deforestation Control and Prevention in the Amazon, which uses Satellite images from INPE's near real-time deforestation detection system. This system has enabled government inspectors to focus their efforts at where deforestation is most critical and to act quickly to prevent new areas from being cleared.

The international view over the Brazilian biodiversity is also very negative, in spite of the fact that Brazil is one of the world's richest megadiversity countries. However, it rarely attracts attention for what it has, rather, it is criticized for what it is losing through deforestation and conversion of natural landscapes into plantations, soybean fields, and pastures (Mittermeier, et al., 2005).

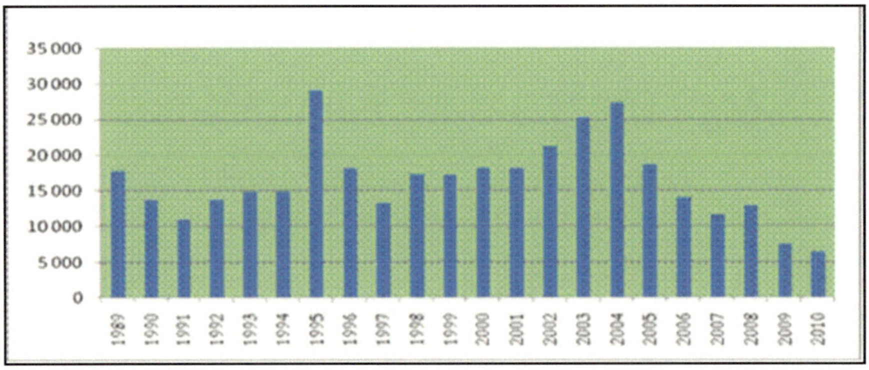

Source: INPE, 2010.

Fig. 5. Deforestation in Legal Amazon per year (km^2) over the period 1989-2010.

[1] Area of the Japan's territory = 378,815 km^2

Concerning the expansion of sugarcane, important Brazilian actors, holders of the *ethanol as green fuel frame,* seem to be more confident about the preservation of the Brazilian Amazon, a confidence that may derive from the history of sugarcane expansion in Brazil. A comparison of rates of deforestation in Amazonia between 1988 and 2009 and of sugarcane expansion in the same period shows that deforestation rates ranged from 1.1 to 2.9 million ha/year (INPE, 2010), while the average rate of sugarcane expansion in the whole country was 0.20 million ha/year (FAOSTAT, 2010). This means that the former was five to fifteen times higher than the latter in the period. Research also shows that the expansion of sugarcane plantations in Brazil is occurring mostly over degraded pastures far from the Amazon (Cerri et al., 2007; Goldemberg, 2008). Most of this expansion to date has occurred in the state of São Paulo. The northern region (which includes almost all of Brazil's Amazon rainforest) has only 21,000 hectares planted with sugarcane, i.e. only 0.3% of Brazil's sugarcane plantations (MAPA, 2007). This means that 99.7% of sugarcane plantations are at least 2,000 km from the Amazon forest, as shown in the map below (Figure 6).

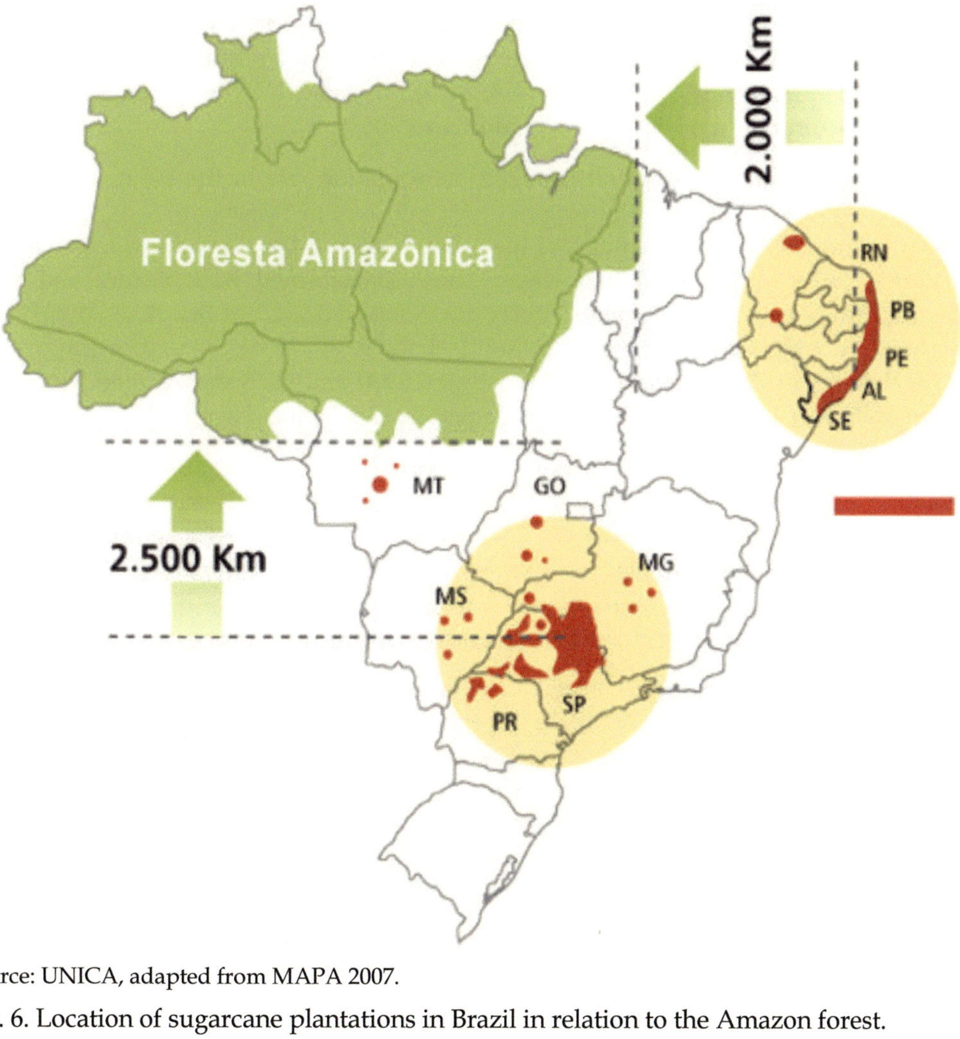

Source: UNICA, adapted from MAPA 2007.

Fig. 6. Location of sugarcane plantations in Brazil in relation to the Amazon forest.

Another position relating to the expansion of sugarcane is that the current sugarcane area represents only 3% of the 264 million ha of agricultural land in use in Brazil, of which nearly 200 million ha are pastoral land. Furthermore, the hotspots of deforestation in the Amazon region have a low suitability for sugarcane production and are not directly threatened by the current sugarcane expansion (Smeet et al., 2008). Fearnside (2005) concluded that Amazon deforestation has been caused mainly by conversion to pastoral land for livestock production and, more recently, for expansion of soybean production.

5.2.2 Risks deriving from expansion of sugarcane to biodiversity in the Cerrado

By 2000, the risks deriving from the expansion of arable and pasture land to the Cerrado biodiversity were already perceived. It was then that the Cerrado was defined as one of the 25 biodiversity hotspots in the world for conservation priorities (Myers et al., 2000). This was due both to its exceptional concentration of endemic species (native to the Cerrado) and to the severe threats to which these species are exposed. The term 'hotspot' was created by the English ecologist Norman Myers in 1988 to identify major terrestrial bio-geographical regions based on the number of endemic species and the degree of threat to biodiversity. It should be noted that at least 130 Cerrado animals are threatened with extinction because of large-scale agricultural expansion (Klink & Machado, 2005).

In 2004, concerns were raised about the high deforestation rate of the biome, which was estimated to be 1-1.5% of the total original area per annum (Machado et al., 2004). This continued up to 2009, when around half the original area of the Cerrado (equal to almost three times the area of Japan) had already been transformed into pasture and other agricultural uses. This is around three times the deforested area of Brazilian Amazon, an area which is double the size of the Cerrado (Machado et al., 2004). Annual clearing in real terms is also higher than in the Amazon: between 1970 and 1978 the average was 40,000 km^2/year, which is 1.8 times the deforestation rate for the Amazon from 1978 to 1988 (Klink & Moreira, 2002).

Devastation of priority areas for biodiversity conservation has also been perceived as a risk to the biome in recent years. For example, 67% of the priority areas in the state of Goiás have already been converted to cropping and pastoral uses (Lobo and Guimarães, 2008). Priority areas for conservation are defined as areas where the biodiversity is very sensitive or vulnerable to anthropogenic pressure (Machado et al., n.d.).

The expansion of sugarcane plantation has tended to occur in the State of São Paulo, very often in areas of the Cerrado. In 2007, the Central South[2] region was responsible for 86% of total sugarcane production in the country. The state of São Paulo alone produced more than 60% (Embrapa, 2007). However, a decrease in the rate of expansion in this state is expected,

[2]Two methods of dividing the country into regions are used in this study depending on the reference, namely geo-political and geo-economic. The official geopolitical method was established by IBGE in 1969 and divides the country into five regions (North, Northeast, South, Southeast and Central-West). The geo-economic method divides the country into three regions (Amazonia, Northeast and Centro-South). The Centro South comprises the states of the South and Southeast regions (except north of Minas Gerais) and the states of Goías, Mato Grosso do Sul and Tocantins. The geo-economic division is intensively used by academics (e.g. Esalq/USP) and government agencies (e.g. EMBRAPA).

mainly due to the lower availability and consequently higher price of land compared with other regions of the country (Torquato, 2006).

The second movement of the expansion of sugarcane production has been towards the Central-Western region. For example, the area of sugarcane increased by 40% between 2007 and 2008 in the states of Goías and Mato Grosso do Sul, which mostly comprise the region of the Cerrado; 16% is the average increase for the Central-Western region as a whole (Canasat, 2008).

In the future, even more significant expansion of sugarcane into the Cerrado biome is expected. This is due mainly to the lower price of land compared with the state of São Paulo, flat landscape suitable for mechanical harvesting and climate conditions suitable for sugarcane plantations (Torquato, 2006). This expansion 'may directly or indirectly affect parts of the Cerrado area with native vegetation and unprotected forest where biophysical, infrastructural and socio-economic conditions are favourable for sugarcane cultivation. Most threatened are those lands adjacent to current production areas. Environmental consequences of sugarcane expansion might range from quite acceptable (conversion of crop land and managed pastures) to very negative where sugarcane expands directly or indirectly in unprotected areas, which still have native vegetation with high bio-diversity, or into unprotected native forests area' (Fisher et al., 2008, p.56). Another consideration is that the cost-benefit of sugarcane production is directly related to the distance between the sugarcane plantation and the mill and distilleries, which is at best up to 30 km and at a maximum 50 km (Gall1, 2011). This means that high biodiversity lands adjacent to current production areas are at risk of destruction. Not so obviously, biodiversity is also reduced when mixed farming systems are replaced by monoculture landscapes (Sawyer, 2008). The result of a study by Feltran-Barbieri (2008) in the Cerrado biome confirms that the use of extensive pasture land (instead of intensive cattle production) helps to protect biodiversity.

The view of the Cerrado as an agricultural resource has also been adopted internationally, as shown by a recent article published by The Economist (2010). It confirms the view of the Cerrado as the Brazilian agricultural frontier and as an example of large-scale agricultural production that should be followed by other countries. The article cites the FAO as stating that Brazil has a lot of spare land, mostly in the Cerrado area: 'It [Brazil] is often accused of levelling the rainforest to create its farms, but hardly any of this new land lies in Amazonia; most is cerrado' (The Economist, 2010).

Although many of the Brazilian actors who are holders of the *ethanol as a green fuel frame* see the expansion of the sugarcane as unthreatening to the conservation of biodiversity in the Cerrado, others point out the uncertainties surrounding how the expansion of ethanol production will occur. The actors in favour argue that the expansion will only occur on degraded land and land of low productivity. These actors ignore (or maybe, do not agree with) the alert issued by Martha Jr (2008) that the dynamics of land use are determined by the profitability of agricultural commodities and by any perceived risks to business. This implies that there is no guarantee that the expansion of sugarcane plantations will occur in low productivity or degraded areas, as suggested in the expansion policy. Furthermore, the adoption of new technologies for intensification of pastoral management (e.g. use of fertilisers, rotational grazing) will increase agro-chemical inputs, production costs (e.g. the final cost of cattle and of meat) and greenhouse gas emissions (Fisher et al., 2008; Galli, 2011).

5.2.3 Risks deriving from public policies and lack of environmental control

Some actors see the high deforestation rates in the Cerrado as deriving from past land use policies, which were often formulated with little attention to their implications for Cerrado conservation, in part because the Amazon Forest was the main focus on the conservation agenda (Klink & Machado, 2005). One example is that the Cerrado is not mentioned in the *Brazilian Magna Carta* from 1988, while the Amazon Forest is considered 'a national patrimony and its use will be in accordance with the law, within the conditions that make sure the preservation of its environment, including the use of its natural resources'. Another example that denotes the scant attention given to Cerrado conservation in public policy in the past, according to Klink & Machado (2005), is to be found in the Forest Code concerning the legal reserve (LR) requirements. While the Code requires that 20% to 35% of a holding in the Cerrado be maintained in its natural state as LR, in the Amazon rainforest this proportion is 80%.

The negative impacts of public policies are also perceived as a risk to natural vegetation. According to Brandon et al., (2005), underlying most of habitat conversion in Brazil is public policies supporting transportation, energy and communications infrastructure that open up areas for conversion, colonisation and other uses.

Lack of effective environmental regulation control is also perceived as a risk to deforestation of the Cerrado biome. An example of the lack of enforcement of the environmental legislation is given by Feltran-Barbieri (2008). He studied the state of preservation of the LRs and APPs in 65 rural properties, 37 dedicated to agriculture and 28 to cattle breeding, located in areas of intense agricultural activities within the Cerrado. The results showed that less than 20% of all properties fulfilled the requirements of the environmental legislation and only 13% of the agricultural properties. This coincides with the results of the study by Chomitz et al. (2005) that related the no compliance of the land owner with the legal reserve obligation.

Fisher et al. (2008) also perceive the risk deriving from lack of environmental control. According to them, increasing demand for livestock products will require replacement of the land converted to sugarcane, leading to substantial shifts of arable and pasture land to other regions, causing pressure on the ecosystems there. This could be avoided by effective environmental regulations and control and by implementation of agricultural policies supporting intensification of production.

5.3 Need for reframing

The availability of agricultural land is an issue that has been debated and discussed in Brazil since the start of the Proálcool programme. From the outset, it has been claimed that the country has sufficient potential land to expand sugarcane plantations (Borges, 1990), a potential that still exists nowadays, as shown in section 4.1. However, more recently, experts have issued warnings about such expansion, which they see as a determining factor in the destruction of the high biodiversity of the Cerrado biome (e.g., Lobo and Guimarães, 2008; Feltran-Barbieri, 2008; Machado et al., n/d; Sawyer, 2008).

Although many authors point to the generally high levels of threat to the biodiversity of the Cerrado biome, Branton et al. (2005, p.598) note that others see 'numerous reasons for cautious optimism provided Brazilian decision makers make choices that favour environmental and economic sustainability over short-term gains'. Lovejoy (2005) highlights the strong conservation science capability, one of the strongest in the world, and major conservation advances that have taken place in Brazil in recent decades. Marina Silva, the former Brazilian Ministry of Environment, believes that protected areas are the major tools available for the conservation of the country's natural resources, and says that experience has shown that 'even when suffering deficiencies in personnel, infrastructure, and management procedures, they are still effective barriers to the disorderly and rapacious occupation of natural environments…' (Silva, 2005, p.611).

In addition to these measures, this chapter identified a need for reframing to guarantee socially and environmentally sustainable expansion of ethanol should the hypothesis of a future large-scale expansion of sugarcane plantations become reality, especially in relation to the preservation of the biodiversity of the Cerrado biome. A great part of the biome has been replaced by arable and pasture land in recent decades and its biodiversity is now being seriously threatened. While some believe that the expansion of sugarcane will only occur on degraded pasture land (e.g. BNDES/CGEE, 2008; Cerqueira Leite et al., 2009; Goldemberg, 2008; 2009; Goldemberg et al., 2008; Macedo et al., 2008; Macedo & Seabra, 2008).

However, others point out that it is not clear how the displacement of crops and cattle for the expansion of sugarcane plantations will occur (Lobo and Guimarães, 2008; Feltran-Barbieri, 2008; Machado et al., n/d; Sawyer, 2008; Martha Jr, 2008). There is a good chance that such expansion will not happen in the Amazon region, as there are so many eyes, nationally and internationally, watching the Amazon forest for signs of deforestation and Brazilian law is attempting to severely limit such devastation. The same cannot be said for the Cerrado, which is seen as an agricultural frontier and not as a biome to be legally protected.

The conclusion in this chapter is that the natural vegetation of the Cerrado is not generally seen as such an important part of nature as the rainforests, especially considering that those defending the latter ignore the former. The dry, open landscapes dotted with the small, twisted trees that make up a large part of the vegetation in the Cerrado biome are remote from the idea of 'natural vegetation' inspired by the images of the luxuriant Amazon rainforest, which is considered to belong to the world.

The sustainability criteria adopted by the EU Directive on highly biodiverse grasslands may still result in the designation of some savannah as 'no-go' areas, according to Zahniser (2010). However, there is no final definition as yet. The EU environmental sustainability criteria have been viewed by ethanol producers and other Brazilian actors as a protectionist restraint on biofuel trade, but from a different standpoint they may be viewed as a genuine attempt to regulate production.

An impetus for reframing the use of the Cerrado biome may be given by the *Nagoya Protocol on Access to Genetic Resources and the Fair and Equitable Sharing of Benefits Arising from their Utilisation*, recently agreed by the parties during the UN Biodiversity Summit in 2010, to which Brazil is a signatory since February 2011. The Protocol creates a framework that balances access to genetic resources on the basis of prior informed consent and mutually agreed terms with the fair and equitable sharing of benefits, while taking into account the role of traditional knowledge. In addition, a plan to guide international and national efforts to save biodiversity and a resource mobilisation strategy to increase the current levels of official development assistance in support of biodiversity were drawn up in the meeting (Normile, 2010).

In addition to these, other incentives should be available to stimulate the conservation of the Cerrado biodiversity. A list of economic incentives to protect the areas of natural vegetation has been proposed to the Brazilian Congress within the scope of the Forest Code by the Instituto de Pesquisa Ambiental da Amazônia, IPAM (2011). Another important aspect of the proposal is the country's need of a better environmental management and control. Without both, the economic incentives and a good and efficient environmental management and control, it will be very difficult for the country to achieve its objectives and targets related to biodiversity and reduction of CO_2 emissions, as well as to consolidate its position as a sustainable agricultural power (IPAM, 2011).

6. Conclusions

During more than 30 years of the Brazilian Ethanol Programme, few conflicts arose in the country. It was not until the first decade of the 21st century that the production of ethanol became an international issue. The perspective of future, large-scale global production of ethanol raised conflicts involving environmental, social and economic factors, and these had an international dimension.

Some Brazilian actors believe that sustainable expansion of ethanol production is possible. They frame *ethanol as green fuel,* arguing that there is enough land available in Brazil for safe expansion of sugarcane plantations. Their arguments are based primarily on the high productivity of sugarcane per hectare and on the fact that an increase in ethanol production will not require much land. Another assumption is that sugarcane expansion will replace degraded pasture areas. Overseas actors, on the other hand, frame *ethanol as brown fuel* and express their concerns about its sustainability, with the focus on deforestation of the Amazon.

Developments during the past five years indicate that the risks to the Amazon forest are being controlled. However, the biodiversity of the Cerrado savannah seems not to be a concern of key actors to date. On the contrary, the Cerrado has been seen as fit for conversion to crops and pasture. Thus, a reframing of the use of Cerrado is needed to protect this fragile biome. The new Nagoya Protocol signed in 2010 may be a first step in placing value on biodiversity, an economic incentive to leave the Cerrado savannah standing that can compete with the economic incentive to tear it down.

7. References

ANFAVEA (2010). *Brazilian Automotive Industry Yearbook*. São Paulo: Anfavea, 2010, Accessed in 9/05/2011. Available from: http://www.anfavea.com.br/anuario.html

BNDES/CGEE (2008). Sugarcane –Based Bioethanol, Energy for Sustainable Development. Coodinated by BNDES and CGEE. Published by BNDES, Rio de Janeiro. Available at <http://www.bioetanoldecana.org/en/index.html>

Berndes, G., Hoogwijk, M. and van den Broek, R. (2003). The contribution of biomass in the future global energy supply: a review of 17 studies. *Biomass and Bioenergy*, 25, pp.1-28.

Borges, J.M.M. (1990). The Brazilian Alcool Program: Foundations, Results, and Perspectives. *Energy Sources*. Vol. 12, pp.451-46.

Brandon, K., Fonseca, G.A.B., Rylands, A.B. & Silva, J.M.C. (2005) Brazilian Conservation: Challenges and opportunities. *Conservation Biology*, Vol. 19, No. 3, pp. 595-600.

Buechler, S.M. (2000). *Social Movements in Advanced Capitalism*, New York: Oxford University Press.

Böhn, A. (2005). Theoretical Coding: Text Analysis in Grounded Theory. *In:* Flick, U., von Kardoff, E. and Steinke, I. (Eds.), *A Companion to Qualitative Research*, London: Sage Publications, pp.270-275.

Canasat (2008) 'Projecto Canasat / INPE/ UNICA/ CTI', Available at: http://www.dsr.inpe.br/canasat/, Accessed on 14 May 2009.

Carter, N. (2001). *The politics of the Environment*, Cambridge University Press.

Carter, N. (n/d). Understanding Sustainable Development. Cambridge University Press, Accessed in: 20/6/2011 Available from: http://www.fathom.com/course/21701763/contributors.html

Cerqueira Leite, R.C., Leal, M.R.L.V., Cortez, L.A.B., Griffin, W.M. and Scandiffio, M.I.G. (2009). Can Brazil replace 5% of the 2025 gasoline world demand with ethanol? *Energy*, vol.34, no.5, pp: 655-661.

Cerri , C.E.P., Easter, M., Paustian, K., Killian, K., Coleman,K., Bernoux, M., Falloon, P., Powlson, D.S., Batjes, N.H., Milne, E. & Cerri, C.C. (2007). Predicted soil organic carbon stocks and changes in the Brazilian Amazon between 2000 and 2030. *Agriculture, Ecosystem and Environment*, 122, pp. 58-72.

Chomitz, K.M., Thomas, T.S. and Brandão. A.S.P. (2005. The economic and environmental impact of trade in forest reserve obligations: a simulation analysis of options for dealing with habitat heterogeneity. *Revista de Economia e Sociologia Rural*, Vol. 43, No. 4, pp.657–682.

Conservation International (n/d). Biodiversity hot spot. Availabel from: <http://www.biodiversityhotspots.org/xp/Hotspots/>

Cox P.M., Betts R.A., Jones C.D., Spall S.A. and Totterdell I.J. (2000) Acceleration of global warming due to carbon-cycle feedbacks in a coupled climate model. *Nature*, Vol. 408, pp. 184-187.

Cvetkovich, G. and Earle, T.C. (1992). Environmental Hazards and the Public. *Journal of Social Issues*, vol. 48, no. 4, pp. 1-20.

Douglas, M. and Wildavsky, A. (1982). *Risk and Culture: An essay on the selection of technological and environmental dangers*. Berkely: University of California Press.

Dryzek, J. (2005). *The Politics of the Earth: Environmental Discourses*, Oxfor University Press: New York, pp. 146.

Elliot, M. (2003). Risk Perception Frames in Environmental Decision Making. *Environmental Practice*, Vol. 5, pp.214-222.

EMBRAPA (2007). *Mercado Interno.* Accessed on 19/5/2009, *Available at: http://www.agencia.cnptia.embrapa.br/gestor/cana-de-acucar/arvore/CONTAG01_112_22122006154842.html*

EMBRAPA (n/d) *Embrapa nos Biomas brasileiros - Atuação Potencial da Embrapa nos Biomas Brasileiros*, Accessed on 21/4/2009, Available from: http://www.embrapa.br/publicacoes/institucionais/laminas-biomas.pdf/view

FAO (2008) United Nations Food and Agriculture Organization, High-level conference on world food security: the challenges of climate change and bioenergy, food security and sustainability - towards an international framework , Available from: ftp://ftp.fao.org/docrep/fao/meeting/013/k2498e.pdf

FAO (2002).World Agriculture: Towards 2015/2030, Rome: Food and Agriculture Organization, 2002.

FAOSTAT (2010) United Nations Food and Agriculture Organization. Access on 29/3/2011, Available from: http://faostat.fao.org/site/567/DesktopDefault.aspx?PageID=567#ancor

Fargione, J., Hill, J., Tilmam., D. Polasky, S. and Hawthorne, P. (2008). Land Clearing and the Biofuel Carbon Debt. *Science*, 319, pp: 1235-1238.

Fearnside, P.M. (2005). Deforestation in Brazilian Amazon: History, Rates, and Consequences. *Conservation Biology*, 19, pp. 680-688.

Feltran-Barbieri, R. (2008). Por que fazendeiros mantêm fragmentos de cerrados em suas propriedades? *In:* Grostein, M.D. (ed.), *Ciência Ambiental – questoões e abordagens*, São Paulo: Fapesp, Annablume, Procam, pp: 186-211.

Financial Times (2008) 'Poor practices taint Brazil's Ethanol exports', published on 20/5/2008.

Fisher, G., Teixeira, E., Hizsnyik, E.T. & Velthuizen, H.van (2008). Land use Dynamics and sugarcane production. In: *Sugarcane Ethanol, Contributions to climate change mitigation and the environment*, Peter Zuurbier & Jos van de Vooren editors, The Netherlands: Wageningen Academic, pp.29-62.

Fischhoff, B., Slovic, P. and Lichtenstein, S. (1978). How Safe is Safe Enough? A Psychometric Study of Attitudes Towards Technological Risks and Benefits. *Policy Sciences*, vol. 9, pp. 127-152.

Galli, E. (2011). Frame Analysis in Environmental Conflicts: The case of ethanol production in Brazil. Doctoral Thesis presented at the KTH-Royal Institute of Technology, Stockholm, Available from: <http://kth.diva-portal.org/smash/record.jsf?pid=diva2:440064>

Galli, E. (2013). Framing land use change in the expansion of Brazilian ethanol production. *International Journal of Sustainable Society*, ,Vol. 5, No. 2, (ISSN 1756-2538).

Galli, E. (2008). *Frame Analysis in Environmental Conflicts*, Saarbrücken: VDM Verlag Dr. Müller.

Goldemberg, J. (2009). The limits of first-generation ethanol, *In:* Official Annals of the Ethanol Summit 2009, Unica: São Paulo, p.33-42.

Goldemberg, J. (2008). Foreword' In: BNDES, *Sugarcane –Based Bioethanol, Energy for Sustainable Development*, Rio de Janeiro: Published by BNDES, pp.13-14.

Goldemberg, J. Coelho, S.T. & Guardabassi, P. (2008). The sustainability of ethanol production from sugarcane. *Energy Policy*, 35, pp.2086-2097.

Goffman, E. (1984). *Frame Analysis – An Essay on the Organization of Experience*, Boston: North-eastern University Press.

Hansen, J., Nazarenko, L., Ruedy, R., Sato, M., Willis, J., Del Genio, A., Koch, D., Lacis, A., Lo, K., Menon, S., Novakov, T., Perlwitz, J., Russell, G., Schmidt, G.A. and Tausnev, N.(2005). Earth's energy imbalance: Confirmation and implications. *Science*, Vol. 308, 1431-1435.

Hoogwijk, M., Faaij, A., Van Den Broek, R., Berndes, G., Gielen, D. and Turkenburg, W. (2003). Exploration of the ranges of the global potential of biomass for energy, *Biomass and Bioenergy,* 25, n. 2, pp.119-133.

Huber, J. (2000, 2005). Towards industrial Ecology: Sustainable Development as a Concept of Ecological Modernization. *Journal of Environmental Policy and Planning,* vol.2, no.4, pp.269-285. Reprinted in: Redclift, Michael R. / Woodgate, Graham (eds): New Developments in Environmental Sociology, Cheltenham: Edward Elger, 2005.

IDB (2010) *European Union and United States Biofuel Mandates, Impacts on World Markets.* Inter-American Development Bank, Technical notes, No. IDB-TN-191, (2010).

INPE (2010). Taxas Anuais de desmatamento 1988-2010, Access on 20/04/2010, Available from: http://www.obt.inpe.br/prodes/prodes_1988_2010.htm

IPAM (2011). *Reforma do Código Florestal: qual o caminho para o consenso?* Contribuições para o Relatório de Meio Ambiente do Senado Federal sobre a reforma do Código Florestal Brasileiro, IPAM: Amazõnia. Acessed in 22/11/2011, Available from <http://www.ipam.org.br/ipam/reforma-do-codigo-florestal:-qual-o-caminho-para-o-consenso>

Jasanoff, S. (1997). The political science of risk perception. *Reliability Engineering and System Safety,* vol. 59, pp. 91-99.

Kahneman, D. and Tversky, A. (1979). Prospect Theory - Analysis of decision under risk. *Econometrica,* 47, pp. 263-291.

Kahneman, D. and Tversky, A. (1984). Choices, values and frames. *American Psychologist,* 39, pp.341-350.

Kahneman, D. (2002). Maps of Bounded Rationality: A Perspective of Intuitive Judgment and Choice. Nobel Prize Lecture, December 8, 2002.

Kaufman, S., Elliott, M. and Shmueli, D., (2003). Frames, framing and reframing. *Beyond Intractability.* Ed. Guy Burgess and Heidi Burgess. Conflict Research Consortium, University of Colorado, Boulder, Colorado, USA.

Klink, C.A. & Machado, R.B. (2005) Conservation of the Brazilian Cerrado. *Conservation Biology.* Vol.19, No. 3, pp.707-713.

Klink, C.A. & Moreira, A. G. (2002). Past and Current human occupation and land use. In: P.S. Oliveira and R.J. Marquis, editors. *The Cerrados of Brazil: ecology and natural history of a Neotropical savanna,* Columbia University Press, New York.

Kvale, S. (2008). *Doing Interviews.* Los Angeles, London, New Delhi, Singapore: Sage Publications. First published in 2007.

Langevin, M.S. (2008). Renewable Cooperation? Reflection on United States - Brazil Cooperation on Biofuels, Access 20/6/10, Available at : http://www.unc.edu/depts/diplomat/item/2008/1012/comm/langevin_biofeul.html

La Rovere, E.L. (2004). The Brazilian Ethanol Program - Biofuels for Transport. International Conference for Renewable Energies, Bonn, Accessed in 23/10/2007, Available from: http://www.renewables2004.de/ppt/Presentation4-SessionIVB(11-12.30h)-LaRovere.ppt

Lélé, S.M. (1991) 'Sustainable development: a critical review', *World Development,* vol. 19, no.6, pp.607-621.

Lewicki, R.J., Gray, B. and Elliot, M. (Eds.) (2003). *Making Sense of Intractable Environmental Conflicts, Concepts and Cases.* Island Press, Washington, Covelo and London.

Licht, F.O. (2008a) 'Ethanol in Asia - Tough Times Getting Tougher', *World Ethanol & Biofuels Report.* 7, 8, pp. 149-154.

Licht, F.O. (2008b) In Renewable Fuels Association - Industry Statistics, 2008. Accessed in 10/2/2011, Available from: http://www.ethanolrfa.org/pages/statistics#E

Lobo, F. and Guimarães, L.F. (2008). Vegetação Remanescente nas Áreas Prioritárias para Conservação da Biodiversidade em Goías: Padrões de Distribuição e Características. Boletim Goiano de Geografia, Vol.28, No.2, pp.89-104.

Lovejoy, T. (2005). Spotlight on Brazil. *Conservation Biolog, Editorial.*Vol. 19, no.3, pp.587-588.

Macedo , I.C., Seabra, J.E.A. and Silva, J.E.A.R. (2008). Green house gases emissions in the production and use of ethanol from sugarcane in Brazil: The 2005/2006 averages and a prediction for 2020. Biomass and Bioenergy, 32, pp. 582-595.

Macedo, I.C. and Seabra, J.E.A. (2008). Mitigation of GHC emissions using sugarcane bioethanol. In: Zuurbier, P. and Vooren J. van de (eds.), Sugarcane Ethanol, Contributions to climate change mitigation and the environment, The Netherlands: Wageningen Academic Publishers: The Netherlands, pp. 95-112

Machado, R.B., Paglia, A.P. & Fonseca, R.L. (n.d.) Priority Areas and Landscapes in the Cerrado and Pantanal. Brasília: International Conservation, Access on 21/10/2011, Available from: www.fbds.org.br/Apresentacoes/6_Areas_Cerrado_Pant_Amaz_Paglia.pdf

Machado, R.B., Ramos Neto, M.B., Pereira, P.G.P., Calsdas, E.F., Goncalves, D.A., Santos, N.S., Tabok, K. and Steininger, M. (2004). *Estimativas de perda da área do Cerrado brasileiro,* Technical Report. Brasilia: Conservation International.

MAPA (2009a) Sugarcane Agroecological Zooning. English version, Access on 14/11/2011, Available from: http://www.unica.com.br/downloads/sugarcane-agroecological-zoning.pdf.

MAPA (2009b). Brasil: Área colhida lavouras temporárias e permanents, available at: < *http://www.agricultura.gov.br/*>

MAPA (2007) *Açúcar e Álcool no Brasil.* Ministério da Agriculture, Pecuária e Abastecimento, Secretaria de Produção e Abastecimento. Departamento da Cana-de-açúcar e Agroenergia, Março de 2007, Access on 8/12/2007, Available from: www.agriculturra.gov.br

MAPA (2008, 2010) *Usinas e destilarias cadastradas,* available at: http://www.agricultura.gov.br/.

Martha Jr., G.B. (2008) 'Expansão da cadeia da cana-de-açúcar e suas implicações para o uso da terra e desenvolvimento do Cerrado', 1° Workshop do Observatório do Setor Sucroalcooleiro, FEA/RP.

Miranda, E.E. (2009). Alcance territorial das áreas protegidas. *Opiniões sobre o setor de Florestas Plantadas,* Access on 14/11/2011, Available from: <*http://www.revistaopinioes.com.br/cp/materia.php?id=615*>

Miranda, E.E. (2007a). *Quando o Amazonas corria para o Pacífico: uma história desconhecida da Amazônia,* Petrópolis, Editora Vozes 2nd Edition, 251 pages.

Miranda, E.E. (2007b) 'Who are the Deforestations Champions?' O Estado de São Paulo, 17/01/2007, Accessed 15/03/2008), Available from:

http://www.desmatamento.cnpm.embrapa.br/artigo/miranda_deforestationchampions.pdf

Mittermeier, R.A., Fonseca, G.A.B., Rylands, A.B. & Brandon, K. (2005). A Brief history of Biodiversity Conservation in Brazil. *Conservation Biology*, Vol.19, No.3, pp.601-607.

Myers, N., Mittermeier. R.S., Mittermeier, C.G., Fonseca, G.A.B. and Kent, J. (2000). Biodiversity hotspots for conservation priorities. *Nature*, 403, no.24, pp. 853-858.

Nassar, A.M., Rudorff, B.F.T, Antoniazzi, L.B., Aguiar, D.A, Bacchi, M.R.P and Adami, M. (2008). Prospects of the sugarcane expansion in Brazil: impacts on direct and indirect land use changes. In: *Sugarcane Ethanol, Contributions to climate change mitigation and the environment*, Peter Zuurbier & Jos van de Vooren editors, The Netherlands: Wageningen Academic, pp.63-93.

Nastari, P.M. (2006). A expansão anunciada: A expansão industrial no setor sucroalcooleiro. *Revista Opiniões*, São Paulo. October 2006.

Natale Netto, J. (2007). *A Saga do Álcool, Fatos e verdades sobre os 100 anos de história do álcool combustível em nosso país*, Osasco, SP: Novo Século Editora.

Normile, D. (2010). Convention on biological diversity: U.N. biodiversity summit yields welcome and unexpected progress, *Science*, Vol. 330, No. 6005, pp.742–743.

Otway, H. (1992). Public Wisdom, Expert Fallibility: Toward a Contextual theory of Risk. In S. Krimsky and D. Golding (Eds.), *Social Theories of Risk*. Westport, Connecticut, London: Praeger Publishers, pp. 215-228.

REN21 (2010) Renewables 2010 Global Status Report, Paris: REN21 Secretariat.

Schmidt, C. (2005). The Analysis of Semi-structured interviews. In: Flick, U., von Kardoff, E. and Steinke, I. (eds.), A Companion to Qualitative Research, London: Sage Publications, pp.253-258.

Schön, D.A. and Rein, M. (1994). *Frame Reflection: Toward the Resolution of Intractable Policy Controversies*, New York: Basic Books.

Searchinger, T., Heimlich, R., Houghton, R.A., Dong, F., Elobeid, A., Fabiosa, J., Tokgoz, S., Hayes, D. and Yu, T.S. (2008). Use of U.S. Croplands for Biofuels Increases Greenhouse Gases Through Emissions from Land-Use Change. *Science*, pp:1238-1319

Silva, M. (2005). The Brazilian Protected Areas Program. *Conservation Biology*. Vol.19, No.3, (June 2005), pp.608-6011.

Slovic, P. (Ed.) (2000). *The Perception of Risk*. Earthscan Publ., London and Sterling, VA.

Slovic, P. (1992). Perceptions of Risk: Reflections on the Psychometric Paradigm. In S. Krimsky and D. Golding (Eds.) *Social Theories of Risk*. Westport, Connecticut, London: Praeger Publishers, pp. 117-152.

Smeets, E., Junginger, M., Faaij, A., Walter, A., Dolzan, P. and Turkenburg, W. (2008). The sustainability of Brazilian ethanol - an assessment of the possibilities of certified production. *Biomass and Bioenergy*, 32, pp.781-813.

Sawyer, D. (2008). Climate change, biofuels, and eco-social impacts in the Brazilian Amazon and Cerrado. *Phil.Trans. R. Sos.B.* Vol. 363, pp.*1747-1752.*

Starr, C. (1969). Social benefit versus technological risk - What is our society willing to pay for safety? *Science*, Vol. 165, pp.1232–1238.

The Economist (2010). 'The miracle of the cerrado, Brazil has revolutionised its own Farms. Can it do the same for others?' 26/8/ 2010.

The Guardian (2008). The sweet hereafter, 28/5/2008.

The Washington Post, 'Latin America - the 'Persian Gulf' of Biofuels?' 23/2/2007.

Thompson, M., Ellis, R. and Wildavsky, R. (1990). *Cultural Theory*. Boulder, CD: Westview.

Torquato, S.A. (2006). Cana de açúcar para Indústrias: O quanto vai precisar Crescer. *Instituto de Economia Aplicada,* IEA, Access on 6/5/2011, Available from: <http://www.iea.sp.gov.br/out/verTexto.php?codTexto=7448>

Zahniser, S. (2010) 'EU environmental sustainability requirements and Brazilian biofuel exports', *EuroChoices*, Vol. 9, pp.28-34.

16

Analysis of Hydrologic Alteration Due to River Diversion

Giancarlo Principato and Giuseppe Viggiani
Department for Soil Conservation,
University of Calabria, Rende
Italy

1. Introduction

Relevant alteration of rivers is most often caused by water diversions for irrigation, hydropower, industry and/or domestic uses. The operation of weirs and reservoirs has resulted in severe modification of the hydrologic regime (Rosenberg et al., 2000; Postel & Richter, 2003; World Commission on Dams [WCD], 2000; World Conservation Union [IUCN], 2000). As consequence, the discharges released into or left in a river for ecological purposes - the so-called "environmental flow requirements" - have received great attention in the last few decades from scientists, technicians and water managers (Richter et al., 2003; Tharme, 2003).

Among ecologists, the importance of flow variability in determining the structure and the function of the river ecosystem has been widely recognized (Richter et al., 1996; Poff et al., 1997; Bunn & Arthington, 2002). Richter et al. (1996) stated the natural flow paradigm as 'the full range of natural intra- and inter-annual variation of hydrological regimes, and associated characteristics of timing, duration, frequency and rate of change, are critical in sustaining the full native biodiversity and integrity of aquatic ecosystems'.

The above "variation of hydrological regime" should be calculated statistically, i.e. as the change in the frequency distribution of a set of hydrologic variables useful to describe the overall flow regime (Arthington et al., 2006) or specific flow events (Stewardson & Gippel, 2003).

According to this environmental flow approach, the first step is to define ecologically-meaningful flow variables that capture natural flow variability. A great number of variables has been suggested for this purpose (Olden & Poff, 2003) and the 33 Indicators of Hydrologic Alteration (IHA) (Richter et al., 1996) are intended to represent each of the major facets of the flow regime (Olden & Poff, 2003; Monk et al., 2007). The IHA method assumes that, if these indicators are significantly altered, this will result in a decline in overall ecosystem health.

Therefore, the second step of the analysis involves the estimation of the alteration due to a water diversion through comparison of the probability distributions of the hydrologic variables (the IHA or other variables) in the pre- and post-impact conditions.

Unfortunately, the probability distributions can be difficult to obtain for several variables. Data may show skewed, scattered or multi-modal distribution (Principato & Viggiani, 2006) and non-normal distribution is recurrent (The Nature Conservancy, 2009).

Consequently, simplified statistical analyses have been proposed.

The Range of Variability Approach (RVA) (Richter et al., 1997) was formulated to quantify the modification of the IHA by comparing the frequencies within three fixed intervals. The RVA is a milestone in the hydrologic alteration assessment and has been widely utilised (e.g.: Galat & Lipkin, 2000; Irwin & Freeman, 2002; Shiau & Wu, 2004, 2007a, 2007b). Zolezzi et al. (2009) combined the RVA with a wavelet transform analysis in order to separate the scales of variability and investigate their alterations independently.

The method of Suen & Eheart (2006) integrated the natural flow paradigm with the "intermediate disturbance hypothesis" (Connel, 1978), which states that species diversity will be highest at sites that have had an intermediate frequency of disturbance and lower at sites that have experienced extremely high or extremely low disturbance frequencies. Based on this assumption, they suggested that the range of intermediate disturbance levels for six eco-hydrological indicators could be used as a management target.

Shiau and Wu (2008) developed a Histogram Matching Approach (HMA) for resolving the limited flow variability resulting from RVA. For each IHA, histograms with a greater number of intervals (to RVA) are first obtained. Then, the histograms are compared using a metric proposed by Niblack et al. (1993).

Botter et al. (2010) avoided the histogram analysis through the estimation of the probability density functions of flow (in the pre- and post-impact conditions) by means of a few climate, soil and vegetation parameters. The relevant parameters of the probability density functions (like mean, mode, skewness, peak probability) are the solely used to estimate the hydrologic alteration.

As final step, a flow release that minimizes hydrologic alteration has to be defined. To this end, great effort has been focused on the use of optimization models (Shiau & Wu, 2004, 2007a, 2007b; Suen & Eheart, 2006).

Despite the accuracy of such optimization models, the robustness and the statistical accuracy of the current methods used for hydrologic alteration analysis have not been extensively evaluated and the second step of the overall analysis, i.e. the alteration assessment, remains vague.

This chapter deals with the second step. First, the drawbacks of few existing methods are highlighted in a conceptual framework. Then, a real case is selected as the test case, in which a 40 years release scenario downstream of a diversion is compared to an unaltered condition. More precisely, the frequency distributions of the hydrologic variables in the pre- and post-impact conditions are compared using two of the methods mentioned above, namely RVA and HMA. Then, the need for a more accurate frequency analysis of hydrologic variables and the importance of capturing the full range of variation of the flow regime are incorporated in a novel Frequency-Based Approach (FBA). The FBA results are analysed and compared to the results of other methods.

To this aim, a reference solution cannot be obtained. Therefore, a proper analysis technique is developed, based on the physical correctness of diversion-alteration relationships. Comparison of methods is performed both for single indicators (IHA) and in terms of overall hydrologic alteration.

2. Existing methods

Once the natural flow regime (or another condition) is chosen as reference condition, the hydrologic alteration has to be evaluated by taking into account the modification of flow descriptors (IHA or other variables) from pre- to post-impact conditions. The variables in pre-impact conditions are usually computed using daily hydrological measurements from a relevant gauge station from a period prior that reflects the unaltered condition. The post-impact condition can be computed using either post-impact measurements (i.e., downstream of an existing diversion) at the same gauging station or synthesised release scenarios (i.e. downstream of a planned diversion).

Said n_1 and n_2 the number of years in the pre-impact and post-impact conditions, each hydrologic variable can be regarded as a stochastic variable for which there are two samples $\mathbf{a}=a_i$ ($i=1,...,n_1$), $\mathbf{b}=b_j$ ($j=1,...,n_2$).

2.1 Range of Variability Approach (RVA)

In the RVA, the alteration is evaluated by comparing the frequency with which observed (pre-impact) and post-impact variables (usually the IHA) fall within three intervals (categories). The standard interval limits are equal to 0, 33th, 67th and 100th percentiles, represented by a histogram with three bins. The same number of values ($n_1/3$) is expected to fall in each interval. However, if some values in the set are equal, the values in each category can be different from $n_1/3$.

Discrepancies can also occur because values that are equal to the category boundaries are placed in in the middle category (The Nature Conservancy, 2009).

Hydrologic alteration is assumed to occur if the number of the post-impact values falling in the central interval (33rd÷67th percentiles) differ from the expected ones (i.e. the number of the pre-impact values). To assist in the evaluation of this, an Hydrologic Alteration Factor (HAF) is calculated for each variable in the central category

$$HAF = \frac{F-F_0}{F_0} \quad (1)$$

where F is the observed frequency and F_0 is the expected frequency. A positive HAF means that the frequency of values in the category has increased from the pre-impact to the post-impact period (with a maximum value of 2), while a negative value means that the frequency of values has decreased (with a minimum value of -1). If the other categories are not taken into account, however, important information can be lost. For example, when dealing with minimum 1-day annual flow, the value HAF=-1 may indicate that all values in the post-impact condition fall in the low category (extreme drought decrease in magnitude) or all values in the post-impact conditions fall in the high category as well (extreme drought increase in magnitude).

Thus, in order to more accurately evaluate the hydrologic alteration, HAF should be computed for the other two categories and all three values of HAF evaluated (latest version of the RVA, reported in The Nature Conservancy, 2009).

Information may also be lost if the distribution of the hydrologic variable is skewed, as the analysis may be reduced to a comparison of histograms in which most values fall in a single bin. This is more likely to arise for variables describing the number of events (number of zero flow days, number of high/low pulses, number of reversals).

Most of the drawbacks of the RVA are due to the low number of intervals (or histogram bins) used to analyse frequency modification, that is unrelated to n_1, n_2. Moreover, since the pre-impact data set determines the intervals size, the resultant intervals can be unsuitable to describe the post-impact data set or to compare the two sets.

Limitations of the RVA were discussed in the work of Shiau & Wu (2008) and are further detailed in the application below.

2.2 Suen and Eheart (2006) method

The method introduced by Suen & Eheart (2006) is based on six eco-hydrological indicators (coefficient of efficiency of the yearly trend of the hydrograph, dry season 10-day minimum, wet season 3-day maximum, number of high-flow events, mean duration of low-flow events and mean of all positive differences between consecutive values in wet season, i.e. the rising rate). Fuzzy theory was applied to represent the degree of disturbance levels and a Gaussian shape membership function has been used to describe indicators variability. In accordance to the intermediate distance hypothesis (Connel, 1978), disturbance occurs if indicators values are far from intermediate level (i.e. membership values are not close to 1). The ecosystem needs objective are coupled with human needs objective and incorporated in an optimization model.

However, the intermediate disturbance hypothesis partially contrasts the conceptual statement of the natural flow paradigm, according to which the *full* range of natural intra- and inter-annual variation of hydrological regimes has to be taken into account. Moreover, biodiversity concepts, initially based only on species diversity, have been progressively extended to integrate biotic and abiotic patterns and processes across scales, encompassing structural and functional processes, which are crucial in riverine context (Ward & Tockner, 2001). In fact, the importance of supra-seasonal extreme events has been assessed (Poff et al., 1997), especially for highly variable flow rivers (Lake, 2003), for which efforts to reduce the flow variability in order to increase biodiversity or to "restore" the river system to one that better fits a perception of a "healthy" river may not be the best ecological option (Boulton et al., 2000).

Finally, as noted above, the Gaussian distribution can be unappropriate for several hydrologic variables.

For these reasons, the Suen & Eheart (2006) method can be weak both from conceptual and statistical point of view.

2.3 Histogram Matching Approach (HMA)

In the HMA (Shiau & Wu, 2008), pre- and post-impact values of each IHA variable are represented with histograms whose number of bins is calculated taking into account all

n_1+n_2 data and the characteristics of data distribution. As is the case of RVA, dissimilarity of histograms (in the pre- and post-impact condition) denotes the extent of the hydrologic alteration. The dissimilarity is evaluated using the quadratic-form distance proposed by Niblack et al. (1993), which accounts for both the class-by-class correspondence (differences of frequencies in each bin, like in the RVA) and cross-class information (distance between central values of bins).

The outcome is a 'degree of histogram dissimilarity', D_Q, for each variable.

Different options can be adopted depending on the value of α, which is the exponent of the similarity function ($1<\alpha<\infty$).

The HMA is incorporated in an aggregated multi-objective optimization genetic algorithm that minimizes both alteration (D_Q) and a shortage ratio that accounts for human needs - see Shiau & Wu (2008) for details.

One drawback of the HMA is that the metric used by Niblack et al. (1993) underestimates the distances between bins because it tends to accentuate the similarity of distributions without a pronounced mode (Rubner et al., 1998). Serratosa & Sanfeliu (2006) observed that most distance measures considers the overlap or intersection between two histograms as a function of the distance value, but they do not take accurately into account the similarity of the non-overlapping parts of the two histograms. The effects of these drawbacks are examined through the test case reported below.

3. Frequency Based Approach (FBA)

The Frequency Based Approach (FBA) is proposed here. It is an analysis of the frequency modification of each hydrologic variable from pre- to post-impact condition based on the Earth Mover's Distance (EMD) algorithm (Rubner et al., 1998, 2000) applied to "extended signatures" (Serratosa & Sanfeliu, 2006).

Three key elements are considered in the FBA, due to the necessity to encompass:

i. *histograms that better represent an empirical estimate of the probability density function*

The histogram of a set of univariate data (n_1 and n_2 data points of each hydrologic variable) provides the basis for an empirical estimate of the probability density function, that is required for comparing pre- and post- impact conditions. The formal definition of a histogram with N bins used for estimating the density function of a stochastic variable x with n data (Silvermann, 1986) is

$$f(x) = \frac{h_j}{n\Delta x}, (x_0 + j\Delta x \leq x_i < x_0 + (j+1)\Delta x) \tag{2}$$

where h_j is the number of data points, x_i, within the jth bin, whose limits are $x_0+j\Delta x$ and $x_0+(j+1)\Delta x$, (in which x_0 is an origin of the range of interest). It is worth noting that the count is also ratioed to the bin width Δx and therefore the definition satisfies the condition of histogram total area equal to 1, that matches the probability definition. In effect, the information conveyed is the area of the plot - not only the height of the column representing the bin.

ii. *histograms that avoids to compute dissimilarity caused by histograms structure*

This is because dissimilarity between histograms can be influenced by arbitrary choose of bin centres and limits of the two histograms, as well as by bin number.

iii. *a different metric that overcomes the weakness of other existing metrics*

If the ordering of the elements in the set is unimportant, a histogram is a lossless representation of the set itself (Serratosa & Sanfeliu, 2006). Thus, the "distance" between two sets can be computed in an efficient way by computing the distance between their histograms. For this reason, a number of measures of similarity between histograms have been proposed and used in other fields, especially in computer vision and pattern recognition (e.g.: Protein classification). These include: the quadratic-form distance proposed by Niblack et al. (1993), the B-distance, the Matusita approach and the "K–L distance" (Cha & Srihari, 2002). As noted above, many methods do not accurately take into account the similarity of the non-overlapping parts of the two histograms (Serratosa & Sanfeliu, 2006). This problem was overcome by Rubner et al. (1998, 2000), who presented a new definition of the distance measure between histograms called Earth Mover's Distance (EMD), that is more robust than histogram matching techniques.

3.1 Histograms construction

The above described points (*a*), (*b*) are taken in account with a proper construction of histograms.

A histogram $\{h_j\}$ is defined as a mapping from a set of *d*-dimensional integer vectors *j* to the set of nonnegative integers. These vectors typically represent bins (or their centres) in a fixed-size partitioning of the relevant region of the underlying space, while the associated integers are a measure of the mass of the distribution that falls into the corresponding bin.

On the other hand, a signature is a *variable-size description* of a distribution, aimed to reach a balance between expressiveness and efficiency of representation, thus overcoming deficiency of fixed-size structure like histograms (Rubner et al., 2000). A element of a signature is defined as

$$\{s_j = (m_j, w_j)\} \tag{3}$$

in which m_j is the mean (or mode) of the *cluster j* and w_j is the number of elements that belong to that *cluster*.

In the present application, fixed-size histograms can be still adequate and can be considered as a special case of signatures, i.e. the histogram $\{h_j\}$ with N bins can be viewed as a signature in which m_j is the central value of the bin *j* of the histogram (j=1,..., N) and w_j is equal to h_j. Therefore, in such case, the signatures

$$\{\mathbf{s}_a=(\mathbf{m}_a,\mathbf{w}_a)\} \tag{4a}$$

$$\{\mathbf{s}_b=(\mathbf{m}_b,\mathbf{w}_b)\} \tag{4b}$$

- in which $\mathbf{w}_a=w_{a,j}$ $(j=1,\dots N_a)$, $\mathbf{w}_b=w_{b,j}$ $(j=1,\dots N_b)$, are the number of elements of **a**, **b** in clusters $\mathbf{m}_a=m_{a,j}$ $(j=1,\dots N_a)$, $\mathbf{m}_b=m_{b,j}$ $(j=1,\dots N_b)$ - are only a formal definition of the histograms of the variables **a**, **b**.

The variables **a**, **b** and the vector $\mathbf{m}_a$ can be adimensionalised by dividing for the mean of a_i (i=1,...,n_1)

$$\mathbf{a}'=\mathbf{a}/\ \text{mean}(a_i) \tag{5a}$$

$$\mathbf{b}'=\mathbf{b}/\ \text{mean}(a_i) \tag{5b}$$

$$\mathbf{m}_a'=\mathbf{m}_a/\text{mean}(a_i) \tag{6}$$

As subsequent step, the histograms can be represented as "extended signatures" (Serratosa & Sanfeliu, 2006), in which the minimum number of empty bins is added to assure that the number of bins for both pre- and post- impact data are the same. To this aim, the vector **m′** is obtained adding equal size bins to $\mathbf{m}_a'$ in order to include all values of **b′**. The size of **m′**, N', is equal or greater than N. As result, the same m_j' (j=1,..., N') are used for both adimensionalised variables **a′**, **b′**.

The extended signatures

$$\{\mathbf{s}_a'=(\mathbf{m}',\mathbf{w}_a')\} \tag{7a}$$

$$\{\mathbf{s}_b'=(\mathbf{m}',\mathbf{w}_b')\} \tag{7b}$$

(in which $\mathbf{w}_a'$, $\mathbf{w}_b'$ are the number of elements of **a′**, **b′** in each cluster of **m′**) are obtained.

Finally, the vectors $\mathbf{W}_a=W_{a,j}$ (j=1,..., N'), $\mathbf{W}_b=W_{b,j}$ (j=1,..., N') are calculated using the equations

$$\mathbf{W}_a = \frac{\mathbf{w}_a'}{n_1 \Delta m'} \tag{8a}$$

$$\mathbf{W}_b = \frac{\mathbf{w}_b'}{n_2 \Delta m'} \tag{8b}$$

(in which $\Delta m'$ is the **m′** cluster size) in order to satisfy the conditions related to frequency

$$\sum \mathbf{W}_a \Delta m' = 1 \tag{9a}$$

$$\sum \mathbf{W}_b \Delta m' = 1 \tag{9b}$$

The final result consists of two signatures

$$\{\mathbf{S}_a=(\mathbf{m}',\mathbf{W}_a)\} \tag{10a}$$

$$\{\mathbf{S}_b=(\mathbf{m}',\mathbf{W}_b)\} \tag{10b}$$

corresponding to histograms with same bins (same centres and bin limits) and total area equal to 1. Each bin includes a non-negative number of elements.

3.2 The Earth Mover's Distance (EMD) for hydrologic variable signatures

The Earth Mover's Distance was proposed by Rubner et al. (1998, 2000). Given two distributions, one can be seen as a mass of earth properly spread in space (*supplier*), the other as a collection of holes in that same space (*consumer*). The EMD measures the least amount of work needed to fill the holes with earth. A unit of work corresponds to the cost of transporting a unit of earth by a unit of ground distance.

The supplier and the consumer can be also two signatures. In such case, the EMD is defined as the minimum amount of work that must be performed to transform one signature into the other by moving distribution mass.

The cost c_{ij} is the ground distance between element i in the first signature and element j in the second signature (in the present case, i=1,...,N', j=1,...,N'). c_{ij} can be any distance, e.g. an Euclidean distance. The matrix **C** of elements c_{ij} is the cost matrix.

For signatures 10a, 10b, the problem is finding a set of flows f_{ij} that minimize the overall cost

$$\sum_{i=1}^{N'}\sum_{j=1}^{N'} c_{ij} f_{ij} = \min \qquad (11)$$

subject to the following constraints

$$f_{ij} \geq 0, \ \ (i = 1, \dots, N', j = 1, \dots, N') \qquad (12)$$

$$\sum_{j=1}^{N'} f_{ij} \leq W_{a,i} \ \ (1 < i < N') \qquad (13)$$

$$\sum_{i=1}^{N'} f_{ij} \leq W_{b,i} \ \ (1 < j < N') \qquad (14)$$

$$\sum_{i=1}^{N'}\sum_{j=1}^{N'} f_{ij} = min\left(\sum_{i=1}^{N'} W_{a,i}, \sum_{j=1}^{N'} W_{b,j}\right) \qquad (15)$$

Constraint (12) allows shipping of supplies from a supplier to a consumer and not vice versa. Constraint (13) limits the amount of supplies that can be sent to the clusters of the consumer to their weight. Constraint (14) limits the clusters of the consumer to receive no more supplies than their weight. Constraint (15) is related to *total flow* and forces to move the maximum amount of supplies possible.

The optimal flow matrix **F**={f_{ij}} (whose size, in the present case, is $N'\times N'$) is obtained solving the classic transportation problem by the simplex algorithm (Dantzig, 1951).

Once the transportation problem has been solved and the optimal flow has been computed, the Earth mover's distance is defined as

$$EMD = \frac{\sum_{i=1}^{N'}\sum_{j=1}^{N'} c_{ij} f_{ij}}{\sum_{i=1}^{N'}\sum_{j=1}^{N'} f_{ij}} \qquad (16)$$

where the denominator is a normalization factor that coincides with the smaller signature, because of constraint (15). This factor is needed when the two signatures have different total weight and it avoids favoring signatures with smaller total weights.

If the ground distance is a metric and the total weights of two signatures are equal, the EMD is a true metric.

For a complete description of the Rubner et al. (1998, 2000) algorithm, see http://www.cs.duke.edu/~tomasi/software/emd.htm.

3.3 The FBA algorithm

FBA comprises the following steps:

i. in order to compare the hydrologic alteration of different variables (whose values can show different order of magnitude and even different units), the values of each variable (in both conditions) are adimensionalised dividing by the mean of each pre-impact variable set **a** (eq. 5a, 5b);
ii. the histograms and their signatures are defined for each variable using the above described procedure, in order to obtain eq. 10a, 10b;
iii. the Earth Mover's Distance (16) is computed for each variable using the algorithm available in C++ at http://www.cs.duke.edu/~tomasi/software/emd.htm, that is incorporated in a Matlab routine.

In the present case, total weight of the two signatures are coincident (eq. 9a, 9b) as well as their cluster number N'. Therefore, constraints 13, 14, 15 become

$$\sum_{j=1}^{N'} f_{ij} = W_{a,i} \quad (1 < i < N') \tag{17}$$

$$\sum_{i=1}^{N'} f_{ij} = W_{b,i} \quad (1 < j < N') \tag{18}$$

$$\sum_{i=1}^{N'} \sum_{j=1}^{N'} f_{ij} = \sum_{i=1}^{N'} W_{a,i} = \sum_{j=1}^{N'} W_{b,j} \tag{19}$$

The number N of bins in the signatures 4a, 4b is chosen according Sturges rule

$$N = 1 + \log_2 n_1 \tag{20}$$

Though this rule is not applicable to sets of data that have strongly non-Gaussian distribution, for moderate n_1 (less than 200) it gives similar results to other alternative rules (Scott, 1992) and thus it is assumed for FBA. The Euclidean distance

$$c_{ij} = |m'_i - m'_j| = |j - i|\Delta m' \tag{21}$$

is chosen as ground distance between central points of i-bin in signature $\{\mathbf{s}_a=(\mathbf{m}',\mathbf{W}_a)\}$ and j-bin in signature $\{\mathbf{s}_b=(\mathbf{m}',\mathbf{W}_b)\}$, in order to obtain the $N'\times N'$ cost matrix **C** of elements c_{ij} (i=1,...,N', j=1,...,N').

C, $\mathbf{W}_a$, $\mathbf{W}_b$ are the input of the algorithm, whose output is the scalar EMD and the $N'\times N'$ flow matrix **F**.

The algorithm is applied to each hydrologic variable, for which the EMD is a dimensionless measure of the dissimilarity of the two histograms, i.e. measures of the frequency modification (and hence the hydrologic alteration).

It should be noted that the FBA is entirely based on sample(s), while statistical modelling aimed to estimate the probability distribution is not performed. In other words, the registered flows are considered the most reliable information.

Finally, it should be stressed that the full range of variation of the flow regime is adequately incorporated in the FBA, while in the RVA extreme events (minimum and maximum values of each variable) are included in histogram bins that contain also ordinary events (first and third bins). In the FBA and HMA, extreme events are more accurately represented using histograms with a greater number of bins. However, as noted above, the similarity of the non-overlapping parts of the two histograms (mainly the external bins) is more accurately computed using the EMD algorithm rather than Niblack et al. (1993) metric adopted in the HMA.

3.4 Case study

The Crati River is the main watercourse of Calabria (South Italy), with catchment area of 2431 km^2, mean altitude of 600 m and length of 81 km. It is a typical Mediterranean river, characterized by an irregular and perennial flow regime - that strongly depend on rainfall. Periods of zero-flow are rare, while severe summer droughts and autumn-winter floods are fairly regular. Crati estuary area is a natural reserve.

A time-series of 40-years records (1927-1966) of mean daily (near) natural discharge is available from "Conca" gauge station. Relevant data for Conca gauge station are:

- distance from estuary: 23 km, altitude: 35 m;
- catchment area: 1332 km^2, with a mean altitude of 664 m and moderate permeability;
- mean annual rainfall: 1260 mm (rainfall concentrated from November to May);
- mean annual flow: 26.0 m^3/s, mean February flow: 56.5 m^3/s, mean August flow: 3.8 m^3/s, annual coefficient of variation (standard deviation of all the daily flow values, divided by the mean annual flow): 1.27.

The alluvial plain of the Crati River (downstream of Conca gage station) is a major agricultural area of Calabria and, as consequence, the Tarsia diversion weir (near Conca gage station) was completed before 1970 to provide water for off-channel irrigation.

In the present application, a variable release scenario was hypothesized to occur for a 40 years period (1967-2006), during which time the flow regime represented by the recorded registered mean daily discharge (Q) in the period 1927-1966 was assumed to occur upstream of the flow diversion, while releases downstream of the diversion (Q_r) were assumed to be in accordance to the operational rule

$$\begin{cases} Q \le Q_{\min} \rightarrow & Q_r = Q \\ Q_{\min} \le Q \le Q_{\min} + Q_d \rightarrow & Q_r = Q_{\min} \\ Q > Q_{\min} + Q_d \rightarrow & Q_r = Q - Q_d \end{cases} \tag{22}$$

where Q is the flow upstream the diversion, Q_{min} is a the value of flow corresponding to some environmental flow prescription and Q_d is the projected diversion. Eq. (22) is the typical flow release rule for an unregulated weir.

In this application, according to prescription of the Basin Authority of Calabria, Q_{min}=4.5 m^3/s for the entire year was assumed (Principato & Viggiani, 2009). Different (monthly or daily) assumptions can be done for Q_{min}.

Q_d was set equal to 2, 5, 10, 15, 20, 25, 30, 35, 40 m^3/s (larger values were not considered because mean annual flow is 26.2 m^3/s).

4. Results analysis

The hydrologic alteration due to different values of Q_d was calculated for the 33 IHAs using RVA (with standard interval limits), HMA (with α=1) and FBA. The 33 IHA are considered, although other variables can also be used. The Suen & Ehart (2006) method has been discussed above, but is based on other variables and thus not used here for comparison.

Thus, the Alteration Indexes AI (HAF_k, $D_{Q,k}$ and EMD_k, k=1,...,33) were calculated.

The **a**=a_i (i=1,...,n_1), **b**=b_j (j=1,...,n_2) for March mean flow (with a diversion of Q_d=20 m^3/s) and the corresponding histograms and signatures (with n_1=n_2=40, N=6, N'=8, mean flow: 40.16 m^3/s) are reported in Fig. 1-2.

The March mean flow signatures of **a** and **b** are first defined by (eq. 4a, 4b)

$$\mathbf{w}_a = (13, 13, 5, 6, 1, 2)$$

$$\mathbf{m}_a = (26.35, 36.65, 46.95, 57.25, 67.55, 77.85)$$

$$\mathbf{w}_b = (1, 10, 5, 5, 1, 2)$$

$$\mathbf{m}_b = (9.37, 19.12, 28.87, 38.62, 48.37, 58.12)$$

Then, the adimensional vector $\mathbf{m}_a'$ is calculated by dividing each element of $\mathbf{m}_a$ for the mean of **a** (eq. 5a) and the vector

$$\mathbf{m}' = (0.14, 0.40, 0.66, 0.91, 1.17, 1.43, 1.68, 1.94)$$

(with N'=8) is obtained adding two equal size bins to $\mathbf{m}_a'$ in order to include all values of $\mathbf{b}'$ (eq. 6). $\mathbf{m}'$ contains the central values of bins in the signatures of adimensional variables $\mathbf{a}'$, $\mathbf{b}'$, for which (eq. 7a, 7b)

$$\mathbf{w}_a' = (0, 1, 12, 13, 5, 6, 1, 2)$$

$$\mathbf{w}_b' = (13, 13, 5, 6, 1, 1, 1, 0)$$

Finally, the vectors

$$\mathbf{W}_a = (0, 0.098, 1.170, 1.267, 0.487, 0.585, 0.098, 0.195)$$

$$\mathbf{W}_b = (1.267, 1.267, 0.487, 0.585, 0.098, 0.098, 0.098, 0)$$

are calculated using equations 8a, 8b. Total area of each signature $\{\mathbf{S}_a=(\mathbf{m}',\mathbf{W}_a)\}$, $\{\mathbf{S}_b=(\mathbf{m}',\mathbf{W}_b)\}$ is 1. The cost matrix **C** of elements c_{ij} is given by eq. 21 (in which $\Delta m'$=0.257).

The optimal flow matrix **F**, whose values are positive (eq. 12), is

F=

	j=1,...,N'=8								$\mathbf{W}_a$:
i=1,...,N'=8	0	0	0	0	0	0	0	0	0
	0.098	0	0	0	0	0	0	0	0.098
	0.682	0	0.487	0	0	0	0	0	1.170
	0.487	0.195	0	0.585	0	0	0	0	1.267
	0	0.487	0	0	0	0	0	0	0.487
	0	0.390	0	0	0.098	0.098	0	0	0.585
	0	0.098	0	0	0	0	0	0	0.098
	0	0.098	0	0	0	0	0.098	0	0.195
$\mathbf{W}_b$:	1.267	1.267	0.487	0.585	0.098	0.098	0.098	0	

The flow matrix indicates the amounts of $\mathbf{W}_a$ to ship to locations specified by $\mathbf{m}'$ to obtain $\mathbf{W}_b$, or *vice versa* (each signature can be a supplier or a consumer because of coincident total weights).

Therefore, sum of rows is equal to $\mathbf{W}_a$, while sum of column is equal to $\mathbf{W}_b$. E.g., the amounts in first column have to be shipped from 2nd, 3rd and 4th elements of $\mathbf{W}_a$ to obtain first element of $\mathbf{W}_b$.

It should be noted that amounts of $\mathbf{W}_a$ are not added to $\mathbf{W}_b$, but to empty locations (clusters) of $\mathbf{W}_b$ (or *vice versa*), while the total cost is minimum (eq. 11).

The correspondence between histograms centres and bin size in the pre- and post-impact conditions has to be stressed. In fact, it ensures that histograms dissimilarity is due to flow alteration rather than histograms structure.

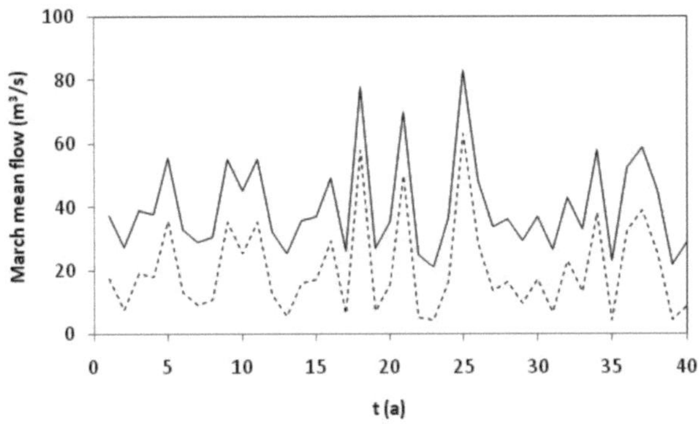

Fig. 1. March mean flow in the pre-impact (—) and post-impact (- - -) conditions in the n_1=n_2=40 year periods (Q_d=20 m³/s).

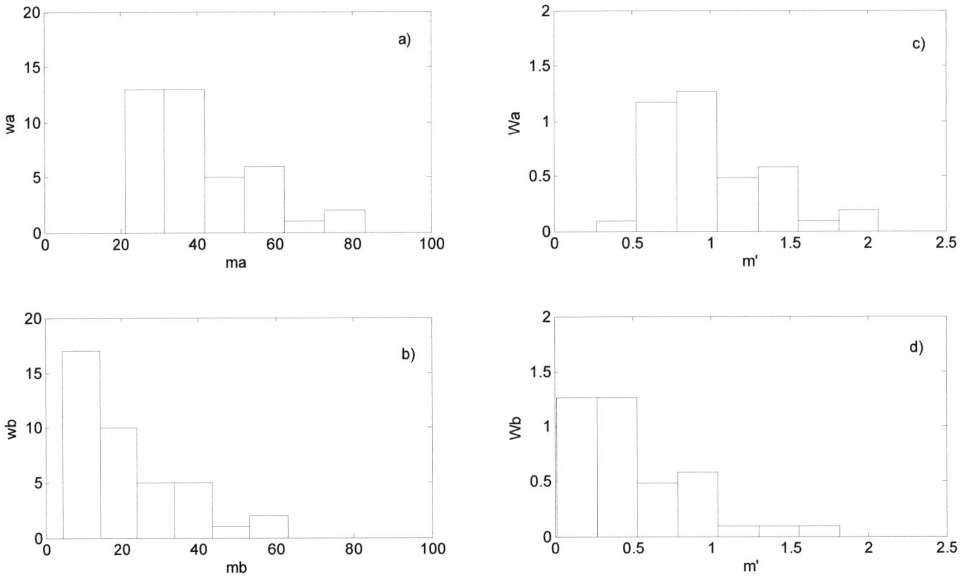

Fig. 2. Histograms for March mean flow in the pre-impact (a) and post-impact (b) conditions and corresponding extended dimensionless signatures (c, d).

4.1 "Expected variations" technique for evaluating methods accuracy

Once the RVA, HMA and FBA results were obtained, the problem to evaluate and compare their accuracy was considered. Preferably, results should be compared to a reference solution, that can usually be an analytical solution or a set of experimental data. Unfortunately, when analysing hydrologic alteration, the reference solution is not known neither analytically nor experimentally.

Moreover, the indexes HAF_k, $D_{Q,k}$ and EMD_k are all influenced by natural *variability* of flow regime, that cannot easily be distinguished from hydrologic *alteration*. In order to overcome this problem and to focus the analysis on methods accuracy, a post-impact flow (upstream of diversion) equal to the pre-impact flow is hypothesized in the present application. This option ensures that all alteration is attributable to the flow diversion. In other words, if $Q_d=0$ (no diversion), all indexes of alteration are equal to zero.

When $Q_d>0$, different values of alteration indexes AI are obtained ($-1<HAF<2$, $0<D_Q<1$, $0<EMD<1$). These values cannot be compared each other. Therefore, an accuracy analysis based on indexes *values* cannot be conducted (what is the *exact* value of alteration index?).

Relevant conclusions on indexes *variation* can be formulated instead. In fact, for many variables, the *expected* variation of AI due to Q_d increase can be qualitatively stated. For the IHA, such expected *variations* are detailed below and summarized in Tab. 1.

Since the operational rule (22) causes the flow to be reduced (if $Q>Q_{min}$) or unaltered (if $Q<Q_{min}$):

- for variables related to the magnitude of flow, alteration is expected to increase (or remain constant) when diverted discharge increases (Not Decreasing condition-ND);
- Number of zero-flow days is not influenced by operational rule (22) – (No alteration condition - N).
- Base flow is computed as 7-day minimum flow/mean flow for year. Mean flow decreases with Q_d, while, according to (22), 7-day minimum flow is not modified if it is less than Q_{min}. Therefore, base flow alteration is generally expected to increase with Q_d (Increasing condition - IN), but the condition 7-day minimum flow$<Q_{min}$ has to be verified in each application.
- Julian date of each annual 1-day maximum is not influenced by operational rule (22), because each annual maximum is reduced by the same quantity (Q_d) in the post-impact condition (N), except in the years when the annual 1-day maximum flow is less than the sum of Q_{min} and Q_d (this condition has to be verified in each application). On the contrary, Julian date of each annual 1-day minimum can be altered because, if there are multiple days in the water year with the same minimum flow value, the earliest date is considered in the computation (The Nature Conservancy, 2009). In this case, alteration should either increase or remain constant (ND).
- Number and duration of high/low pulses, fall/rise rate can increase or decrease owing to (22), depending on the specific sequence of natural flow. General conclusions cannot be drawn.
- Number of reversals decreases when $Q_{min}\leq Q\leq Q_{min}+Q_d$ and $Q_r=Q_{min}$; otherwise it is not modified. Being periods with $Q_{min}\leq Q\leq Q_{min}+Q_d$ more long and frequent for large Q_d, alteration increases with Q_d (IN).

As result, the variation of AI due to Q_d increase/decrease is known for 25 to 27 IHA (depending on the conditions to be tested), while for remaining 6 variables conclusions cannot be drawn. Furthermore, three kinds of variation can occur (IN, ND, N). Therefore, the technique here developed allows to evaluate the accuracy of RVA, HMA and FBA through the analysis of 25-27 IHA.

4.2 Comparison of results for single variables

The results for indexes HAF, D_Q, EMD - expressed in terms of variation due to Q_d increase - are summarized in Tab. 1.

If a minimum exists in the AI(Q_d) relationship, results are not physically correct, since alteration cannot decrease if diverted flow Q_d increases. In such case, Decrease (D) is indicated in Tab. 1 (it means that the corresponding method fails). In summary:

- Monthly mean flows (ND expected; Fig. 3): ND condition does not occur both for HAF (February, March, April, October, December) and D_Q (February, August, September). Moreover, relationships HAF(Q_d) and D_Q(Q_d) are irregular also for other monthly flow. Such conditions are not observed for index EMD, that shows a regularly not decreasing relationship to Q_d.
- Annual minima – 1, 3, 7, 30, 90 days mean (ND expected; Fig. 4): HAF is not sensitive to alteration of 1, 3, 7 days mean, while D_Q values for 30, 90 days mean are irregularly related to Q_d. ND condition is satisfied only using EMD.
- Annual maxima – 1, 3, 7, 30, 90 days mean (ND expected; Fig. 5): HAF(Q_d) relationship is very irregular (two minima are observed for all but one variables). D_Q(Q_d) irregularity is less evident, but slight decrease occurs for 1, 3, 7, 30 days mean. ND condition is satisfied using EMD.
- N. of zero-flow days (N expected): all indexes are equal to zero for all values of Q_d.
- Base flow (IN expected; Fig. 6a): a highly irregular relationship is observed both for HAF and D_Q, while EMD satisfies IN condition.
- Julian dates of each annual 1-day minimum (ND expected; Fig. 6b): most of alteration is due to 1954 winter drought, that preceded the rainy summer of the same year (in which $Q>Q_{min}$). So, in 1993, minimum flow release ($Q=Q_{min}$) would occur 1st January. HAF is not sensitive to such alteration, while D_Q does not satisfy the ND condition. On the contrary, index EMD satisfies the ND condition.
- Julian dates of each annual 1-day maximum (N expected): alteration is not expected nor observed for the three indexes.
- Number and duration of low/high pulses: conclusions cannot be drawn.
- Rise/fall rate: conclusions cannot be drawn.
- Number of reversals (IN expected): alteration decreases with Q_d both for HAF and D_Q, while it correctly increases when using EMD (Fig. 8).

As result, only FBA matches IN, N and ND conditions for the 27 variables for which qualitative relationships are known. Discrepancies are significant both for RVA (17 mismatches) and HMA (13 mismatches) and therefore a fictitious alteration (due to method accuracy) is added to (or subtracted from) real alteration.

In accordance with Shiau & Wu (2008), it should be concluded that the RVA can led to misleading outcome. However, also HMA reveals unphysical response to diverted flow increase/decrease and should be discarded or carefully evaluated as hydrologic alteration assessment method.

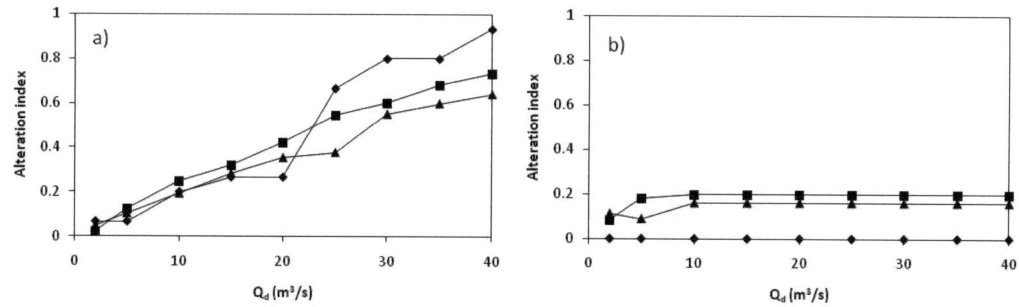

Fig. 3. Q_d - Alteration indexes HAF (♦), D_Q (▲), EMD (■) relationships for January mean flow (a), August mean flow (b), September mean flow (c), October mean flow (d)

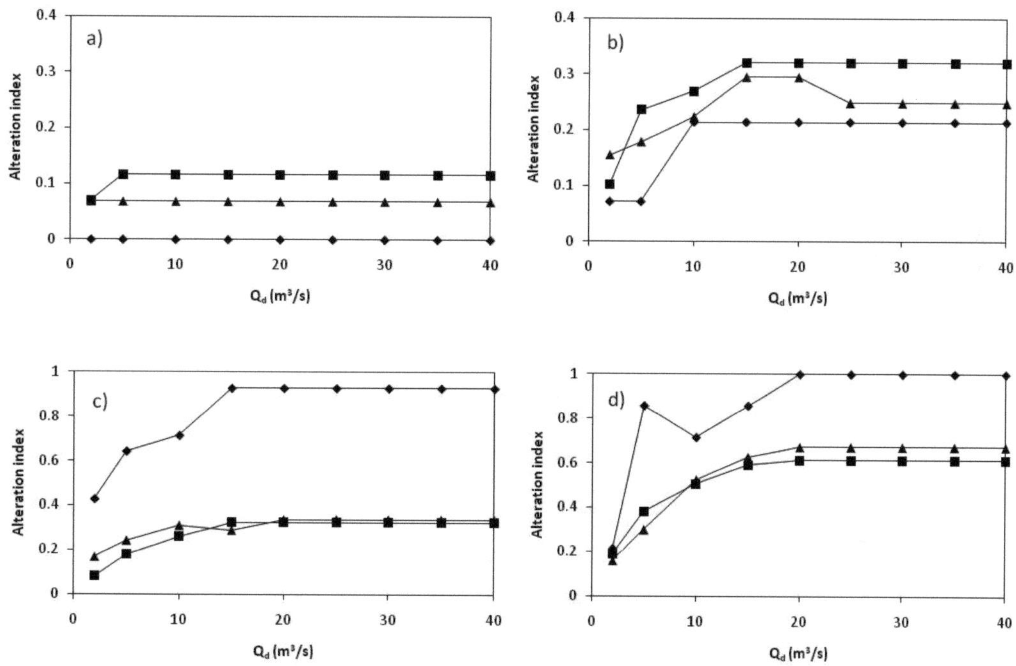

Fig. 4. Q_d - Alteration indexes HAF (♦), D_Q (▲), EMD (■) relationships for annual minima-7 days mean (a), annual minima-90 days mean (b)

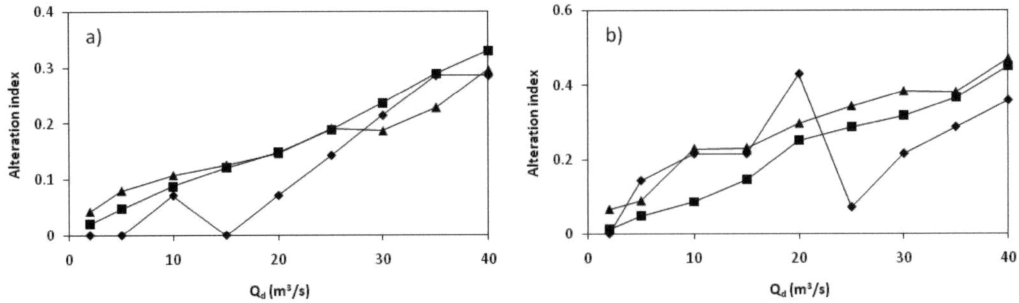

Fig. 5. Qd - Alteration indexes HAF (♦), DQ (▲), EMD (■) relationships for annual maxima-7 days mean (a), annual minima-30 days mean (b)

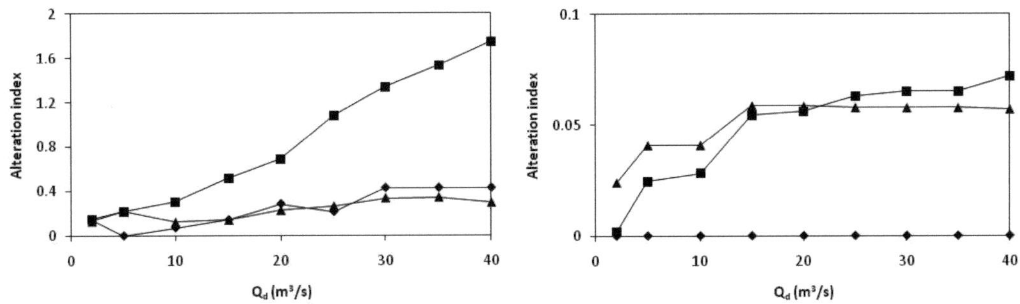

Fig. 6. Q_d - Alteration indexes HAF (♦), D_Q (▲), EMD (■) relationships for base flow (a) and Julian date of each annual one-day minimum (b)

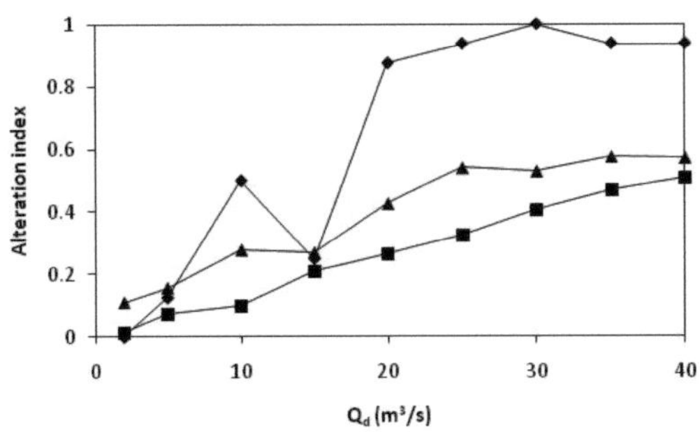

Fig. 7. Q_d - Alteration indexes HAF (♦), D_Q (▲), EMD (■) relationships for number of reversals

	Expected results	RVA	HMA	FBA
IHA Group 1: Discharge for each cal. month (m^3/s)				
January	ND	ND	ND	ND
February	ND	D	D	ND
March	ND	D	ND	ND
April	ND	D	ND	ND
May	ND	ND	ND	ND
June	ND	ND	ND	ND
July	ND	N	ND	ND
August	ND	N	D	ND
September	ND	ND	D	ND
October	ND	D	ND	ND
November	ND	ND	ND	ND
December	ND	D	ND	ND
IHA Group 2: Discharge (m^3/s)				
Annual min. - 1 day mean	ND	N	ND	ND
Annual min. - 3 days mean	ND	N	ND	ND
Annual min. - 7 days mean	ND	N	D	ND
Annual min. - 30 days mean	ND	ND	D	ND
Annual min. - 90 days mean	ND	ND	D	ND
Annual max. - 1 day mean	ND	D	D	ND
Annual max. - 3 days mean	ND	D	D	ND
Annual max. - 7 days mean	ND	D	D	ND
Annual max. - 30 days mean	ND	D	D	ND
Annual max. - 90 days mean	ND	D	ND	ND
N. of zero-flow days	N	N	N	N
Base flow	IN	D	D	IN
IHA Group 3: timing of annual extreme water conditions				
Julian date of each annual 1-day minimum	ND	N	D	ND
Julian date of each annual 1-day maximum	N	N	N	N
IHA Group 4: frequency and duration of high/low pulses				
Number of low pulses each year	-	IN	D	D
Duration of low pulses within each year (days)	-	D	D	D
Number of high pulses each year	-	D	D	IN
Duration of high pulses within each year (days)	-	D	D	D
IHA Group 5: rate/frequency of water condition changes				
Rise rate	-	ND	D	ND
Fall rate	-	ND	D	IN
Number of hydrologic reversals	IN	D	D	IN

Table 1. Analysis of AI-Q_d relationships: computed and expected results for RVA, HMA and FBA (ND: not decreasing, IN: increasing, D: decreasing, N: no variation)

4.3 Comparison of results for overall hydrologic alteration

Once alteration indexes AI are computed for each N variables, overall hydrologic alteration can be estimated using the index proposed in Shiau & Wu (2007a). The Index

$$L_2 = \left(\frac{1}{N}\sum_{k=1}^{N} AI_k^2\right)^{0.5} \tag{23}$$

is adopted here for the purpose of obtaining overall alteration-Q_d relationships for HAF, D_Q and EMD (Fig. 8). The suitability of L_2 is not discussed here. Other indexes could be used for comparison.

Adopting HAF, the pattern is quite irregular and L_2 for Q_d=35 m³/s is less than L_2 for Q_d=30 m³/s. For D_Q, pattern is less irregular, while more regular variations of overall alteration are observed for EMD. These patterns are obviously due to irregularities in many of the 33 AI(Q_d) relationships.

It should be concluded that the scarce accuracy of RVA and HMA (for single variables) also affects the estimation of overall hydrologic alteration.

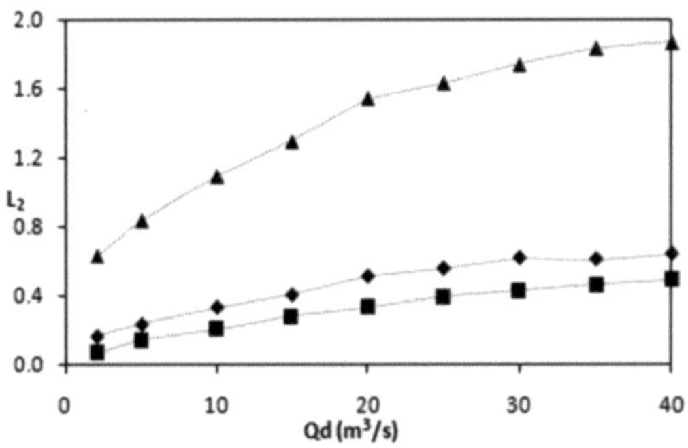

Fig. 8. L_2-Q_d relationships from RVA (♦), HMA (▲) and FBA (■)

5. Conclusions

The main characteristics and the accuracy of few methods for hydrologic alteration assessment have been analysed.

First, few theoretical drawbacks are highlighted for few existing methods, which are due to poor statistical analysis (RVA), inadequate metric (HMA) and weak ecological hypothesis (Suen & Eheart, 2006 method).

Then, quantitative evaluations have been conducted for a test case study using RVA and HMA, revealing weakness of both methods.

As consequence, a more robust method has been proposed here for the first time, namely the FBA. It considers the full range of flow regime variations (in accordance to natural flow paradigm) and ensures that a statistically-based analysis is conducted (comparison of the frequency distributions of a proper set of hydrologic variables), as required in the ecological context.

The FBA allows to obtain physically correct results (AI(Q_d) relationships) for all hydrologic variables (IHA) for which qualitative relationships are known. Consequently, also overall alteration index are more accurately computed.

The hydrologic alteration analysis performed with FBA has many potential applications, like definition of water release plans, evaluation of river status (and correlation to ecological status), analysis of flow variability (natural variability or the effects of climate change).

However, hydrologic alteration analysis is far from being conducted with the same accuracy in river reaches for which adequate registered flow series are not available. Accurate applications in non-gauged river reaches remain to be defined both for predictive purposes (water release plans for future diversions) and existing diversions.

It is worth noting that, if alteration-diverted flow relationships are not physically and statistically meaningful, optimization efforts aimed to define release plans are not correctly addressed. Consequently, the integration of FBA in the existing optimization models - aimed to perform the third step of the overall analysis - is encouraged.

Notations

a_i	values of **a**
a	vector of hydrologic variable in the pre-impact condition
AI	alteration index
b_i	values of **b**
b	vector of hydrologic variable in the post-impact condition
c_{ij}	cost to ship a unit of supply
C	matrix of elements c_{ij} (cost matrix)
D	decrease
D_Q	degree of histogram dissimilarity in the HMA
EMD	Earth Mover's Distance
f_{ij}	set of flows from supplier to consumer
F	matrix of elements f_{ij} (flow matrix)
F	observed frequency in the RVA
F_0	expected frequency in the RVA
h_j	number of data within the j_{th} bin of the histogram
HAF	Hydrologic Alteration Factor in the RVA
IC	Increasing Condition
L_2	overall hydrologic alteration index
m_j	central value of the bin j of the histogram or signature
m_j'	central value of the bin j of the adimensionalised histogram or signature
m, m′	vector of m_j, m_j'
n_1, n_2	number of year in the pre-impact and post-impact conditions
N	number of histogram bins
N'	number of extended signature clusters
ND	Not Decreasing condition
Q	flow upstream the diversion (m^3/s)
Q_{min}	flow corresponding to some environmental flow prescription (m^3/s)
Q_d	projected diversion (m^3/s)
Q_r	released discharges (m^3/s)

$\{s_j\}$	element of a signature
$\{\mathbf{s}_a'\}$	signature of **a′**
$\{\mathbf{s}_b'\}$	signature of **b′**
$\{\mathbf{s}_a'\}$	extended signature of **a′**
$\{\mathbf{s}_b'\}$	extended signature of **b′**
$\{\mathbf{S}_a\}$	normalized extended signature of **a′**
$\{\mathbf{S}_b\}$	normalized extended signature of **b′**
t	time (years)
w_j	number of data within the j_{th} cluster of the signature
w_j'	number of data within the j_{th} cluster of the adimensionalised signature
$W_{a,j}$	normalized value of $w_{a,j}$
$W_{b,j}$	normalized value of $w_{b,j}$
$\mathbf{W}, \mathbf{w}'$	vector of w_j, w_j'
$\mathbf{W}_a$	vector of $W_{a,j}'$
$\mathbf{W}_b$	vector of $W_{b,j}'$
α	parameter of the similarity function

6. Acknowledgements

The authors acknowledge Yossi Rubner, Carlo Tomasi, and Leonidas J. Guibas for letting their C code available on the web.

7. References

Arthington, A.H. ; Bunn, S.E.; Poff, N.L. & Naiman R.J. (2006). The challenge of providing environmental flow rules to sustain river ecosystems. *Ecological Applications*, Vol. 16, No. 4, pp. 1311–1318.

Botter, G.; Basso, S.; Porporato, A.; Rodriguez-Iturbe, I. & Rinaldo, A. (2010). Natural streamflow regime alterations: Damming of the Piave river basin (Italy). *Water Resources Research*, Vol. 46, W06522, doi:10.1029/2009WR008523.

Boulton, A.J.; Sheldon, F.; Thoms, M.C. & Stanley, E.H. (2000). Problems and constraints in managing rivers with variable flow regimes. Chapter 18, In : *Global Perspectives on River Conservation: Science, Policy and Practice*, P. J. Boon, B. R. Davies & G. E. Petts (Ed.), 415-430, John Wiley and Sons Ltd.

Bunn, S.E. & Arthington, A.H. (2002). Basic principles and ecological consequences of altered flow regimes for aquatic biodiversity. *Environmental Management*, Vol. 30, pp. 492–507.

Cha, S.H. & Srihari, S.N. (2002). On measuring the distance between histograms. *Pattern Recognition*, Vol. 35, pp. 1355–1370.

Connell, J.H. (1978). Diversity in tropical rain forests and coral reefs. *Science*, Vol. 199, pp. 1302–1310.

Dantzig, G.B. (1951). Application of the simplex method to a transportation problem, In: *Activity Analysis of Production and Allocation*, 359–373, JohnWiley and Sons.

Galat, D.L. & Lipkin, R. (2000). Restoring ecological integrity of great rivers: historical hydrographs aid in defining reference conditions for the Missouri River. *Hydrobiologia*, Vol. 422/423, No. 0, pp. 29-48.

Irwin, E.R. & Freeman, M.C. (2002). Proposal for Adaptive Management to Conserve Biotic Integrity in a Regulated Segment of the Tallapoosa River, Alabama, USA. *Conservation Biology*, Vol. 16, No. 5, pp. 1212-1222.

Lake, P.S. (2003). Ecological effects of perturbation by drought in flowing waters. *Freshwater Biology*, Vol. 48, pp. 1161–1172.

Monk, W.A.; Wood, P.J.; Hannah, D.M. & Wilson D.A. (2007). Selection of river flow indices for the assessment of hydroecological change. *River Research and Applications*, Vol. 23, pp. 113-122.

Niblack, W.; Barber, R.; Equitz, W., Flickner, M.D. ; Glasman, E.H., Petkovic, D.; Yanker, P.; Faloutsos, C.; Taubin, G.; Heights, Y. (1993). Querying images by content, using color texture, and shape, In : *SPIE Conference on Storage and Retrieval for Image and Video Databases*, 1908, 173–187, San Jose, CA, USA. DOI: 10.1117/12.143648.

Olden, J.D. & Poff N.L. (2003). Redundancy and the choice of hydrologic indices for characterizing streamflow regimes. *River Research and Applications, Vol.* 19, pp. 101–121.

Poff, N.L.; Allan, D.; Bain, M.B.; Karr, J.R.; Prestegaard, K.L.; Richter, B.D.; Sparks, R.E. & Stromberg, J.C. (1997). The Natural Flow Regime. A paradigm for river conservation and restoration. *BioScience*, Vol. 47, No. 11, pp. 769-784.

Postel, S.; Richter, B. 2003. *Rivers for Life: Managing Water for People and Nature*, Island Press, Washington, USA.

Principato, G. & Viggiani, G. (2006). Condizioni idrologiche di riferimento per la tutela dei corsi d'acqua. *L'Acqua*, Vol. 4, pp. 31-46.

Principato, G. & Viggiani, G. (2009). Hydrologic alteration assessment in Mediterranean rivers: perspective from a South Italy case study. *International Journal of River Basin Management*, Vol. 7, No. 4.

Richter, B.D.; Baumgartner, J.V.; Powell, J. & Braun, D.P. (1996). A method for assessing hydrologic alteration within ecosystems. *Conservation Biology, Vol.* 10, No. 4, pp. 1163-1174.

Richter, B.D.; Baumgartner, J.V.; Wigington, R. & Braun, D.P. (1997). How much water does a river need? *Freshwater Biology*, Vol. 37, No.1, pp. 1–249.

Richter, B.D.; Mathews, R.; Harrison, D.L. & Wigington, R. (2003). Ecologically sustainable water management: Managing river flows for ecological integrity. *Ecological Applications*, Vol. 13, No. 1, pp. 206–224.

Rosenberg, D.M.; McCully, P.& Pringle, C.M. (2000). Global-scale environmental effects of hydrological alterations: introduction. *BioScience*, Vol. 50, No. 9, pp. 746–751.

Rubner, Y.; Tomasi & C.; Guibas, L.J. (1998). A Metric for Distributions with Applications to Image Databases, In : *Proceedings of the 1998 IEEE International Conference on Computer Vision*, Bombay, India, 59–66.

Rubner, Y.; Tomasi, C.; Guibas, L.J. (2000). The Earth Mover's Distance as a Metric for Image Retrieval. *International Journal of Computer Vision*, Vol. 40, No. 2, pp. 99–121.

Scott, D.W. 1992. *Multivariate Density Estimation: Theory, Practice, and Visualization*, Wiley Series in Probability and Statistics.

Serratosa, F. & Sanfeliu, A. (2006). Signatures versus histograms: Definitions, distances and algorithms. *Pattern Recognition*, Vol. 39, pp. 921–934.

Shiau, J.T. & Wu, F.C. (2004). Assessment of hydrologic alterations caused by Chi-Chi diversion weir in Chou-Shui Creek, Taiwan. *River Research and Applications*, Vol. 20, No. 4, pp. 401–412.

Shiau, J.T. & Wu F.C. (2007a). Pareto-optimal solutions for environmental flow schemes incorporating the intra-annual and interannual variability of the natural flow regime. *Water Resources Research*, Vol. 43: W06433. DOI: 10.1029/2006WR005523.

Shiau, J.T. & Wu, F.C. (2007b). A dynamic corridor-searching algorithm to seek time-varying instream flow releases for optimal weir operation: comparing three indices of overall hydrologic alteration. *River Research and Applications*, Vol. 23, pp. 35–53. DOI: 10.1002/rra.962

Shiau, J.T. & Wu, F.C. (2008). A histogram matching approach for assessment of flow regime alteration: application to environmental flow optimization. *River Research and Applications*, Vol. 24, pp. 914–928. DOI: 10.1002/rra.1102.

Silverman, B.W. (1986). *Density Estimation for Statistics and Data Analysis, Monographs on Statistics and Applied Probability*, Chapman & Hall, London, UK.

Stewardson, M.J. & Gippel, C.J. (2003). Incorporating flow variability into environmental flow regimes using the Flow Events Method. *River Research and Applications*, Vol. 19, pp. 459–472.

Suen, J.P. & Eheart, J.W. (2006). Reservoir management to balance ecosystem and human needs: Incorporating the paradigm of the ecological flow regime. *Water Resources Research*, Vol. 42: W03417. DOI:10.1029/2005WR004314.

Tharme, R.E. (2003). A global perspective on environmental flow assessment: emerging trends in the development and application environmental flow methodologies for rivers. *River Research and Applications*, Vol. 19, pp. 397–441.

The Nature Conservancy (2009). Indicators of Hydrologic Alteration Version 7.1 User's Manual, Available from http://www.nature.org

Ward, J.V. & Tockner, K. (2001). Biodiversity: towards a unifying theme for river ecology. *Freshwater Biology*, Vol. 46, pp. 807–819.

World Commission on Dams (WCD) (2000). *Dams and Development. A New Framework for Decision-making. The report of the World Commission on Dam*, Earthscan Publications, London, UK.

World Conservation Union (IUCN) (2000). *Vision for Water and Nature. A World Strategy for Conservation and Sustainable Management of Water Resources in the 21st Century*, IUCN, Gland, Switzerland and Cambridge, UK.

Zolezzi, G.; Bellin, A.; Bruno, M.C.; Maiolini, B. & Siviglia, A. (2009), Assessing hydrological alterations at multiple temporal scales: Adige River, Italy. *Water Resources Research*, Vol. 45, 12421, doi:10.1029/2008WR007266.